Speculation by Commodity Index Funds

The Impact on Food and Energy Prices

Speculation by Commodity Index Funds

The Impact on Food and Energy Prices

Scott H. Irwin

Laurence J. Norton Chair of Agricultural Marketing
Department of Agricultural and Consumer Economics
University of Illinois at Urbana-Champaign, USA

Dwight R. Sanders

Professor
Department of Agribusiness Economics
Southern Illinois University at Carbondale, USA

CABI

"Speculation by Commodity Index Funds: The Impact on Food and Energy Prices" reveals how leaders in empirical finance and in international commodity policy analysis led industry participants, investors and academics and governments astray early in the twenty-first century. Highly cited analyses by leaders in empirical finance and global commodity policy touted in turn the benefits of commodity index investing, the evils of "financialization" of the markets and the role of speculative "bubbles" in commodity price turmoil. A reader will search the top finance journals in vain for a clear exposition of how leading academics got each of these issues so wrong. But it is all laid out in the sequence of eleven papers in this book. Each preceded by a truly delightful retrospective foreword which young academics will find highly informative about the importance, and the challenges, of publishing unfashionable economic truths."

Brian Wright, Distinguished Professor, Department of Agricultural and Resource Economics, University of California-Berkeley

"Many observers and policy makers are convinced that participation in commodity markets by speculators and financial investors contributed to the volatility in commodity prices over the last two decades. Irwin and Sanders have been at the forefront of careful academic investigation of the evidence that could support this conclusion. The studies in this book use data on the positions of index fund traders in commodity futures markets, roll yields and rollover effects, and order-flow costs together with the economic theory and practical understanding of how futures markets function. This evidence is must reading for anyone who has concluded that we need to tighten limits on participation in these markets."

James Hamilton, Robert F. Engle Professor of Economics, University of California at San Diego

"When food prices began spiking around the world in late 2007 – and multiple times since then – observers offered a range of competing explanations. One popular explanation was the 'financialization' of agricultural commodity markets. The idea that financial speculators were diversifying from more traditional financial securities into commodity futures, including through new commodity index fund structures, and that this innovation was enriching a few while immiserating millions captured popular and policymakers' imaginations. The correlational evidence indeed seemed consistent with that hypothesis. Two experts in agricultural finance, Scott Irwin and Dwight Sanders, took those concerns seriously. Using the best available theory, methods and data they rigorously tested a range of hypotheses around the central claim: financial speculation increasingly drives food prices globally. Their findings are the most convincing evidence available on the topic, largely showing the correlation is largely spurious, not causal. Food price spikes do not seem attributes to financial markets. This book pulls together various papers they wrote on the topic into a coherent, compelling narrative. It is a must read for anyone interested in this topic at a serious level."

Chris Barrett, Stephen B. and Janice G. Ashley Professor of Applied Economics and Management, Cornell University

"This collection of papers by Irwin and Sanders, spanning their decade plus of collaborations on commodity futures markets and the impacts of index fund investing on those markets, is greater than the sum of its parts. The authors have included additional insights and backstories in new forewords to each of the eleven papers collected here. These alone and worth the price of admission. If you study, invest in, or regulate commodity futures markets, this book is essential reading."

Jeff Dorfman, Professor of Agricultural and Applied Economics, University of Georgia

"Professors Irwin and Sanders analyze and discuss in a rigorous and clear way the role of speculation in commodity and energy markets and the impact on food and energy prices. They show a balanced review of various – sometimes conflicting - opinions about the role of speculators in commodity futures markets. They are not afraid to ask difficult questions: challenging the work done in this domain by others and themselves and subsequently synthesizing and improving on that work. A book that will help better understand the structure of markets, their characteristics (volatilities) and their participants. A timely, must-read book for readers interested in commodity markets and the role of financial institutions."

Joost Pennings, Professor of Finance and Marketing, and Director of the Marketing-Finance Research Lab, Maastricht University School of Business and Economics

"Irwin and Sanders' work provides the backbone for our understanding of the impacts of index investment in commodity markets. I view the papers collected here as essential reading for any policy maker looking to understand this issue. Their combination of rigorous econometric analysis and clear exposition also serves as an excellent example for researchers seeking to develop an empirical agenda in any area of applied economics."

Rob Ready, Associate Professor and Robert J. and Leona M. De Armond Research Scholar, University of Oregon

"Irwin and Sanders' book on "Speculation by Commodity Index Funds" provides a wonderful historical record, covering their 15 years of focused research on this topic. Irwin and Sanders' book shows that they have ably picked up the baton from their agricultural economist predecessors: Holbook Working, Roger Gray, and Thomas Hieronymous, in carrying out empirical studies on the role of market participants in commodity price formation. This work will guide future economists in how to objectively respond to the inevitable skepticism on commodity futures trading that has existed since the 1880s. Importantly, the authors educate readers on allowing the data to provide answers to important policy questions, no matter where these conclusions may lead, which can variously include friction with market participants, politicians, other academics, and/or regulatory staff. While this approach is scientific thinking at its best, it is only for the brave!"

Hilary Till, Principal, Premia Research LLC, and Co-Editor, Commodity Insights Digest (Bayes Business School)

"What has been the role of index investors in driving commodity futures prices? And what are the effects of financialization in commodity futures markets for the economy at large? Irwin and Sanders dive into these issues, searching for the tell-tale footprints of speculation and illuminating its role in modern commodity markets. They offer compelling answers to these big questions which run counter to the easy solutions often peddled by politicians and the public while also providing insights into markets which collectively move trillions of dollars and help feed billions of people every year."

David Jacks, J.Y. Pillay Professor of Social Sciences, National University of Singapore

"... the book by Scott Irwin and Dwight Sanders addresses key issues in the more than 15-year old controversy about the impact of index fund investment on commodity futures prices. It is a must-read book for scientists, politicians, regulators and interested public needing well-founded information on the role of index funds in commodity futures markets. The collection of articles provides an important contribution to our knowledge of the functioning of commodity markets and shows convincingly how scientific discourse contributes to the understanding the complicated way of working of financial markets. Given the heated debate on speculation by commodity index funds the reader is a backbone for fact-based discussions on the role of index fund investment."

Martin Bohl, Professor, Westfälische Wilhelms-University Münster, Germany

"Commodity markets have experienced structural changes unprecedented levels of activity in the past two decade. Part of the considerable growth in the commodity derivatives open interest, and major shifts in volume dynamics, during this period are tied to financial investors' desire for exposure to commodity prices in general and index trading in particular. Irwin and Sanders have written some of the most cited scientific articles on whether the influx of commodity index money into futures markets has, directly or indirectly, artificially boosted commodity price levels and impacted commodity risk or liquidity premia. This book, by presenting the authors' own selection among their prolific output, and by spelling out their views of where each paper in the collection fits within the literature and the ongoing policy debate, is an essential read for scholars and practitioners alike. The sharing of numerous anecdotes about the genesis and the reception of each paper is the cherry on a very tasty cake."

Michel Robe The Clearing Corporation Foundation Professor in Derivatives Trading, University of Illinois at Urbana-Champaign

"During the global price spikes of 2007/08 and 2010/11, Scott Irwin and Dwight Sanders were beacons of light in their detailed analyses of the role of speculation in agricultural commodity markets. Following in a long tradition dating back to the work of Holbrook Working, Roger Gray, Tom Hieronymus, Lester Telser and others, Irwin and Sanders brought clarity to the impact of commodity index funds on commodity prices and price volatility. This compendium of their work could not come at better time. The war in Ukraine has sent prices to record highs and price volatility has exceeded even 2007/08 levels. This book will be a welcome reminder to critics of futures markets to look first to market fundamentals to understand price volatility."

Joe Glauber, Senior Research Fellow, International Food Policy Research Institute, and Former Chief Economist of the USDA

In "Speculation by Commodity Index Funds," Professors Irwin and Sanders delve into the controversial topic of whether financialization of commodities distorts price discover in futures markets. With the reoccurring debates about whether price spikes occur outside normal expected ranges, this book provides a compelling case for why there is not empirical proof to suggest that index investors' positions cause large change in commodity futures prices. Through a collection of essays, the authors make a valuable contribution to the understanding of this complex subject, providing a comprehensive and in-depth treatment of the topic, presenting evidence that index speculation was not the primary driver of the great price spikes that occurred during the 2007-2013 cycle. This book is highly recommended for academics, investors, traders, and regulators seeking a deep understanding of market dynamics. It is a must-read for anyone interested in gaining insight into the efficiency of the price discovery process in futures markets."

Ivo Sarjanovic, Professor of Agricultural Commodities, Torcuato Di Tella University, and Lecturer in Market Intelligence, University of Geneva

CABI is a trading name of CAB International

CABI
Nosworthy Way
Wallingford
Oxfordshire OX10 8DE
UK

Tel: +44 (0)1491 832111
E-mail: info@cabi.org
Website: www.cabi.org

CABI
200 Portland Street
Boston
MA 02114
USA

Tel: +1 (617)682-9015
E-mail: cabi-nao@cabi.org

A catalogue record for this book is available from the British Library, London, UK.

ISBN-13: 9781800622081 (hardback)
9781800622098 (ePDF)
9781800622104 (ePub)

DOI: 10.1079/9781800622104.0000

Commissioning Editor: Ward Cooper
Editorial Assistant: Emma McCann
Production Editor: Marta Patiño

Typeset by SPI, Pondicherry, India
Printed and bound in the UK by Severn, Gloucester

Contents

About the Authors

Scott H. Irwin holds the Laurence J. Norton Chair of Agricultural Marketing in the Department of Agricultural and Consumer Economics at the University of Illinois at Urbana-Champaign. He earned a BS in agricultural business from Iowa State University and both an MS and a PhD in agricultural economics from Purdue University. He has published hundreds of scholarly articles and is best known for his work on commodity speculation and biofuels. He has written articles and op-eds for the *New York Times*, *Washington Times*, and *Time* magazine. He also directs the award-winning *farmdoc* program at the University of Illinois.

Contact information: Scott H. Irwin, Department of Agricultural and Consumer Economics, 344 Mumford Hall, 1301 W. Gregory Drive, University of Illinois at Urbana-Champaign, Urbana, Illinois, 61801, USA. Email: sirwin@illinois.edu, scotthirwin.com

Dwight R. Sanders is Professor of Agribusiness Economics at Southern Illinois University at Carbondale where he teaches and conducts research in price analysis and commodity risk management. He was raised in southwest Missouri and attended undergraduate school at Missouri State University earning BS degrees in economics and agricultural economics. Dwight received his MS degree in finance and PhD in agricultural economics from the University of Illinois. His business experience includes production agriculture, futures brokerage, and positions with The Pillsbury Company and Darden Restaurants, Inc.

Contact information: Dwight R. Sanders, Agribusiness Economics, Southern Illinois University, 1205 Lincoln Drive, Agriculture Building, Room 226E, MC 4410, Carbondale, Illinois, 62901, USA. Email: dwights@siu.edu

1

Intersections

We met in 1993 at the University of Illinois, where one of us (Scott) was on a sabbatical leave from Ohio State and the other (Dwight) was a PhD student in agricultural economics. We had the good fortune of intersecting in several graduate classes, but two stand out. The first was a graduate seminar in the business college on the newly emerging field of behavioral finance taught by Professor Jay Ritter. The second was a graduate course on time-series econometrics taught by Professor Paul Newbold. In addition to being a brilliant 'near' Nobel-Prize-winning econometrician, Professor Newbold's mannerisms and dry British humor were seemingly pulled from a Monty Python skit. This delighted us no end, even though few others in the class seemed to appreciate the daily entertainment. I think it is safe to say that this is where our personal friendship and professional partnership began.

It was clear from the outset that both of us were fascinated by commodity futures markets, which provide both price discovery and risk management opportunities for commodity producers and consumers. These markets are central to the operation of much of the global commodity system. Naturally, being agricultural economists, our main interest was in agricultural futures markets, but we were also interested in other commodity futures markets, such as the relatively new crude oil futures market.

The next highly fortuitous intersection occurred in July 2005, when the roles were reversed, and Dwight contacted Scott about spending his sabbatical leave from Southern Illinois University back at the University of Illinois. Dwight was by then a faculty member at Southern Illinois and Scott had moved to Illinois from Ohio State in 1997. Commodity prices were just starting to take off and we thought this might be a good opportunity to dig into issues surrounding speculation in commodity futures markets. We would not be starting at ground zero because Dwight examined 'noise trader' issues in futures markets for his dissertation research and Scott had done a couple of papers on speculation and price volatility in futures. So we were fortunate to have that base to build upon.

What's the old saying, 'It's better to be lucky than good'? It was incredibly good fortune that Dwight ended up spending much of spring semester 2006 on the Champaign-Urbana campus for his sabbatical leave. Scott was advising an MS student, Robert Merrin, who was also interested in speculation issues. So an idea was born. We would jointly supervise Robert's thesis on the impact of hedging and speculative positions on agricultural futures prices. We would use the time-series statistical tests that Dwight employed in his dissertation. Who could have imagined what followed?

Commodity futures prices exploded in 2007–2008 just as we were finishing our work

DOI:10.1079/9781800622104.0001

with Robert. At the same time, concerns about a new type of participant in commodity futures markets began to emerge. In particular, market participants, regulators, and civic organizations began raising concerns that inflows from new 'commodity index' funds were driving the increases in commodity prices instead of economic fundamentals. The main argument was that unprecedented buying pressure from these speculative long-only futures traders created massive bubbles that resulted in prices substantially exceeding fundamental value, as much as 80% by some accounts. If true, this would raise major questions about the efficiency of price discovery in commodity futures markets and the usefulness of the markets for managing risk. Numerous proposals were offered to restrict speculation in commodity futures markets around the globe, including the creation of a 'virtual reserve' whereby a public agency would take futures positions opposite speculators during periods of high market volatility, a tax on futures transactions, and tighter limits on speculative positions. During this period, it was not uncommon to link concerns about speculation to world hunger, food crises, and civil unrest.

The initial empirical analysis presented by those raising concerns about commodity speculation consisted of simple graphs that showed a concurrent increase in long-only index futures positions and price levels. These analyses were quite effective in catching the eye of politicians and the public. But they clearly failed to establish a rigorous statistical link between actual trader positions and futures prices.

We realized right away that the problem had to be well defined from an empirical perspective. This led us to argue for the importance of establishing a causal link strictly between futures positions and futures prices. Once the relevant empirical problem was defined, the proper commodity futures position data had to be utilized. We then used exhaustive empirical tests across numerous markets, time frames, and data sets to show that there was no consistent evidence that positions held by index investors caused large changes in commodity futures prices. Batteries of time-series and cross-sectional tests failed to find consistent temporal causality between index positions and futures prices. This body of work conclusively demonstrated that index speculation was not the main driver of the great commodity price spikes that occurred between 2007 and 2013.

While we and others have written review articles on the role of index funds in commodity futures markets, there is no single resource that provides a comprehensive and in-depth treatment of this important subject. In our own case, we have written more than two dozen articles and reports on this controversy since 2008. These publications have appeared in various journals over a more than 15-year span of time. We believe that there is value in collecting the most important of these articles in a single volume and organizing the articles in a manner that reflects how and why our work evolved as it did.

Hence, the purpose of this book is to present a curated selection of articles from our body of work on the impact of index funds on commodity futures prices. It is important to note at the outset that the selected articles do not simply represent a 'greatest hits' list based on citation totals. Instead, the selections roughly follow the chronology of our involvement in the worldwide debate about commodity speculation as it evolved after 2007. The 11 articles selected for inclusion in this volume highlight key issues that we addressed as the debate evolved. Some of the articles ended up being highly cited and some did not.

In addition to the articles in their original published form, we include new author forewords for each article that provide context and interesting backstories about the development of the research. The finished product functions as a guided tour through more than 15 years of work on index funds and the behavior of commodity futures prices.

A synopsis of each chapter in the book follows.

Chapter 2. Devil or Angel? The Role of Speculation in the Recent Commodity Price Boom (and Bust). This is the first paper that we wrote on the speculation controversy that erupted in 2007–2008. The article itself was largely a synthesis of the arguments we had been making about the role of index funds in the commodity price spike of 2007–2008 in presentations and other reports. We argued in this 2009 article that the charge of index funds creating a massive bubble simply did not stand up to close scrutiny. The charges were inconsistent with some basic facts, such as the observation that price movements in commodity futures markets with substantial index investment were not uniformly upward in 2007–2008.

Chapter 3. New Evidence on the Impact of Index Funds in US Grain Futures Markets. We thought that the speculation controversy would die out as prices crashed in the second half of 2008. We quickly realized that we were wrong and set out to do our first econometric analysis of the relationship between index positions and price movements in grain futures markets. We obtained some interesting new data on commodity index trader (CIT) positions and showed that the big growth in index positions actually occurred before the massive grain price spike of 2007–2008. Hence, it was no surprise when our Granger causality tests did not find consistent evidence of a relationship between CIT positions and grain futures price movements.

Chapter 4. The Impact of Index and Swap Funds in Commodity Futures Markets. Not only did the commodity speculation debate fail to flame out as we expected, but it actually picked up steam heading into the early 2010s. Civic organizations such as Oxfam jumped into the speculation debate and tended to react in a fiercely negative manner. We were approached in the summer of 2009 by the Organisation for Economic Co operation and Development (OECD) to produce a report on the controversy. When the report was published in June 2010 it started a global firestorm that spilled into the pages of major financial publications such as *The Economist*. The analysis was actually quite straightforward, but the results went against the grain of conventional wisdom in many places and organizations.

Chapter 5. Testing the Masters Hypothesis in Commodity Futures Markets. Critics of our OECD report focused on both data and methodological issues. The main data concern was the lack of accurate data on index positions in energy futures markets, particularly West Texas Intermediate (WTI) crude oil. The principal methodology issue was a supposed lack of power of Granger causality time-series tests. This 2012 paper was our response to those criticisms. It was the first to use positions from the new Commodity Futures Trading Commission (CFTC) *Index Investment Data* (IID) report and we also employed cross-sectional tests in addition to time-series statistical tests. The results were pretty much the same as before – no consistent relationship between index positions and commodity futures price movements. The article is probably most influential for having introduced the term 'Masters Hypothesis.'

Chapter 6. Financialization and Structural Change in Commodity Futures Markets. While working on the OECD report, it also became clear to us that there was a great deal of confusion about the nature of 'financialization' and the types of market impacts associated with it. In this 2012 article, we began by defining financialization as large-scale buying by financial index investors in commodity futures markets. A major complication in any analysis of the impact of financialization in commodity futures markets is that a number of historically large and important structural changes were taking place at roughly the same time as the rise of commodity index investment. For example, the switch from open outcry to electronic trading basically ran in parallel to financialization. This can make it difficult to disentangle market impacts due to financialization and other structural changes.

Chapter 7. A Reappraisal of Investing in Commodity Futures Markets. Another idea occurred to us while working on the OECD report: Did all this commodity index investment really make economic sense in the first place? A famous 2004 article by Gorton and Rouwenhorst was crucial in kick-starting the boom in commodity index investment, with its conclusion that commodity futures offered 'equity-like' returns. This ran directly counter to the evidence in classic commodity futures market studies by Telser, Rockwell, Dusak, and Hartzmark. In this 2012 article, we collected over five decades of daily futures prices and found that the return to individual futures markets was zero, consistent with the classics. This was also the first academic study to argue that 'roll yields' could not drive returns in commodity futures markets, which was considered conventional wisdom at the time. In some ways, this article was our most original and pre-dated the conclusions in other papers by nearly a decade.

Chapter 8. The 'Necessity' of New Position Limits in Agricultural Futures Markets: the Verdict from Daily Firm-level Position Data. CFTC proposals to extend speculative position limits to all futures markets for physical commodities became a focal point of the global controversy surrounding index trading in commodity futures markets. At the heart of the political and legal battle was the question of whether the CFTC had to meet the 'necessity' test before expanding position limits. Simply put, new regulations on trading had to be justified based on empirical evidence. For this 2016 article, we had access to daily data for a major private index fund and used them to bring new evidence to bear on the necessity question. The results were similar to our previous work, and we argued that the CFTC had flunked the necessity test.

Chapter 9. Bubbles, Froth and Facts: Another Look at the Masters Hypothesis in Commodity Futures Markets. This 2017 article is important for two reasons. First, it reflects the evolution of our understanding of the policy question at the heart of the controversy surrounding index funds in commodity futures markets. Second, we address the major criticisms that had appeared in the literature about the statistical methods we had used in previous studies. As a result, this article contains the most comprehensive set of time-series and cross-sectional tests of any of our published articles. We find once again that the Masters Hypothesis comes up short on its most basic market predictions.

Chapter 10. Mapping Algorithms, Agricultural Futures, and the Relationship between Commodity Investment Flows and Crude Oil Futures Prices. The 2014 study by Singleton is one of the most influential and widely cited in the financialization literature. He reports an economically large and statistically significant influence of index positions on crude oil futures prices. This truly puzzled us because it was completely at odds with virtually all of our own work. We discovered that Singleton (and others) inferred index positions in non-agricultural markets from index positions in agricultural markets. This is based on the seemingly sensible idea that there is an approximately fixed relationship among commodity index positions, reflecting the fixed nature of weights for the underlying target indexes. It turns out that Singleton's results can be directly traced to a surge of index investment in, of all things, feeder cattle futures during 2007–2008 that were not matched in crude oil futures. The implication is that Singleton's original results really are spurious.

Chapter 11. Sunshine versus Predatory Trading Effects in Commodity Futures Markets: New Evidence from Index Rebalancing. Many people do not appreciate that the failure of the Masters Hypothesis does not mean that we should end the search for price impacts of financialization in commodity futures markets. Rather, the search should focus on smaller price impacts associated with more rational market dynamics. The annual rebalancing of major commodity market indexes is tailor-made for just this type of investigation. In this article, we studied the annual rebalancing of the Standard and Poor's Goldman Sachs Commodity Index (S&P GSCI), which is by far the most widely tracked commodity index. We found that the price impact of S&P GSCI rebalancing reaches a peak of 72 basis points in the middle of the week following the rebalancing period, but the impact is temporary as it declines to near-zero within the next week. The findings showed that the impact of rebalancing order flows in commodity futures prices is modest and temporary, consistent with the prediction of sunshine trading theory.

Chapter 12. The Order Flow Cost of Index Rolling in Commodity Futures Markets. Investments that track the S&P GSCI roll positions forward from the nearby contract to the next deferred contract over a fixed 5-day window from the fifth to the ninth business day of every month. This is an especially interesting event to test theories of the market impact of financialization because the entire position of index investors in the commodity futures market must be rolled every month. We estimated that commodity index investors paid a total of $29 billion in order flow costs during monthly rolls over 1991–2019 and this was heavily concentrated in the growth period of financialization over 2004–2011. A careful examination of the yearly estimates revealed that order flow costs nosedived after 2006. This coincided almost perfectly with the transition to electronic trading in commodity futures markets. We concluded that a dramatic increase in the supply of liquidity brought on by the transition to electronic trading is primarily responsible for the remarkable decline of roll order flow costs.

Chapter 13. Lessons Learned. The controversy over commodity index funds contains important lessons for the future. We examine those lessons in this final chapter and discuss useful directions for future research in the area.

Note that we reproduced the articles in Chapters 2–12 based on the final Word and Excel files submitted to publishers. To the extent possible, edits made in the galley proof stage for each article were also incorporated. While not necessarily exact reproductions of the original published articles, the versions included in this book are extremely close to the published versions. We also made a few minor editorial corrections that were missed in the original publication process. All articles were also reformatted to have a consistent style throughout the book.

We hope you enjoy reading this book as much as we did in putting it together.

Scott H. Irwin
University of Illinois
Dwight R. Sanders
Southern Illinois University
October 2022

2

Devil or Angel? The Role of Speculation in the Recent Commodity Price Boom (and Bust)[1]

New Author Foreword

This article was the direct result of a phone call from Alfons Weersink of the University of Guelph in the summer of 2008, right at the height of the commodity price boom. The Department of Food, Agriculture, and Resource Economics at Guelph offered the annual Farrell Distinguished Public Policy Lecture on campus and asked if one of us (Scott) would be interested in delivering the talk later that fall. Alfons had seen some of our work on commodity market speculation and thought it would be a good topic for the lecture given everything going on in the markets at the time. Alfons also suggested that the lecture needed a really attention-getting title. This is what Scott came up with and it certainly turned out to be attention grabbing.

The Farrell Lecture was held in November 2008 and it went off quite well, although there was a good bit of skepticism by some in the audience about the conclusions reached. One of the requirements was to write up a paper to go along with the talk. This was fortuitous because sometime in the fall of 2008 we were contacted by an old friend, Hector Zapata, of Louisiana State, about writing up and presenting an invited paper for a session on commodity market volatility that he was organizing for the Annual Meeting of the Southern Agricultural Economics Association to be held in Atlanta during February 2009. Since we already had a draft of a paper, this seemed like a no-brainer. We even used the same title. However, the version of the paper presented in Atlanta was considerably expanded and polished. We also added Robert Merrin as a co-author because we used material from his MS thesis. Scott presented the paper in what was a very cold hotel because the heat was not working properly during the meeting, and it was unusually cold in Atlanta at the time. (Funny the things that stick in your memory.)

The article itself was largely a synthesis of the arguments we had been making about the role of index funds in the commodity price spike of 2007–2008 in earlier presentations, reports, articles, and even an op-ed (an opposite the editorial page) article we had published in *The New York Times*. It was commonly argued at the time that large-scale buying by commodity index investors in futures markets created a massive bubble and this was one of the main explanations for the commodity price spike. We argued in the article that the charge of index funds creating a massive bubble simply did not stand up to close scrutiny. One of the most fundamental errors made by bubble proponents was equating money flows into futures markets with demand. Investment dollars flowing into either the long or short side of futures markets is not the same thing as demand for physical commodities. Futures markets are zero-sum games where all money flows must by definition net to zero. It makes as much logical sense to call the long positions of index investors new 'demand' as it does to call the positions on the short side of the same contracts new 'supply.'

We also pointed out some basic facts that were blatantly inconsistent with the bubble arguments. These included the fact that price movements in commodity futures markets with substantial index investment were not uniformly upward in 2007–2008, and the lack of a notable build-up in commodity inventories during the alleged

DOI:10.1079/9781800622104.0002

bubbles. Finally, we included a section on 'Lessons from History' that laid out our view of the controversy from a historical perspective. Simply put, we viewed the political storm over index buying as just the latest in a very long line of attacks on speculation in commodity futures markets. This was our favorite section of the article and we referenced it frequently in later presentations and papers.

The article was published in the August 2009 issue of the *Journal of Agricultural and Applied Economics*, and it was an instant hit. According to Google Scholar, the article has been cited more than 450 times since it was published, a notable achievement considering that it was not published in a mainline economics or finance journal. Field (2017) provides important documentation of the far-reaching influence of the article. He interviewed 28 key participants at the center of commodity index trading and regulation and found that: 'Several key informants mention Gorton and Rouwenhorst's (2004) or Irwin *et al.*'s (2009) work as evidence of the benefits of index speculation and to support the view that financial speculation is unrelated to price volatility.' Most importantly, the arguments in the article have held up remarkably well. Any reasonable reading of the large literature that followed would have to conclude that, in the main, we got it right the first time around.

Abstract

It is commonly asserted that speculative buying by index funds in commodity futures and over-the-counter (OTC) derivatives markets created a 'bubble' in commodity prices, with the result that prices, and crude oil prices in particular, far exceeded fundamental values at the peak. The purpose of this paper is to show that the bubble argument simply does not withstand close scrutiny. Four main points are explored. First, the arguments of bubble proponents are conceptually flawed and reflect fundamental and basic misunderstandings of how commodity futures markets actually work. Second, a number of facts about the situation in commodity markets are inconsistent with the existence of a substantial bubble in commodity prices. Third, available statistical evidence does not indicate that positions for any group in commodity futures markets, including long-only index funds, consistently lead futures price changes. Fourth, there is a historical pattern of attacks upon speculation during periods of extreme market volatility.

Key words: commodity, futures, index fund, market, speculation

Journal of Economic Literature **(JEL) categories:** Q11, Q13

2.1 Introduction

Led by crude oil, commodity prices reached dizzying heights during mid-2008 and then subsequently declined with breathtaking speed (see Fig. 2.1). The impact of speculation, principally by long-only index funds, on the boom and bust in commodity prices has been hotly debated.[2] It is commonly asserted that speculative buying by index funds in commodity futures and over-the-counter (OTC) derivatives markets created a 'bubble,' with the result that commodity prices, and crude oil prices in particular, far exceeded fundamental values at the peak (e.g. Gheit, 2008; Masters, 2008; Masters and White, 2008). The main thrust of bubble arguments is that: (i) a large amount of speculative money was invested in different types of commodity derivatives over the last several years; (ii) this 'titanic' wave of money resulted in significant and unwarranted upward pressure on commodity prices; and (iii) when the flow of speculative money reversed, the bubble burst. Based on the bubble argument, a number of bills have been introduced in the US Congress with the purpose of prohibiting or limiting index fund speculation in commodity futures and OTC derivative markets.

The purpose of this paper is to show that the bubble argument simply does not withstand close scrutiny. Four main points are explored. First, the arguments of bubble proponents are conceptually flawed and reflect fundamental and basic misunderstandings of how commodity futures markets actually work. Second, a number of facts about the situation in commodity markets are inconsistent with the existence of a substantial bubble in commodity prices. Third, available statistical evidence does not indicate that positions for any group in commodity futures markets, including long-only index funds, consistently lead futures price changes. Fourth, there is a historical pattern of attacks

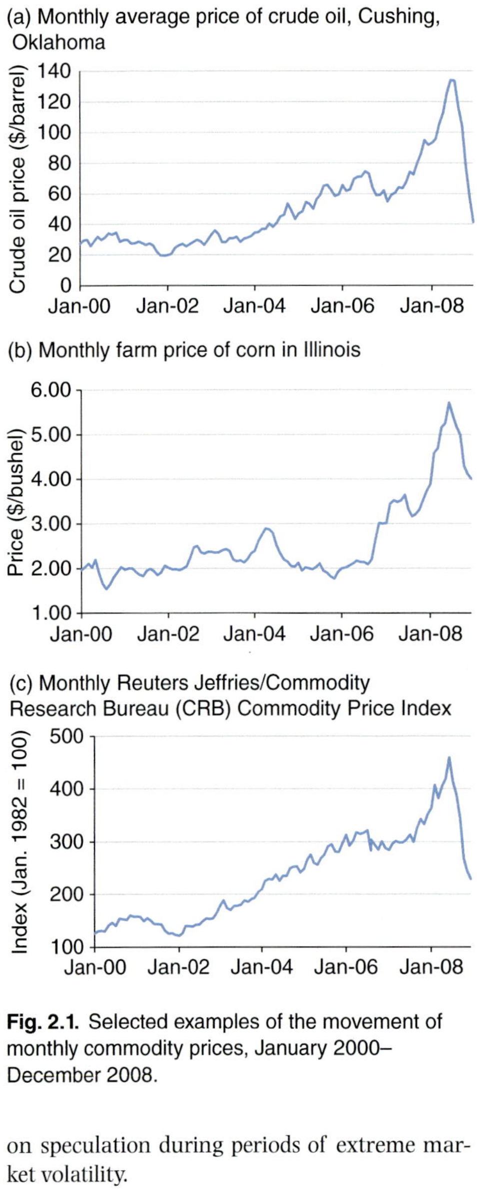

Fig. 2.1. Selected examples of the movement of monthly commodity prices, January 2000–December 2008.

on speculation during periods of extreme market volatility.

2.2 Conceptual Errors

As noted in the introduction, bubble proponents argue that large investment flows, through index-type investments, resulted in unjustified upward pressure on commodity prices. Not only was the pressure unjustified according to bubble proponents, but it also caused very large overvaluations of commodities. For example, Fadel Gheit, Managing Director and Senior Oil Analyst for Oppenheimer & Co. Inc., made the following statement while testifying before the US House of Representatives in June 2008:

> I firmly believe that the current record oil price in excess of $135 per barrel is inflated. I believe, based on supply and demand fundamentals, crude oil prices should not be above $60 per barrel ... There were no unexpected changes in industry fundamentals in the last 12 months, when crude oil prices were below $65 per barrel. I cannot think of any reason that explains the run-up in crude oil price, beside excessive speculation.
>
> (Gheit, 2008)

While bubble arguments may seem sensible on the surface, they contain conceptual errors that reflect a fundamental and basic misunderstanding of how commodity futures and OTC derivative markets actually work.

The first and most fundamental error is to equate money flows into futures and derivatives markets with demand, at least as economists define the term. Investment dollars flowing into either the long or short side of futures or derivative markets is not the same thing as demand for physical commodities. Our esteemed predecessor at the University of Illinois, Tom Hieronymus, put it this way: 'For every long there is a short, for everyone who thinks the price is going up there is someone who thinks it is going down, and for everyone who trades with the flow of the market, there is someone trading against it' (Hieronymus, 1977, p. 302). These are zero-sum markets where all money flows must by definition net to zero. It makes as much logical sense to call the long positions of index funds new 'demand' as it does to call the positions on the short side of the same contracts new 'supply.'

An important and related point is that a very large number of futures and derivative contracts can be created at a given price level. In theory, there is no limit. This is another way of saying that flows of money, no matter how large, do not necessarily affect the futures price of a commodity at a given point in time. Prices will change if new information emerges that causes market participants to revise their estimates of physical supply and/or demand. Note that a contemporaneous correlation can exist between money flows (position changes) and price changes if information on fundamentals is changing at the same time. Simply observing that large investment has flowed into the long side of commodity futures markets at the same time as prices have risen substantially (or the reverse) does not necessarily prove anything. This is more than likely the classical statistical mistake of confusing correlation with causation. One needs a test that accounts for changes in money flow and fundamentals before a conclusion can be reached about the impact of speculation.

It should be said that the previous argument assumes all market participants are equally informed. When this is not the case, it is rational for participants to condition demands on both their own information and information about other participants' demands that can be inferred ('inverted') from the futures price (Grossman, 1986). The trades of uninformed participants can impact prices in this more complex model if informed traders mistakenly believe that trades by uninformed participants reflect valuable information. An argument along these lines can be applied to the rise of index funds in commodity markets. It is possible that traders interpreted the large order flow of index funds on the long side of the market as a reflection of valuable private information about commodity price prospects, which would have had the effect of driving prices higher as these traders subsequently revised their own demands upward. Given the publicity that accompanied index fund entry into commodity futures markets and the transparency of their trading methods, it is highly doubtful that this happened on a wide-enough scale in recent years to consistently drive price movements (more on this in a later discussion of noise trading).

The second conceptual error is to argue that index fund investors artificially raise both futures and cash commodity prices when they only participate in futures and related derivatives markets. In the short-run, from minutes to a few days, commodity prices typically are discovered in futures markets and price changes are passed from futures to cash markets (e.g. Garbade and Silber, 1983). This is sensible because trading can be conducted more quickly and cheaply in futures compared to cash markets. However, longer-term equilibrium prices are ultimately determined in cash markets where buying and selling of physical commodities must reflect fundamental supply-and-demand forces. This is precisely why all commodity futures contracts have some type of delivery or cash settlement system to tie futures and cash market prices together. Of course, delivery systems do not always work as well as one would hope (Irwin *et al.*, 2008).

It is crucial to understand that there is no change of ownership (title) of physical quantities until delivery occurs at or just before expiration of a commodity futures contract. These contracts are financial transactions that only rarely involve the actual delivery of physical commodities. In order to impact the equilibrium price of commodities in the cash market, index investors would have to take delivery and/or buy quantities in the cash market and hold these inventories off the market. There is absolutely no evidence of index fund investors taking delivery and owning stocks of commodities. Furthermore, the scale of this effort would have had to have been immense to manipulate a worldwide cash market as large as the crude oil market, and there simply is no evidence that index funds were engaged in the necessary cash market activities.

This discussion should make it clear that it is wrong to draw a parallel (e.g. Masters and White, 2008) between index fund positions and past efforts to 'corner' commodity markets, such as the Hunt brothers' effort to manipulate the silver market in 1979–1980. The Hunt brothers spent tens of millions of dollars buying silver in the cash market, as well as accumulating and financing huge positions in the silver futures market (Williams, 1995). All attempts at such corners eventually have to buy large, and usually increasing, quantities in the cash market. As Tom Hieronymus noted so colorfully, there is always a corpse (inventory) that has to be disposed of eventually. Since there is no evidence that index funds had any participation in the delivery process of commodity futures markets or the cash market in general, there is no obvious reason to expect their trading to have impacted equilibrium cash prices.

A third conceptual error made by many bubble proponents and, unfortunately, many other observers of futures and derivatives markets is an unrealistic understanding of the trading activities of hedgers and speculators. In the standard story, hedgers are benign risk-avoiders and speculators are active risk-seekers. This ignores nearly a century of research by Holbrook Working, Roger Gray, Tom Hieronymus, Lester Telser, Anne Peck, and others, showing that the behavior of hedgers and speculators is actually better described as a continuum between pure risk avoidance and pure speculation. Nearly all commercial firms labeled as 'hedgers' speculate on price direction and/or relative price movements, some frequently, others not as frequently. In the parlance of modern financial economics, this is described as hedgers 'taking a view on the market' (e.g. Stulz, 1996). Apparently, there is also some contamination in the non-commercial category, with 'speculators' engaged in hedging activities. This problem is highlighted in the recent Commodity Futures Trading Commission (CFTC) report on swap dealers and index traders, which included the statement that, 'The current data received by the CFTC classifies positions by entity (commercial versus noncommercial) and not by trading activity (speculation versus hedging). These trader classifications have grown less precise over time, as both groups may be engaging in hedging and speculative activity' (CFTC, 2008b, p. 2).

What all this means is that the entry of index funds into commodity futures markets did not disturb a sterile textbook equilibrium of pure risk-avoiding hedgers and pure risk-seeking speculators, but instead the funds entered a dynamic and ever-changing 'game' between commercial firms and speculators with various motivations and strategies. Since large commercial firms can take advantage of information gleaned from their far-flung cash market operations, it is not unreasonable to expect that these firms have a trading advantage compared to all but a few very large speculators.[3] The following passage from a recent article on Cargill, Inc. (Davis, 2009) corroborates this view of the operation of commodity futures markets:

> Wearing multiple hats gives Cargill an unusually detailed view of the industries it bets on, as well as the ability to trade on its knowledge in ways few others can match. Cargill freely acknowledges it strives to profit from that information. "When we do a good job of assimilating all those seemingly unrelated facts," says Greg Page, Cargill's chief executive, in a rare interview, "it provides us an opportunity to make money ... without necessarily having to make directional trades, i.e. outguess the weather, outguess individual governments."
>
> (Davis, 2009)

This sheds an entirely different light on the entry of large index fund speculators into commodity futures and derivatives markets. Large hedgers are no innocents in this game and their economic interests are not easily harmed by new entrants.

2.3 Inconsistent Facts

In addition to logical errors, a number of facts about the situation in commodity markets are inconsistent with the arguments of bubble proponents. To begin, if speculation drove futures prices consistently above fundamental values, the available data indicate it was not obvious in the relative level of speculation to hedging. The statistics on long-only index fund trading reported in the media and discussed at Congressional hearings tend to view speculation in a vacuum – focusing on absolute position size and activity. As first pointed out by Working (1960), an objective analysis of futures market activity must consider the balance between speculators and commercial firms hedging market risks. A key insight from this framework is that speculation can only be considered 'excessive' relative to the level of hedging activity in the market.[4]

Weekly *Commitments of Traders* (COT) data provided by the CFTC are enlightening in this regard. Table 2.1 shows the division of open interest for nine commodity futures markets, averaged for the first 3 months of 2006 and 2008.[5] The four basic hedging and speculative positions are: (i) HL = hedging long = commercial long positions; (ii) HS = hedging short = commercial short positions; (iii) SL = speculation long = non-commercial long + index trader long positions; and (iv) SS = speculation short = non-commercial short + index trader short positions. Note that index fund traders are allocated almost exclusively to the SL category in Table 2.1 and that HL + SL = HS + SS.[6]

As expected, Table 2.1 reveals that long speculation – driven by index funds – increased sharply in all but one of the nine commodity futures markets over January 2006 through April 2008.[7] In four of the eight markets with an increase in long speculation (corn, soybeans, soybean oil, and cotton), the increase in short hedging actually exceeded the increase in long

Table 2.1. Speculative and hedging positions (number of contracts) in agricultural futures markets, first quarter of 2006 and 2008.[a] (From Sanders *et al.*, 2008a.)

Market[b]	HL	HS	SL	SS
Corn				
2006	328,362	654,461	558,600	208,043
2008	598,790	1,179,932	792,368	182,291
Change	270,428	525,471	233,768	–25,752
Soybeans				
2006	126,832	192,218	183,105	107,221
2008	175,973	440,793	351,379	74,844
Change	49,141	248,575	168,274	–32,377
Soybean oil				
2006	66,636	124,134	92,515	35,599
2008	121,196	228,515	128,546	25,844
Change	54,560	104,381	36,032	–9,755
CBOT wheat				
2006	57,942	213,278	251,926	92,148
2008	70,084	240,864	300,880	121,578
Change	12,141	27,585	48,954	29,430
KCBT wheat				
2006	43,993	110,601	80,158	13,560
2008	46,459	96,556	67,827	15,767
Change	2,466	–14,045	–12,330	2,207
Cotton				
2006	41,582	108,085	86,777	21,824
2008	107,826	296,434	200,773	18,918
Change	66,244	188,349	113,995	–2,906
Live cattle				
2006	54,549	128,951	129,786	45,305
2008	34,970	144,549	198,211	80,303
Change	–19,579	15,599	68,425	34,998
Feeder cattle				
2006	10,707	17,725	20,769	10,632
2008	6,310	13,435	28,284	18,111
Change	–4,397	–4,290	7,515	7,479
Lean hogs				
2006	15,949	65,438	93,522	40,036
2008	36,825	113,971	149,415	69,055
Change	20,876	48,533	55,893	29,019

[a]HL, hedging, long; HS, hedging, short; SL, speculating, long; SS, speculating, short. The data reflect average positions in the first calendar quarter of 2006 and 2008, respectively. Open interest is aggregated across futures and options, with options open interest delta-adjusted to a futures equivalent basis.
[b]CBOT, Chicago Board of Trade; KCBOT, Kansas City Board of Trade.

speculation. Corn provides a pertinent example. Speculative buying in corn, which includes commodity index funds for this analysis, increased by nearly 250,000 contracts; but selling by commercial firms involved in the production and processing of corn increased by an even greater amount, around 500,000 contracts. What this means is that long speculators (as a group) must have been trading with short hedgers. Working (1960) argued that this was beneficial to overall market performance since speculators provide liquidity and risk-bearing capacity for hedgers.

In the other four markets with an increase in long speculation (Chicago Board of Trade (CBOT) wheat, live cattle, feeder cattle, and lean hogs), the increase in short hedging was less than the increase in long speculation. Live cattle provides a pertinent example here. Speculative buying in cattle, again including commodity index funds, increased by nearly 70,000 contracts;

whereas selling by commercial firms increased by only about 16,000 contracts. In this situation the bulk of the increase in long speculation had to be absorbed by an increase in short speculation. Working (1960, p. 210) argued that trading between speculators generally was 'unneeded' and reflected either 'entry into the market of a considerable group of inexpert or ill-informed speculators' or 'recognition by one group of speculators of significant economic conditions or prospects that are currently being ignored by other, equally expert and generally well-informed, speculators.' Either case could result in a deterioration of market performance. However, Sanders *et al.* (2008a) show that the observed increase in speculation for these markets was still well within historical bounds for commodity futures markets. Even higher levels of speculation have been observed in the past without adverse consequences for market performance.

In sum, observed speculative levels in commodity futures markets since early 2006, even after accounting for index trader positions, either did not exceed the hedging needs of commercial firms or did not exceed historical norms for the level of speculation relative to hedging needs. Simply put, there is no compelling evidence that speculation was 'excessive.'

The second inconsistent fact is that price movements in futures markets with substantial index fund investment were not uniformly upward through the spring of 2008. Panel A in Table 2.2 shows the increase in commodity futures prices over January 2006–April 2008 for the same nine markets as in Table 2.1. The spectacular price increases were concentrated in grain and oilseed markets, while prices in other markets either increased moderately or declined. It is especially interesting to note that prices either dropped or rose only slightly in the markets with the highest level of speculation relative to hedging (Table 2.1: live cattle, feeder cattle, and lean hogs). Figure 2.2 reveals the same pattern in a different form. Here the position of commodity index traders over time is plotted as a percentage of total market open interest. The highest concentration of index fund positions was often in livestock markets, the very markets without large price increases through the spring of 2008. It is difficult to rationalize why index fund speculation would have little or no impact in commodity futures markets with the highest concentration of index positions, relative to either hedging positions or total open interest, yet have a large impact in the markets with the lowest concentration.

The third inconsistent fact is that high prices were also observed in commodity markets not connected to index fund investment. Panels B and C in Table 2.2 provide four examples.[8]

Table 2.2. Change in commodity prices, January 3 2006–April 15 2008.[a]

Commodity	January 2006	April 2008	Change (%)
Panel A: Futures markets included in popular indexes			
Corn	$2.20/bushel	$6.06/bushel	175
Soybeans	$6.28/bushel	$13.80/bushel	120
Soybean oil	22.96¢/lb	62.52¢/lb	172
CBOT wheat	$3.46/bushel	$8.96/bushel	159
KCBOT wheat	$3.90/bushel	$9.50/bushel	136
Cotton	55.24¢/lb	75.23¢/lb	36
Live cattle	$96.37/cwt	$91.57/cwt	–5
Feeder cattle	$114.00/cwt	$103.95/cwt	–9
Lean hogs	$64.65/cwt	$71.65/cwt	11
Panel B: Futures markets not included in popular indexes			
Rough rice	$8.27/lb	$22.17/lb	168
Fluid milk	$12.65/cwt	$17.29/cwt	37
Panel C: No futures markets			
Apples fresh use	$0.26/lb	$0.41/lb	58
Edible beans	$19.30/cwt	$34.40/cwt	78

[a]CBOT, Chicago Board of Trade; ¢, cents; cwt, 100 pounds; KCBOT, Kansas City Board of Trade; lb, pounds. All prices refer to the relevant nearby futures price except apples and edible beans, which are monthly prices received by farmers.

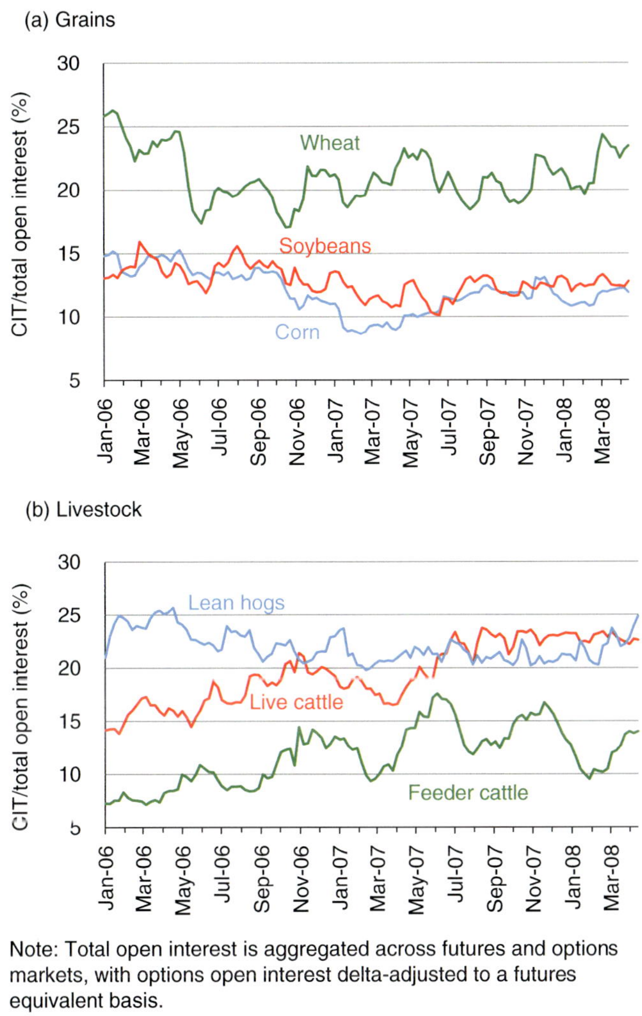

Fig. 2.2. Proportion of open interest held by commodity index traders (CITs) in grain and livestock futures markets, January 2006–June 2008. (From Sanders *et al.*, 2008a.)

Rough rice futures and fluid milk futures are not included in popular commodity indices tracked by index funds, but prices in these two markets increased 162% and 37%, respectively, over January 2006–April 2008. Apples for fresh use and edible beans do not have futures markets, and thus no index fund investment, yet prices in these markets increased 58% and 78%, respectively, over the same time interval. If index fund speculation caused a bubble in commodity prices, why then did prices increase substantially in commodity markets without any index fund activity?

A fourth inconsistent fact has to do with inventories for storable commodities. Following Krugman (2008), Fig. 2.3 illustrates market equilibrium for a storable commodity with and without a price bubble. The standard equilibrium occurs at the intersection of the supply-and-demand curves and results in a P_E. Now assume there is a bubble in the market that pushes price above equilibrium to P_B. At this inflated price the quantity supplied exceeds quantity demanded and the excess shows up as a rise in inventories. We should therefore observe an increase in inventories when a bubble is present in storable commodity markets. In fact, inventories for corn, wheat, and soybeans fell sharply from 2005 through 2007. Inventories of other commodities, such as

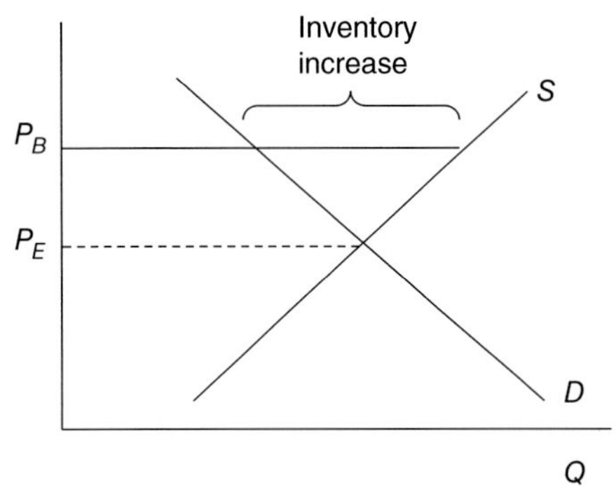

Fig. 2.3. Theoretical impact of a price bubble in a storable commodity market. *D*, demand; P_B, price bubble; P_E, price equilibrium; *Q*, quantity; *S*, supply.

crude oil, stayed relatively flat or declined modestly until very recently. The lack of a notable build-up in commodity inventories is one more reason to be skeptical that a large bubble developed in commodity futures prices.

A fifth inconsistent fact is the nature of commodity index trading. The literature on 'noise traders' shows that a group of uninformed traders can consistently push prices away from fundamental value only if their market opinions are unpredictable, with the unpredictability serving as a deterrent to arbitrage (e.g. De Long *et al.*, 1990). This notion seems unlikely given the ease with which other large traders can trade against index fund positions. Index funds do not attempt to hide their current position or their next move. Generally, funds that track a popular commodity index (e.g. Goldman Sachs Commodity Index) publish their mechanical procedures for rolling to new contract months. Moreover, they usually indicate desired market weightings when the index is rebalanced. So the main uncertainty in their trading patterns usually stems from overall in-flow or out-flow of monies associated with the underlying investment vehicle.

The problems created by the mechanical trading of index funds is well illustrated by a recent story (Meyer and Cui, 2009) on problems experienced by the US Oil Fund, LP, the largest exchange-traded crude oil index fund, when rolling positions from one nearby contract to the next:

> "It's like taking candy from a baby," said Nauman Barakat, senior vice president at Macquarie Futures USA in New York. That candy comes out of the returns of investors in the fund. Take Feb. 6, when US Oil moved its 80,000 contracts from March to April at the end of the trading day, selling the March contract and buying April. Because US Oil publishes the dates of its roll in advance, traders knew the switch was coming. At 2 p.m., 30 minutes before closing, trading in New York Mercantile Exchange oil contracts soared, and the price of the April contract narrowed to $4 more than the March contract. Within minutes, that gap had widened and closed at $5.98, according to trading records. As the fund's managers were about to roll their contracts, "suddenly came the awfully extreme move," said one manager. Some said the move is a sign that big trades were placed ahead of US Oil's roll. The price move instantly made it more expensive for US Oil to roll into the April contract and cost the fund about $120 million more than it would have a day earlier.
>
> (Meyer and Cui, 2009)

As the above passage so amply highlights, it is highly unlikely that other well-capitalized speculators, such as commodity trading advisors, hedge funds, and large floor traders, would allow index funds to push futures prices away from fundamental values when index trades are so easily anticipated.

A related point is that large and long-lasting bubbles are less likely in markets where deviations from fundamental value can be readily arbitraged away (easily 'poached' in the terminology of Patel *et al.*, 1991). There are few limitations to arbitrage in commodity futures markets because the cost of trading is relatively low, trades can be executed literally by the minute, and gains and losses are marked-to-the-market daily. Moreover, the finite horizon of futures contracts further diminishes the likelihood that speculative arbitrage is limited (Shleifer and Summers, 1990). This stands in contrast to markets where arbitrage is more difficult, such as residential housing. The low likelihood of bubbles is also supported by numerous empirical studies on the efficiency of price discovery in commodity futures markets (e.g. Zulauf and Irwin, 1998). Where pricing problems have been documented, they are typically associated with the delivery period of particular commodity futures contracts. However, as noted by the CFTC in a recent background memorandum on the application of its emergency powers, even this type of problem has only risen to an 'emergency' level three times since the Commission was founded in 1974 (CFTC, 2008a).

2.4 Empirical Tests

The preceding discussion focuses on empirical facts that are inconsistent with substantial bubbles in commodity futures prices. When considered as a whole, these facts build a persuasive case against bubbles. However, the facts are largely circumstantial, since they tend to rely on indirect evidence. Bubble proponents can then argue that 'this time is different' even if the links between commodity money flows and bubbles are not fully understood. This is an especially difficult argument to settle because the one variable that can provide definitive evidence about the level of commodity prices – fundamental value – is unobservable. It is like politics; everyone has an opinion.

While fundamental value is unobservable, all is not lost. It is still possible to conduct empirical tests of the hypothesis that money flows from index funds aided and abetted the recent boom and bust in commodity prices. This can be done by running standard Granger causality tests between futures price changes and position changes in commodity futures markets. These tests establish whether lagged position changes help to forecast current futures price changes.[9] Sanders *et al.* (2004), Bryant *et al.* (2006), Gorton *et al.* (2007), and Sanders *et al.* (2008b) conduct Granger causality tests using publicly available data on positions of commercial, non-commercial, and non-reporting trader groups from the weekly COT report published by the CFTC.[10] A typical set of results, drawn from Sanders *et al.* (2008b), is presented in Table 2.3. A statistically significant relationship between the movement of commodity futures prices and measures of position change is found in only five out of 30 cases. In other words, position changes by COT trader groups help forecast futures price movements in only 16% of the cases, hardly more than what one would expect based on pure randomness. And the evidence is even slimmer if results are limited to non-commercial traders (speculators).

The previously cited studies cast considerable doubt on the value of position changes for any group in consistently forecasting futures price movements. However, these studies also use publicly reported COT data, which are aggregated across all contracts and reported only on a weekly or monthly basis. This may limit the power of Granger causality tests because positions cannot be matched precisely to contract maturity months and positions cannot be tracked over daily intervals. Some have argued that if speculator positions do impact returns it is most likely over time horizons shorter than a week (Streeter and Tomek, 1992).

The Interagency Task Force on Commodity Markets led by the CFTC recently conducted thorough Granger causality tests for the crude oil futures market using non-public data on the daily positions of commercial and non-commercial

Table 2.3. Granger causality test results for Commodity Futures Trading Commission (CFTC) trader categories; positions do not lead returns, 1995–2006. (From Sanders *et al.*, 2008b.)

	p-values for hypothesis test		
Market[a]	Commercials	Non-commercials	Non-reporting
Wheat CBOT	0.01	0.18	0.54
Wheat KCBOT	0.03	0.24	0.71
Wheat MGE	0.63	0.15	0.76
Corn	0.35	0.79	0.33
Soybeans	0.83	0.05	0.78
Soybean oil	0.24	0.30	0.94
Soybean meal	0.70	0.93	0.61
Lean hogs	0.05	0.34	0.08
Live cattle	0.75	0.83	0.48
Feeder cattle	0.10	0.16	0.23

[a]CBOT, Chicago Board of Trade; KCBOT, Kansas City Board of Trade; MGE, Minneapolis Grain Exchange.

traders (ITFCM, 2008). Daily price changes and position changes for commercial and non-commercial traders, as well as various sub-groups of traders, were examined over January 2003–June 2008. Consistent with the findings in other studies, there was no evidence that daily position changes by any of the trader sub-categories systematically led crude oil futures price changes over the full sample period. This result held for all categories of speculators tracked by the CFTC: non-commercial traders in total, hedge funds, swap dealers, and non-commercial traders combined with swap dealers. At least in the crude oil futures markets, Granger causality test results are unaffected by the use of daily versus weekly data or position changes for sub-groups of traders. This bolsters the findings from other studies that did not have access to such detailed data on trader positions.

Bubble proponents can still point out that none of the above-referenced studies tested specifically whether commodity index trader positions help to forecast price movements over the last several years. Aulerich *et al.* (2009) provide just this type of evidence for 12 commodity futures markets. They conducted Granger causality tests using non-public data from the CFTC on the daily positions of commodity index traders over January 2000 through July 2008. A unique feature of this study is that the authors were able to extend the series on commodity index positions back through the entire sample under study for each of the 12 markets. Aulerich *et al.* found only a few cases where index trader position changes helped to forecast price changes in commodity futures markets. When significance was found, the size of the estimated price impact was small. These findings also held when the sample was broken into sub-periods.

While it is always possible to dither over the power of Granger causality tests or whether specifications adequately control for changing fundamentals, the evidence to date leads to a high degree of skepticism that positions for any group in commodity futures markets, including index traders, consistently forecast futures price changes (this will not be true for skilled individual traders within a group).

2.5 Lessons from History

A pervasive theme running through the history of US futures markets is skepticism or out-and-out hostility towards speculators (Jacks, 2007).[11] Rapidly increasing or decreasing commodity prices at various times over the last 125 years have been accompanied by assorted attempts to curtail speculation or control prices. For example, just after World War II, soaring grain futures prices, especially for wheat, attracted political attention. President Truman proclaimed that, 'The cost of living in this country must not be a football to be kicked around by grain gamblers,' and ordered the Commodity Exchange Authority (precursor to today's CFTC) to require futures exchanges to raise margins to 33% on all speculative positions, a truly extraordinary level. In a statement that echoes those being made today, President Truman added, 'If the grain exchanges refuse, the government may find it necessary to limit the amount of trading.'[12]

In the boldest move against speculators in US commodity futures, trade in onion futures was banned by the US Congress in 1958. The ban, actually still in place, was due to the widespread belief that speculative activity created excessive price variation (Working, 1963). Again, in language very similar to that heard today, a Congressional report stated that 'speculative activity in the futures markets causes such severe and unwarranted fluctuations in the price of cash onions as to require complete prohibition of onion futures trading in order to assure the orderly flow of onions in interstate commerce.'[13]

The experience of the last time period with a comparable level of structural change in commodity markets, 1972–1975, is particularly instructive. US and international commodity markets experienced a period of rapid price increases from 1972 to 1975, setting new all-time highs across a broad range of markets. These price increases were often blamed on speculative behavior associated with the 'tremendous expansion of trading in futures in a wide range of commodities' (Cooper and Lawrence, 1975, p. 702).[14] Following these price increases, public and political pressure to curb speculation resulted in a number of regulatory proposals and the upward adjustment of futures margin requirements (Hieronymus, 1977; Rainbolt, 1977; Tomek, 1985). These changes were accompanied by even more drastic measures – such as federal price controls and an embargo against soybean exports – aimed at lowering commodity price levels.

The actions used to reign in supposedly damaging speculation in the past run the gamut from requiring futures exchanges to raise margins to an outright ban on futures trading. There is little historical evidence that measures to curtail speculation had the desired effect on market prices. For instance, there is no historical evidence that directives to increase futures margins were effective at lowering overall price levels. The only consistently documented impact of the higher margin requirements is a decline in futures trading volume due to the increased cost of trading (Fishe and Goldberg, 1986; Peck and Budge, 1987; Hardouvelis and Kim, 1996).

Finally, it is important to note the historical pattern of attacks upon speculation. Petzel (1981, p. 117) commented that:

> In periods of rising prices (e.g. the early 1920s, the Korean War, inflation, and the 1970s) grain speculators have been accused of increasing the prices of agricultural commodities artificially. During the early 1930s when agricultural prices were low, grain speculators were accused of depressing prices.
>
> (Petzel, 1981, p. 117)

Market cycles seem to be accompanied by a predictable pattern of speculative complaints: when prices are exceptionally low, natural sellers in the market, such as farmers, complain that speculators are the problem and when prices are exceptionally high, natural buyers in the market – consumers and processors – complain about speculators. While his focus was a relatively obscure episode in the 1925 wheat market, the conclusion reached by Petzel (1981, p. 126) applies with equal force today: 'It is all too easy after suffering an economic loss to look for the villain in the piece. In 1925 the public found its villains and conspirators in the large speculators.'

2.6 Conclusions

There is little evidence that the recent boom and bust in commodity prices was driven by a speculative bubble. If speculation by long-only index funds did impact commodity futures prices, it is not evident in the empirical evidence available to date. Economic fundamentals, as usual, provide a better explanation for the movements in commodity prices. The main factors driving prices up in the energy markets included strong demand from China, India, and other developing nations, a leveling out of crude oil production, a decrease in the responsiveness of consumers to price increases, and US monetary policy (Hamilton, 2008). In the grain markets, factors driving up prices also included demand growth from developing nations and US monetary policy, as well as the diversion of row crops to biofuel production and weather-related production shortfalls (Trostle, 2008). The favorable demand factors were reversed in quick order due to the recent financial market meltdown and burgeoning worldwide recession, leading to large price drops across the board in commodity futures markets (Good and Irwin, 2008). The complex interplay between these factors and how they impact commodity prices is often difficult to grasp in real time and speculators have historically provided a convenient scapegoat for frustration with rapidly rising and falling prices.[15]

Legislative proposals currently being considered may in fact curtail speculation – through reduced volume of trade – but the initiatives could severely compromise the ability of commodity markets to accommodate the needs of firms to manage price risks. In particular, limiting the participation of index fund investors would rob the markets of an important source of liquidity and risk-bearing capacity at a time when both are in high demand. The net result is that commodity futures markets will become less-efficient mechanisms for transferring risk from parties who do not want to bear it to those that do, creating added costs that ultimately get passed back to producers in the form of lower prices and back to consumers as higher prices.

The recent attacks on speculation in commodity markets hark back to an earlier era. For most of the past 30 years a consensus seemed to have been reached among policy makers that speculation played a valuable and important role in commodity futures markets. Writing in the 1970s, Tom Hieronymus had this to say about the matter:

> For many years the anti-futures trading arguments tended to prevail so that speculation was treated as a necessary evil that accompanied the desirable hedging process. During the last decade the balance appears to have shifted so that a favorable view is more widely held. It is doubtful that the favorable view is yet in the majority but it is generally held by students of futures markets and increasingly held by members of Congress and the CFTC.
>
> (Hieronymus, 1977, p. 298)

Much to the surprise of agricultural economists, there is little doubt after the political uproar of the last year that a majority of the public still does not hold a favorable view of speculation. It is yet to be determined whether members of the US Congress hold the same view and whether this portends a return to the anti-futures trading environment of an earlier era.

Acknowledgments

Todd Petzel and seminar participants at the University of Illinois, Guelph University, and the Economic Research Service of the USDA (United States Department of Agriculture) provided valuable comments on earlier versions of the paper. Funding support from the Aurene T. Norton Trust is gratefully acknowledged.

Notes

[1] Original citation: Irwin, S.H., Sanders, D.R. and Merrin, R.P. (2009) Devil or angel? The role of speculation in the recent commodity price boom (and bust). *Journal of Agricultural and Applied Economics* 41, 377–391. Reprinted by permission of Cambridge University Press and the Southern Agricultural Economics Association.

[2] In reality, a variety of investment instruments are lumped under the heading 'commodity index fund.' Individuals may enter directly into over-the-counter (OTC) contracts with swap dealers to gain the desired exposure to returns from a particular index of commodity prices. Some firms also offer investment funds whose returns are tied to a commodity index. Exchange-traded funds (ETFs) and structured notes (ETNs) have also recently been developed to make it even easier to gain commodity exposure. ETFs and ETNs trade on securities exchanges in the same manner as stocks on individual companies. See Engelke and Yuen (2008) and CFTC (2008b) for further details.

[3] Hieronymus (1977) argued that large commercial firms dominated commodity futures markets and speculators tended to be at a disadvantage. Based on his theoretical analysis, Grossman (1986, p. S140) asserted, 'It should come as no surprise if a study of trading profit finds that traders representing large firms involved in the spot commodity (i.e. commercial traders) make large trading profits on futures markets.' In the classic empirical study on this subject, Hartzmark (1987) showed that large commercial firms in six of seven futures markets make substantial profits on their futures trades.

[4] Peck (1979–1980, p. 339) provides a succinct re-statement of Working's argument:

> Taken together, these analyses reaffirm the fundamental importance of hedging to futures markets and dependence of total activity upon hedging needs. The results also lend support to the Working definition of an appropriate measure of hedger demands upon a market. Net hedging is not the most useful view of the demands commercial users make on a market. Speculation is needed to offset both long hedging and short hedging. Only coincidentally are long and short hedgers sufficiently alike in date and amount to be offsetting, although increased balance increases the probability of such correspondence and differences in seasonal needs between long and short hedgers decreases this probability. The appropriate measure of minimum required speculation must at least begin with total hedging demand.

[5] Note that total open interest consists of futures open interest and delta-adjusted options open interest.

[6] Non-reporting trader positions are allocated to the commercial, non-commercial, and index trader categories in the same proportion as that which is observed for reporting traders (see Sanders *et al.*, 2008a).

[7] There is an important omission from Table 2.1 – crude oil futures. As the CFTC noted when it first began publishing data on index fund positions, it is difficult to separate out index fund transactions in energy markets because of the degree to which many firms in these markets engage in multiple trading activities that fall into different classifications and the degree to which firms engage in internal netting of these activities. The special swap dealer survey (CFTC, 2008b) does provide an estimate of index trader positions in the crude oil futures market; however, the data are limited to a 6-month period from December 31 2007 to June 30 2008 and reported only on a net long basis. Computations for crude oil that parallel those reported in Table 2.1 can be made only by assuming that short positions for index funds are zero.

[8] The four markets were not selected at random, but instead represent markets that generally have low-cross price elasticities relative to the nine markets in Panel A. If the selected markets had high cross-price

elasticities, then observed price increases could have been due to linkages with the markets in Panel A (and possibly bubble effects in these markets) rather than fundamental factors specific to the selected markets or fundamental factors common to all the markets.

[9] Granger causality tests reflect the basic idea that if event *X* causes event *Y*, then event *X* should precede event *Y* in time. These tests require careful interpretation if the null hypothesis of no causality (no statistical prediction) is rejected (Hamilton, 1994). A statistical correlation may be observed between *X* and *Y* when in reality an omitted variable *Z* is the true cause of both *X* and *Y*. Hamilton (1994, p. 308) suggests it is better to describe 'Granger causality' tests between *X* and *Y* as tests of whether *X* helps forecast *Y* rather than whether *X* causes *Y*. He notes that the tests may have implications for causality in the conventional sense, but only in conjunction with other assumptions.

[10] In a work well ahead of its time, Petzel (1981) conducted Granger causality tests between the daily position changes of three groups of speculators and price changes for the May 1925 wheat futures contract at the Chicago Board of Trade. Foreshadowing later results, he did not find any evidence that lagged position changes helped to forecast current price changes.

[11] See Stout (1999) for an in-depth discussion of the legal and regulatory history of opposition to speculation in the USA.

[12] Quoted in Peck and Budge (1987, p. 172).

[13] Quoted in Working (1963, p. 18).

[14] It is fascinating to observe the similarity of the current public debate about speculation and the one that followed the mid-70s commodity boom. For instance, Labys and Thomas (1975, p. 287) motivate their paper with words that could have been written in 2008 instead of 1975:

> This paper analyses the instability of primary commodity prices during the recent period of economic upheaval, and determines the extent to which this instability was amplified by the substantial increase in futures speculation which also occurred. Of particular interest is the degree to which this speculation rose and fell with the switch of speculative funds away from traditional asset placements and towards commodity futures contracts.

[15] The origin of the word 'scapegoat' is of more than passing interesting in the present context. In ancient Israel, the high priest confessed all the sins of the children of Israel on the Day of Atonement over the head of a live goat. As a symbol of their sins, the goat was then sent into the wilderness to perish.

References

Aulerich, N.M., Irwin, S.H. and Garcia, P. (2009) The price impact of index funds in commodity futures markets: evidence from the CFTC's daily large trader reporting system. In: *Proceedings of the NCCC-134 Conference on Applied Commodity Price Analysis, Forecasting, and Market Risk Management, St Louis, Missouri, April 20–21*. Available at: https://legacy.farmdoc.illinois.edu/nccc134/conf_2009/pdf/confp19-09.pdf (accessed January 14 2022).

Bryant, H., Bessler, D.A. and Haigh, M.S. (2006) Causality in futures markets. *Journal of Futures Markets* 26, 1039–1057.

CFTC (Commodity Futures Trading Commission) (2008a) *CFTC Emergency Authority Background*. CFTC, Washington, DC.

CFTC (Commodity Futures Trading Commission) (2008b) Staff Report on Commodity Swap Dealers & Index Traders with Commission Recommendations. Available at: https://www.cftc.gov/sites/default/files/idc/groups/public/@newsroom/documents/file/cftcstaffreportonswapdealers09.pdf (accessed January 14 2022).

Cooper, R.N. and Lawrence, R.Z. (1975) The 1972–75 commodity boom. *Brookings Papers on Economic Activity* 3, 671–723.

Davis, A. (2009) Cargill's inside view helps it buck downturn. *The Wall Street Journal*, January 14. Available at: http://online.wsj.com/article/SB123189501407679581.html?mod=rss_whats_news_us (accessed January 14 2022).

De Long, J.B., Shleifer, A., Summers, L.H. and Waldmann, R.J. (1990) Noise trader risk in financial markets. *Journal of Political Economy* 98, 703–738.

Engelke, L. and Yuen, J.C. (2008) Types of commodity investments. In: Fabozzi, F.J., Fuss, R. and Kaiser, D. (eds) *The Handbook of Commodity Investing*. Wiley, Hoboken, New Jersey, pp. 549–569.

Field, S. (2017) Examining the influence of intellectuals on commodity speculation. *Geoforum* 83, 71–79.

Fishe, R.P.H. and Goldberg, L.G. (1986) The effects of margins on trading in futures markets. *Journal of Futures Markets* 6, 261–271.

Garbade, K.D. and Silber, W.L. (1983) Price movements and price discovery in futures and cash markets. *Review of Economics and Statistics* 65, 289–297.

Gheit, F. (2008) Testimony before the Subcommittee on Oversight and Investigations of the Committee on Energy and Commerce, US House of Representatives. June 23. Available at: https://www.google.com/books/edition/Speculative_Investment_in_Energy_Markets/mCo6_xKYnQMC?hl=en&gbpv=1&dq=fadel+gheit+testimony&pg=PA50&printsec=frontcover (accessed January 14 2022).

Good, D. and Irwin, S. (2008) *Implications of Credit Market Problems for Crop Prices. Illinois Farm Economics Update 2008-03*. Department of Agricultural and Consumer Economics, University of Illinois at Urbana-Champaign, Illinois.

Gorton, G.B. and Rouwenhorst, K.G. (2004) Facts and fantasies about commodity futures. National Bureau of Economic Research (NBER) Working Paper No. 10595. Available at: https://www.nber.org/system/files/working_papers/w10595/w10595.pdf (accessed January 14 2022).

Gorton, G.B., Hayashi, F. and Rouwenhorst, K.G. (2007) The fundamentals of commodity futures returns. National Bureau of Economic Research (NBER) Working Paper No. 13249. Available at: https://www.nber.org/papers/w13249 (accessed January 14 2022).

Grossman, S.J. (1986) An analysis of 'insider trading' on futures markets. *Journal of Business* 59, S129–S146.

Hamilton, J.D. (1994) *Time Series Analysis*. Princeton University Press, Princeton, New Jersey.

Hamilton, J.D. (2008) Understanding crude oil prices. Working Paper, Department of Economics, University of California-San Diego, California. Available at: http://dss.ucsd.edu/~jhamilto/understand_oil.pdf (accessed January 14 2022).

Hardouvelis, G.A. and Kim, D. (1996) Price volatility and futures margins. *Journal of Futures Markets* 16, 81–111.

Hartzmark, M.L. (1987) Returns to individual traders of futures: aggregate results. *Journal of Political Economy* 95, 1292–1306.

Hieronymus, T.A. (1977) *Economics of Futures Trading for Commercial and Personal Profit*, 2nd edn. Commodity Research Bureau, New York.

ITFCM (Interagency Task Force on Commodity Markets) (2008) *Interim Report on Crude Oil*. ITFCM, Washington, DC. Available at: https://www.cftc.gov/sites/default/files/idc/groups/public/@newsroom/documents/file/itfinterimreportoncrudeoil0708.pdf (accessed January 14 2022).

Irwin, S.H., Garcia, P., Good, D.L. and Kunda, E.L. (2008) Recent convergence performance of CBOT corn, soybean, and wheat futures contracts. *Choices*, 2nd Quarter 23, 16–21. Available at: http://www.choicesmagazine.org/magazine/article.php?article=26 (accessed January 14 2022).

Irwin, S.H., Sanders, D.R. and Merrin, R.P. (2009) Devil or angel? The role of speculation in the recent commodity price boom (and bust). *Journal of Agricultural and Applied Economics* 41, 77–391.

Jacks, D.S. (2007) Populists versus theorists: futures markets and the volatility of prices. *Explorations in Economic History* 44, 342–362.

Krugman, P. (2008) More on oil and speculation. *The New York Times*, May 13. Available at: http://krugman.blogs.nytimes.com/2008/05/13/more-on-oil-and-speculation/ (accessed January 14 2022).

Labys, W.C. and Thomas, H.C. (1975) Speculation, hedging and commodity price behavior: an international comparison. *Applied Economics* 7, 287–301.

Masters, M.W. (2008) Testimony before the Committee on Homeland Security and Governmental Affairs, United States Senate, May 20. Available at: http://hsgac.senate.gov/public/_files/052008Masters.pdf (accessed January 14 2022).

Masters, M.W. and White, A.K. (2008) The accidental Hunt brothers: how institutional investors are driving up food and energy prices. Available at: https://www.cftc.gov/sites/default/files/idc/groups/public/@swaps/documents/file/plstudy_31_ahb.pdf (accessed January 14 2022).

Meyer, G. and Cui, C. (2009) U.S. Oil Fund finds itself at the mercy of traders. *The Wall Street Journal*, March 6. Available at: http://online.wsj.com/article/SB123629874701846317.html (accessed January 14 2022).

Patel, J., Zeckhauser, R. and Hendricks, D. (1991) The rationality struggle: illustrations from financial markets. *American Economic Review* 81, 232–236.

Peck, A.E. (1979–1980) Reflections of hedging on futures markets. *Food Research Institute Studies* 17, 327–349.

Peck, A.E. and Budge, C. (1987) The effects of extraordinary speculative margins in the 1947–48 grain futures markets. *Food Research Institute Studies* 20, 165–180.

Petzel, T.E. (1981) A new look at some old evidence: the wheat market scandal of 1925. *Food Research Institute Studies* 18, 117–128.

Rainbolt, J.V. (1977) Regulating the grain gambler and his successors. *Hofstra Law Review* 6, 1–25.

Sanders, D.R., Boris, K. and Manfredo, M. (2004) Hedgers, funds, and small speculators in the energy futures markets: an analysis of the CFTC's *Commitments of Traders* reports. *Energy Economics* 26, 425–445.

Sanders, D.R., Irwin, S.H. and Merrin, R.P. (2008a) The adequacy of speculation in agricultural futures markets: too much of a good thing? *Marketing and Outlook Research Report 2008-02*, Department of Agricultural and Consumer Economics, University of Illinois at Urbana-Champaign, Illinois. Available at: https://farmdoc.illinois.edu/publications/the-adequacy-of-speculation-in-agricultural-futures-markets-too-much-of-a-good-thing (accessed January 14 2022).

Sanders, D.R., Irwin, S.H. and Merrin, R.P. (2008b) Smart money? The forecasting ability of CFTC large traders. Working Paper, Department of Agricultural and Consumer Economics, University of Illinois at Urbana-Champaign, Illinois.

Shleifer, A. and Summers, L.H. (1990) The noise trader approach to finance. *Journal of Economic Perspectives* 4, 19–33.

Stout, L.A. (1999) Why the law hates speculators: regulation and private ordering in the market for OTC derivatives. *Duke Law Journal* 48, 701–786.

Streeter, D.H. and Tomek, W.G. (1992) Variability in soybean futures prices: an integrated framework. *Journal of Futures Markets* 12, 705–728.

Stulz, R.M. (1996) Rethinking risk management. *Journal of Applied Corporate Finance* 9, 8–24.

Tomek, W.G. (1985) Margins on futures contracts: their economic roles and regulation. In: Peck, A.E. (ed.) *Futures Markets: Regulatory Issues*. American Enterprise Institute for Public Policy Research, Washington, DC, pp. 143–209.

Trostle, R. (2008) Global agricultural supply and demand: factors contributing to the recent increase in food commodity prices. *Outlook Report* No. WRS-0801, Economic Research Service, US Department of Agriculture. Available at: https://www.ers.usda.gov/publications/pub-details/?pubid=40464 (accessed January 14 2022).

Williams, J. (1995) *Manipulation on Trial: Economic Analysis and the Hunt Silver Case*. Cambridge University Press, Cambridge, UK.

Working, H. (1960) Speculation on hedging markets. *Food Research Institute Studies* 1, 185–220.

Working, H. (1963) Futures markets under renewed attack. *Food Research Institute Studies* 4, 13–24.

Zulauf, C.R. and Irwin, S.H. (1998) Market efficiency and marketing to enhance income of crop producers. *Review of Agricultural Economics* 20, 308–331.

3

New Evidence on the Impact of Index Funds in US Grain Futures Markets[1]

New Author Foreword

We were honestly surprised that the furor over speculation and index funds did not die out after the crash in commodity prices during the second half of 2008. An important signal that we had underestimated the intensity of the controversy was the work of the US Senate Permanent Subcommittee on Investigations led by Senator Carl Levin of Michigan. This committee published one of the very first reports pointing the finger at index investment as a driver of surging crude oil futures prices (USS/PSI, 2006). After investigating energy markets, the committee began looking into wheat prices, which not only spiked to record levels in 2008, but also developed unprecedented levels of non-convergence between cash and futures prices (Henriques, 2008). Committee staff members reached out to Scott in February 2009 and began extensive conversations about the impact of speculation in grain futures markets, and more specifically, about non-convergence problems in wheat. It was probably more than a little naïve given the politics of the situation, but it appeared that the staff were open to the kind of arguments we put forth in our just published 'Devil or Angel' article (see Chapter 2, this volume).

We were taken aback when the committee published its final report in June 2009. It was a blistering attack on index investors, accusing them of causing 'unwarranted changes' in wheat futures prices relative to the cash price of wheat (USS/PSI, 2009). This supercharged the debate about speculative position limits in US commodity futures markets, which had been simmering since the price spike of 2007–2008.

A few days before the Senate report was released, the staff kindly provided an embargoed copy of the report to Scott. He was struck by data that the committee had obtained on commodity index trader (CIT) positions in grain futures markets for 2004–2005 and phoned Dwight to discuss the matter. A significant limitation of the public CIT position data from the US Commodity Futures Trading Commission (CFTC) was the lack of data prior to 2006. We knew this was likely an important constraint because we had other evidence that the build-up in CIT positions was concentrated in the 2 or 3 years preceding 2006 (Sanders *et al.*, 2008). However, this observation was based on obscure CFTC Bank Participation reports that contained swap positions for banks. Even though it turned out that the Bank Participation data were accurate about the timing of the build-up in positions, we were uncertain how much confidence to place in those data.

We decided to quickly get in touch with the Senate staff and ask if we could use the weekly CIT position data over 2004–2005 for corn, soybeans, Chicago wheat, and Kansas City wheat for research purposes. Much to our surprise, we received a spreadsheet file with the data a few weeks later. This was a real coup because the CIT position data were derived from the non-public Large Trader Reporting System of the CFTC. Historically, it has been very hard for academics to get access to this kind of data due to legal restrictions. Obtaining the data through the Senate committee was like a (legal) end-run around the usual process. It was a good thing that we received the data as quickly as we did. Scott and other colleagues at the University of Illinois wrote a scathing commentary on the Senate wheat report that was released in late July 2009. End of favors from the committee staffers!

DOI:10.1079/9781800622104.0003

Another important motivation for this study was a report from the International Food and Policy Research Institute (IFPRI) released in February 2009. The report was entitled 'When Speculation Matters,' and it was authored by Miguel Robles, Maximo Torero, and Joachim von Braun, who was the Director of IFPRI. This study was among the first after the 2007–2008 price spike to use Granger causality tests to examine the relationship between measures of futures market activity and price movements. The authors concluded that speculation might have been influential in driving price movements during 2007–2008 and called for the creation of a major new 'virtual reserve' to offset the influence of excessive speculation. This report created quite a stir when it was released.

With this as background, we set out to do our first econometric analysis of the relationship between CIT positions and price movements in grain futures markets. However, the most important contribution in the article was to show that CIT positions in grain futures markets grew most rapidly in 2004–2005, before the great price spike of 2007–2008. The key empirical facts were: (i) grain futures prices were relatively flat during the massive build-up of index positions in 2004–2005; and (ii) index positions were relatively flat during the grain futures price spike of 2007–2008. These two facts were very difficult to reconcile with the assertion that index trader buying created a massive bubble in grain futures prices in 2007–2008. The timing was simply way off. Hence, it was no surprise when our Granger causality tests did not find consistent evidence of a relationship between index trader positions and grain futures price movements.

We actually completed a first draft of the paper in November 2009 and presented it at a Food and Agriculture Organization of the United Nations (FAO) conference in Rome. We thought that the analysis was strong, and with the new data being used in a study for the first time, figured it would be a cinch to get it published quickly. Wrong again. We tried first at *Economic Letters* and were rejected out of hand. We then tried the *Australian Journal of Agricultural and Resource Economics*, which was a very frustrating process. The first-round reviews were quite positive and it looked like it would sail through to acceptance with minimal changes. Then the second-round comments hit, and hit hard. It was as if one of the reviewers completely changed their mind about the paper between the first- and second-round reviews. The changes demanded by the negative reviewer would have so altered the paper that we gave up and decided to start over at a different journal. We then sent it to the *Canadian Journal of Agricultural Economics*, where it was accepted after two relatively painless rounds of reviews.

The article did not appear in print until 2011, despite the straightforward nature of the analysis and the fact that it was basically completed more than 2 years earlier. Furthermore, this was a very hot publishing area at the time. Some of it was just bad luck but we also began to sense that politics was seeping into the review process. The article turned out to be quite well received and began generating citations almost immediately. To date, the article has received over 150 citations according to Google Scholar. The basic empirical findings have also held up rather nicely in the face of numerous follow-on studies. As an added bonus, we used the index trader positions from the Senate committee in a number of other studies. Not bad!

Abstract

Commodity index trader position data are examined for the years prior to the 2007–2008 commodity price increase. New data from 2004–2005 show that a large increase in commodity index positions occurred in select grain futures markets. However, the increased index participation took place well in advance of the 2007–2008 boom in prices. Granger causality tests fail to find any causal link between commodity index activity and grain futures prices. Furthermore, there is little evidence of an index-induced price bubble using long-horizon regressions.

3.1 Introduction

A worldwide controversy has erupted about the role of 'long-only' index funds in the recent commodity price boom.[2] A variety of commodity investment instruments typically are lumped under the heading 'index fund' (Engelke and Yuen, 2008). Regardless of the form, these instruments have a common goal – provide investors with buy-side exposure to returns from a particular index of commodity prices. Several influential research reports in recent years purport that investors can capture substantial risk premiums and reduce portfolio risk through relatively modest investment in commodity futures positions (e.g. Gorton and Rouwenhorst, 2006).

There are a few undisputable facts about commodity futures markets over 2006–2008. First, inflows into long-only commodity index funds increased throughout 2006–2008 (see Fig. 3.1). According to the most widely quoted industry source (Barclays) index fund investment increased from $90 billion at the beginning of 2006 to a peak of just under $200 billion at the end of 2007. Second, commodity prices increased rather dramatically – 71% as measured by the Commodity Research Bureau (CRB) index – from January 2006 through June of 2008 (see Fig. 3.2). Third, prices declined almost equally dramatically from June 2008 through early 2009 (see Fig. 3.2). The data are clear and not in dispute; it is the interpretation of the interaction among these facts that is controversial.

On one side, some hedge fund managers, commodity end-users and policy makers assert that speculative buying by index funds on such a wide scale created a 'bubble,' with the result that commodity futures prices far exceeded fundamental values during the boom (e.g. Gheit, 2008; Masters, 2008; Masters and White, 2008; USS/PSI, 2009). This has led to new regulatory initiatives to limit speculative positions in commodity futures markets (Acworth 2009a, b). On the other side, a number of economists have expressed skepticism about the bubble theory, citing contrary facts and a lack of rigorous empirical evidence linking index positions to commodity futures price movements (e.g. Krugman, 2008; Pirrong, 2008; Sanders and Irwin, 2008). These economists argue that commodity markets were driven by fundamental factors that

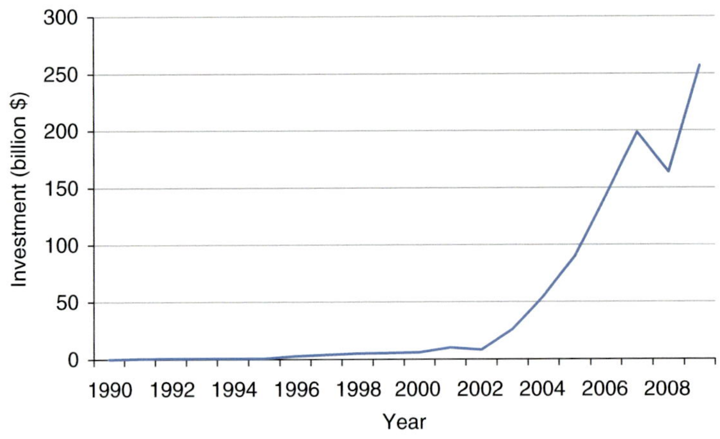

Fig. 3.1. Commodity index fund investment (year end), 1990–2009. (Data from Barclays.)

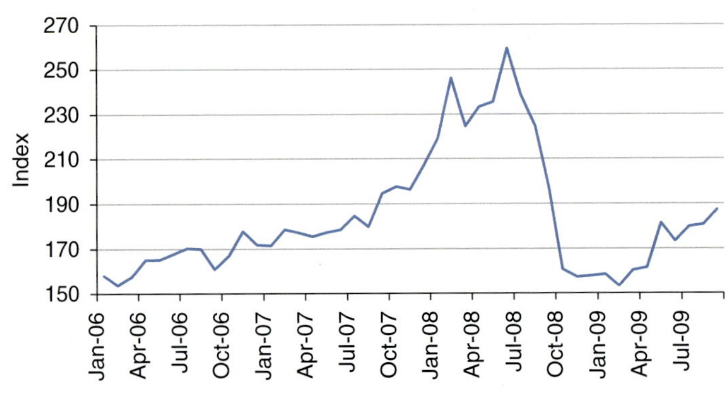

Fig. 3.2. Commodity Research Bureau (CRB) commodity index, January 1990 = 100, January 2006–September 2009.

pushed prices higher. Indeed, McCalla (2009) outlines a number of alternative explanations for rising commodity prices – including macroeconomic factors, supply shocks, and permanent structural shifts in world supply-and-demand conditions. While these fundamental factors may be convincing to economists, they seemed not to have swayed the regulatory and political debate regarding index funds.

The outcome of this debate has potentially broad economic consequences for the marketing, distribution, and pricing of commodity products. Childs and Kiawu (2009) suggest that the increased participation in futures markets by 'non-traditional' investors was one of the causes of the increase in global rice prices and had a detrimental impact on the well-being of rice consumers. More pointedly, Robles *et al.* (2009, p. 7) offer a stark reminder that the efficient pricing of staple food commodities can have wide-ranging implications: 'excess price surges caused by speculation and possible hoarding … could result in unreasonable or unwanted price fluctuations that can harm the poor and result in long-term, irreversible nutritional damage, especially among children.' Along these same lines, an FAO news headline from 2009 reads, 'Financial speculation in basic food commodities played a key role in the 2007–2008 food price crisis which pushed millions of people deeper into hunger …' (FAO, 2009). Given the stakes, it is imperative that policy makers have access to the best possible array of empirical evidence.

In this article, we provide new evidence on the growth and impact of index fund investments in US commodity futures markets using previously unavailable data for the years 2004 and 2005. Data compiled by the US Commodity Futures Trading Commission (CFTC) over 2004–2009 are used to investigate the impact of index traders in US grain futures markets. Previous research suggests that the build-up in index positions was most rapid during 2004–2005 (Sanders *et al.*, 2008), hence the period most likely to show the impact of index traders, if any. Other studies, including Stoll and Whaley (2010) and Sanders and Irwin (2010) have been limited by having access only to 2006 and later observations on index trader positions. The data for 2004–2005 provide valuable new evidence in regard to the ongoing 'bubble debate.'

3.2 Review of Debate

3.2.1 The bubble story

Masters (2008) has interwoven the position and price data to create the oft-told bubble story in US Congressional hearings, painting the activity of index funds as akin to the infamous Hunt brothers' cornering of the silver market. He blames the rapid increase in overall commodity prices from 2006 to 2008 on institutional investors' embrace of commodities as an investable asset class. As noted in the introduction, it is clear that considerable dollars flowed into commodity index funds over this period. The evidence provided by such bubble proponents is limited to anecdotes and the temporal correlation between money flows and prices (e.g. Masters and White, 2008).

Other authors seem to work under the null hypothesis that speculators have an undesirable and somewhat unexplainable impact on market prices. For instance, Robles *et al.* (2009, p. 2) simply claim that 'Changes in supply and demand fundamentals cannot fully explain the recent drastic increase in food prices.' Similarly, a study by the Agricultural and Food Policy Center (2008, p. 32) declares that the 'large influx of money into the markets, typically long positions, has pushed commodities to extremely high levels' and also shows a graphical depiction of investment funds in the Goldman Sachs Commodity Index (GSCI). Alternatively, some analysts have tended to use speculation as a catch-all for those market movements that cannot otherwise be easily explained. For example, Childs and Kiawu (2009, p. 1) report that 'the primary cause of the rise in prices for these commodities from 2006–2008 was rising global incomes, dietary changes, increased use of biofuels, tight grain supplies, and increased participation in futures markets by nontraditional investors.'

While the bubble story and impact of financial speculation is applauded by some US Congressional members and easily absorbed by the public, it glosses over the inherent complexity of price determination in commodity futures markets and the dynamics of trading. Moreover, the lack of rigorous statistical methods generally brings out skepticism in academic circles and arguments against a commodity price bubble.

3.2.2 Arguments against the bubble

While casual bubble arguments are deceptively appealing, they do not generally withstand close examination. Irwin *et al.* (2009) present three logical inconsistencies in the arguments made by bubble proponents as well as five instances where the bubble story is not consistent with observed facts. Here we review these arguments as well as some counter-arguments provided by market observers.

The first logical inconsistency within the bubble argument is that money flows are the same as demand. With equally informed market participants, there is no limit to the number of futures contracts that can be created at a given price level. For each buyer there must also be a seller: index fund buying is no more 'new demand' than the corresponding selling is 'new supply.' Thus, money flows in themselves do not necessarily impact prices. Second, there is no compelling evidence to date that index investors distort cash markets. Index investors are purely involved in a financial transaction using the futures markets; they do not engage in the purchase or hoarding of the cash commodity and any causal linkages between their futures market activity and cash prices is unclear (Headey and Fan, 2008). Index investors do not participate in the futures delivery process or the cash market where long-term equilibrium prices are discovered. Hence, to draw a parallel between index purchases of futures contracts with the Hunt brothers' corner of the silver market – which involved the hoarding of actual physical supplies – is flawed. Lastly, the blanket categorization of speculators as wrongdoers and hedgers as victims of their actions is mistaken. Many market participants are both speculators and hedgers. Seldom is there a clear delineation between hedging and speculating, which makes it difficult to identify the 'victims' (if any) of index participation. Trading dynamics are complex, and it is not easy to understand the interplay between the varied market participants and their motivations for trading, especially in real time. Hence, statements about 'damaging speculation' are at best vague and potentially misguided.

In their rebuttal of the bubble theory, Irwin *et al.* (2009) also identify five areas where the bubble story is not consistent with the observed facts. First, as Krugman (2008) asserts, if a bubble raises the market price of a storable commodity above the true equilibrium price, then stocks of that commodity should increase (much like a government-imposed price floor can create a surplus). Gilbert (2010, p. 408) counters that:

> in the short-to-medium term, inventories are largely predetermined at a level implied by the carryover from previous crop years. With the inventory supply curve near vertical, the increased demand can only be met by an increase in the cash price.
>
> (Gilbert, 2010, p. 408)

Still, stocks were declining, not building, in most grain markets over the multiple-year period of 2006–2008, which is inconsistent with the depiction of a price bubble in these markets.

Second, theoretical models that show uninformed or noise traders impacting market prices rely on the unpredictable trading patterns of these traders to make arbitrage risky (De Long *et al.*, 1990). Because the arbitrage – needed to drive prices to fundamental value – is not riskless, noise traders can drive a wedge between market prices and fundamental values. Importantly, index fund buying is very predictable. That is, index funds widely publish their portfolio (market) weights and roll-over periods. Thus, it seems highly unlikely that other large rational traders would hesitate to trade against an index fund if they were driving prices away from fundamental values.

Third, if index fund buying drove commodity prices higher, then markets without index funds should not have seen prices advance. Again, the observed facts are inconsistent with this notion. Irwin *et al.* (2009) show that markets without index fund participation (fluid milk and rice) and commodities without futures markets (apples and edible beans) also showed price increases over the 2006–2008 period. Headey and Fan (2008) warn against directly comparing commodity markets selected for futures contracts – because they may have characteristics that exacerbate volatility such as relatively inelastic supply and demand – to those commodities without futures markets. Still, Headey and Fan (2008) also cite the rapid increases in the prices for non-securitized commodities such as rubber, onions, and iron ore as evidence that rapid inflation occurred in commodities without futures markets. This would suggest that there were other macroeconomic factors potentially influencing commodity prices. Alternatively, other fundamental shocks – such as rapidly escalating exports and trade shocks – may have driven a number of commodity prices higher over this period (Heady, 2010).

Fourth, speculation is not excessive when correctly compared to hedging demands. Working (1960) argued that speculation must be gauged relative to hedging needs. Utilizing Working's speculative 'T-index,' Sanders *et al.* (2008) demonstrate that the level of speculation in nine commodity futures markets from 2006 to 2008 (adjusting for index fund positions) was not excessive. Indeed, the levels of speculation in all markets examined were within the realm of historical norms. Across most markets, the rise in index buying was more than offset by commercial (hedger) selling.

The fifth observable fact revolves around the impact of index funds across markets. *A priori*, there is no reason to expect index funds to have a differential impact across markets given similar position sizes. That is, if index funds can inflate prices, they should have a uniform impact across markets for the same relative position size. It is therefore difficult to rationalize why index fund speculation would impact one market but not another. Further, one would expect markets with the highest concentration of index fund positions to show the largest price increases. Using cross-sectional tests, Sanders and Irwin (2010) demonstrate that the cross-section of futures market returns have no causal relation with the size of index trader positions. This finding is difficult to rectify with the assertion that index buying represents demand.

Considerable evidence has been gathered that suggests the relationship between price behavior and index fund activity is weak. However, empirical results can vary depending on the time interval used, prices examined, and empirical methods. One common limitation is that the data available for direct empirical tests are largely limited to post-2006 when the CFTC began compiling their commodity index trader reports. While more general tests for bubble-like price behavior (e.g. McQueen and Thorley, 1994; Gilbert, 2009) can examine longer time intervals, these tests use only price data with no ability to establish direct links (if any) between trader groups and price behavior. Here we expand the search for direct links between long-only index traders and futures prices by using an extended data set on index trader positions that include non-public data for 2004–2005, and thereby bring new empirical evidence to the debate.

3.3 Data

Starting in 2007 – in response to complaints by traditional traders about the rapid increase in long-only index money flowing into the markets – the CFTC began reporting the positions held by index traders in 12 commodity futures markets in the *Supplemental Commitments of Traders* (SCOT) report, as an extension to the traditional *Commitments of Traders* (COT) report. According to the CFTC, index trader positions reflect both pension funds that would have previously been classified as non-commercials (speculators) as well as swap dealers who would have previously been classified as commercials (hedgers). The SCOT data are released each Friday in conjunction with the traditional COT report and show the combined futures and options positions as of Tuesday's market close. The index trader positions are simply removed from their prior categories and presented as a new category of reporting traders. The SCOT data include the long and short positions held by commercials (less index traders), non-commercials (less index traders), index traders, and non-reporting traders.

While the CFTC classification procedure has flaws (CFTC, 2008), it is an improvement over the trader classifications in the standard COT reports and is generally thought to represent the activity of index traders. A significant limitation of the public SCOT data is the lack of data prior to 2006. This is an important constraint because other data suggest that the build-up in commodity index positions was concentrated in the preceding two years (Sanders *et al.*, 2008). The CFTC did collect additional data for selected grain futures markets over 2004–2005 at the request of the US Senate Permanent Subcommittee on Investigations (USS/PSI, 2009) and these data are used in the present analysis. The selected markets are: (i) Chicago Board of Trade (CBOT) corn; (ii) CBOT wheat; (iii) CBOT soybeans; and (iv) Kansas City Board of Trade (KCBT) wheat.

To correspond with the release dates for the SCOT data, the Tuesday-to-Tuesday log-relative returns are collected for nearby futures contracts. The futures and SCOT data are available from January 6 2004 through September 1 2009 (296 weekly observations) for each of the four markets.

Table 3.1 presents summary statistics for various position measures, average nearby futures prices, and the cumulative weekly log-relative nearby futures returns by year for 2004–2009. Several interesting trends are apparent. First, the rapid increase in commodity index positions occurred from 2004 to 2006. Over this interval, long positions held by index traders nearly tripled in both corn and CBOT wheat. Likewise, index funds percentage of total open interest nearly doubled in corn and soybeans and increased 40% in CBOT wheat. It is clear that the build-up in commodity index fund positions was concentrated in the 2004–2006 period, not the 2007–2008 period associated with the alleged commodity bubble.

A more complete picture of the index fund build-up in 2004–2006 can be demonstrated graphically. The common association between index fund positions and prices is illustrated with selected data from 2007 to 2008. As shown in Fig. 3.3, for markets such as wheat, the correlation over this period makes a convincing illustration. However, when a larger picture is taken, using data from 2004 to 2009, the perceived association between prices and commodity index trader (CIT) positions breaks down substantially. Indeed, as illustrated in Fig. 3.4, the major increase in CIT net long positions occurred from January 2004 through May 2006. During this period wheat prices were largely unchanged. Similar patterns are observed for the other three markets.

If index trader buying were to have a market impact, surely it would have been during 2004–2006 when their market holdings increased dramatically. It is very difficult to reconcile the build-up of index positions in 2004–2006 with relatively flat prices and the assertion that index trader buying represents new demand. The relationships observed in the 2007–2008 period seem to be a coincidence. However, visual evidence can be deceptive. Therefore, it is important to test for formal statistical links between index positions and prices. Granger causality tests are one such method of determining if there is a causal link between index fund positions and changes in futures price

3.4 Tests for Price Impacts

Granger causality is a standard linear technique for determining whether one time series is useful in forecasting another. The two time series variables we use are log-relative nearby futures returns (R_t) and measures of index fund positions ($Position_t$) to test if there is causality – in a Granger sense – running from index trader positions to futures returns. The following model is estimated,

$$R_t = \alpha_t + \sum_{i=1}^{m} \gamma_i R_{t-i} + \sum_{j=1}^{n} \beta_j \Delta Position_{t-j} + \varepsilon_t \qquad (3.1)$$

The null hypothesis of no causal link between index trader positions and returns is tested using an F-test of the linear restriction that $\beta_j = 0 \forall j$. In an efficient futures market, the null hypothesis of no autocorrelation $(\gamma_i = 0 \forall i)$ will also fail to be rejected.

Each model is estimated for lag lengths of 1–4 weeks $(i = 1 \text{ to } 4 \text{ and } j = 1 \text{ to } 4)$ for a total of 16 possible model specifications and the lag structure of the most efficient model is selected using the Schwartz criterion. The relatively short lag search and Schwartz criterion are used to minimize any data-mining tendencies associated with the model selection procedure. Models are estimated with ordinary least squares (OLS). If the residuals demonstrate serial correlation (Breusch-Godfrey Lagrange multiplier test), additional lags of the dependent variable are added until the null of no serial correlation cannot be rejected. White's test for heteroskedasticity is applied and robust standard errors are used to correct the standard errors when necessary.

Commodity index trader positions in Eqn 3.1 are measured in two ways. First, the position variable is calculated using the net long position of index traders (long contracts minus short contracts). This measure most directly captures the essence of the complaints leveled against index funds: they have become too big. The second position measure is a normalized measure: percentage of long positions. So, the index trader long positions (contracts) are divided by the total long positions in the market

Table 3.1. Summary statistics, commodity index trader positions and futures prices, 2004–2009.

Year/market[a]	Long position (contracts)	Short position (contracts)	Percentage of total open interest	Percentage of total long open interest	Nearby futures price (cents)	Nearby futures return
2004						
CBOT corn	118,286	455	7	14	255	–31.90
CBOT soybeans	36, 862	1,717	6	12	748	–15.60
CBOT wheat	57,187	744	15	30	349	–33.10
KCBT wheat	14,792	4	10	19	369	–16.90
2005						
CBOT corn	236,424	4,135	14	27	211	–22.30
CBOT soybeans	78,740	1,973	11	22	610	4.00
CBOT wheat	138,821	1,851	24	48	321	–8.50
KCBT wheat	18,307	4	10	19	346	12.10
2006						
CBOT corn	408,138	7,662	13	26	262	33.40
CBOT soybeans	119,287	3,679	14	26	594	–4.60
CBOT wheat	201,605	4,883	21	42	405	21.70
KCBT wheat	25,954	115	8	17	469	18.40
2007						
CBOT corn	370,682	12,020	11	21	375	–2.60
CBOT soybeans	155,864	4,766	12	23	866	45.90
CBOT wheat	197,338	11,179	21	39	639	40.20
KCBT wheat	31,560	519	11	22	644	49.70
2008						
CBOT corn	405,241	44,122	12	21	528	–28.60
CBOT soybeans	162,233	12,765	14	26	1228	–29.20
CBOT wheat	198,485	27,644	24	43	797	–49.50
KCBT wheat	26,687	1,054	13	24	836	–46.40
2009						
CBOT corn	316,896	45,133	14	25	374	–29.80
CBOT soybeans	138,406	17,230	15	27	1037	27.10
CBOT wheat	168,117	23,220	24	42	543	–37.30
KCBT wheat	26,508	1,243	15	29	585	–27.40

[a]CBOT, Chicago Board of Trade; KCBOT, Kansas City Board of Trade. Data for 2009 ends on September 1 2009.

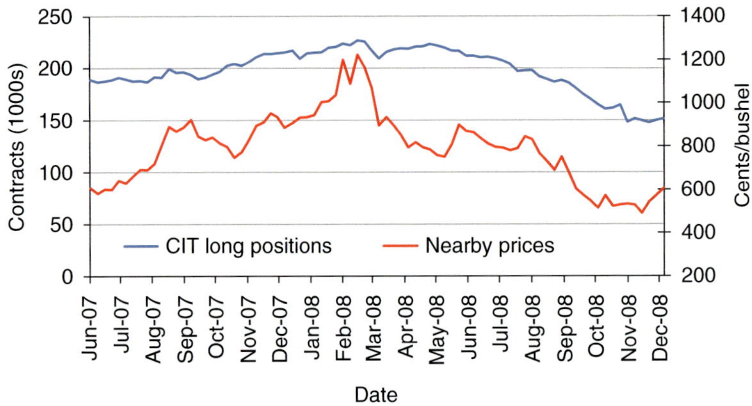

Fig. 3.3. Long-only index fund positions and Chicago Board of Trade (CBOT) wheat prices, June 2007–December 2008. (CIT, commodity index trader.)

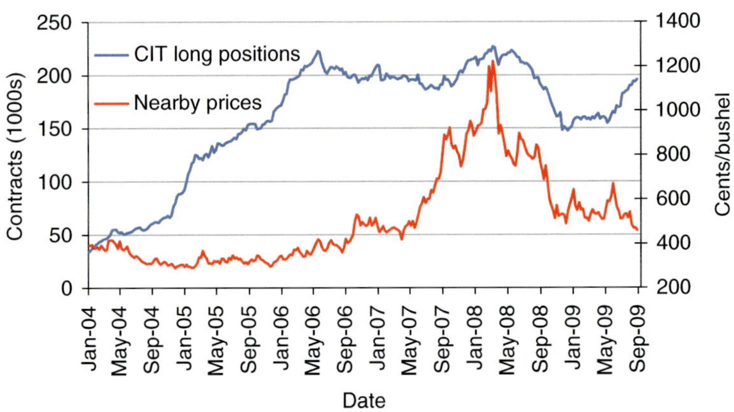

Fig. 3.4. Long-only index fund positions and Chicago Board of Trade (CBOT) wheat prices, January 2004–September 2009. (CIT, commodity index trader.)

(contracts) to get the percentage of long positions within that market held by index traders. Augmented Dickey-Fuller tests indicate both variables were non-stationary in levels and stationary in first differences over 2004–2009; therefore, first differences of the position measures are used to estimate Eqn 3.1.

As shown in Table 3.2, the model selection procedure chooses a simple ($m = 1, n = 1$) model for each market and position measure where just a one period lag of both returns and positions minimized the Schwartz criterion. Given this model selection, it is not surprising that the null hypothesis of no causality from positions to returns $\left(H_0 : \beta_j = 0\right)$ cannot be rejected at the 5% level for any market or position measure. Indeed, the only marginally statistically significant result (p-value = 0.103) is for corn using the percentage of long positions as the independent variable. In this sample, full rationality of the markets $\left(\gamma_i = \beta_j = 0 \forall i, j\right)$ and no autocorrelation $\left(\gamma_i = 0 \forall i\right)$ also cannot be rejected.

Based on these results, there is no evidence that commodity index trader positions 'cause' price changes or returns. However, it is possible that the causal relationship shifted after the initial build-up of index positions in the first half of the sample. To test this, the originally specified models are re-estimated incorporating a 2004–2006 slope-shift variable for the estimated β coefficients. As shown in the final column of Table 3.2, the shift variable is not statistically different from zero. This suggests that impact of lagged positions on returns was

Table 3.2. Granger causality test results for Commodity Futures Trading Commission (CFTC) commodity index traders, positions do not lead returns, 2004–2009.[a]

$$R_t = \alpha_t + \sum_{i=1}^{m} \gamma_i R_{t-i} + \sum_{j=1}^{n} \beta_j \Delta Position_{t-j} + \varepsilon_t$$

Market	m,n	$\beta_j = 0, \forall j$	$\gamma_i = 0, \forall i$	$\gamma_i = \beta_j = 0, \forall i,j$	2004–2006 β_j shift
		p-values for Hypothesis Tests			
Panel A: Positions measure in net long contracts					
CBOT corn	1,1	0.413	0.998	0.713	0.2994
CBOT soybeans	1,1	0.446	0.468	0.430	0.6737
CBOT wheat	1,1	0.841	0.741	0.916	0.4387
KCBT wheat	1,1	0.895	0.462	0.757	0.3419
Panel B: Positions measured in percentage of long positions					
CBOT corn	1,1	0.103	0.710	0.263	0.6287
CBOT soybeans	1,1	0.171	0.256	0.225	0.3155
CBOT wheat	1,1	0.402	0.864	0.618	0.6152
KCBT wheat	1,1	0.384	0.481	0.473	0.7200

[a]CBOT, Chicago Board of Trade; KCBOT, Kansas City Board of Trade. Data for 2009 ends on September 1 2009.

equally unimportant in both the 2004–2006 and the 2007–2009 sub-samples.

The Granger causality tests are designed to detect the relationship, if any, between weekly positions and returns. Such tests may have low power to detect relationships over longer horizons (e.g. Summers, 1986). Index trader positions may flow in 'waves' that build slowly – pushing prices higher – and then fade slowly. In this scenario, horizons longer than a week may be necessary to capture the predictive component of index trader positions. So, we implement the long-horizon regression 'fads' models of Jegadeesh (1991).

$$R_t = \alpha_t + \sum_{i=1}^{m} \gamma_i R_{t-i} + \beta \sum_{j=1}^{n} \frac{\Delta Position_{t-j}}{n} + \varepsilon_t. \quad (3.2)$$

In essence, Eqn 3.2 is analogous to Eqn 3.1, except that instead of positions entering the model at alternative lags, it enters the model as a moving average calculated over the most recent n observations. Jegadeesh shows that letting the independent variable enter the equation as an average over the most recent n observations provides the highest power against a fads-type alternative hypothesis using standard OLS estimation and testing procedures. If the estimated β is positive (negative), then it indicates a fads-style model where prices tend to increase (decrease) slowly over a relatively long time period after widespread buying. The 'fads' stylization captured in Eqn 3.2 is consistent with the popular notion of speculative pressures creating a 'bubble' in commodity prices.

To adequately capture any long-horizon impacts, i and j are determined using a search procedure of the last 12 weeks (i =1 to 12 and j = 1 to 12) and choosing the m, n that minimize the Schwartz criterion. The estimated β coefficients for Eqn 3.2 are shown in Table 3.3. As it happens, for all of the markets and positions – except for soybeans in Panel A – the estimation of Eqn 3.2 is trivial as the selection criteria chooses models equivalent to those estimated for Eqn 3.1. That is, $m = n = 1$ which causes Eqn 3.2 to be functionally equivalent to Eqn 3.1. Only the estimated β coefficient for soybeans in Panel A is statistically different from zero at the 5% significance level. For that model, the estimated β is positive suggesting that increases in long contracts were associated with positive market returns. However, this result has to be viewed with skepticism given the lack of rejections in other markets and the total number of models estimated. Generally, the results of the long-horizon regressions are consistent with the Granger causality results. There is almost no evidence linking commodity index positions with grain futures market prices.

3.5 Conclusions

Data from 2004 to 2009 are used to examine the overall size and impact of commodity index trader positions in the CBOT corn, CBOT soybeans,

Table 3.3. Long-horizon Granger causality test results for Commodity Futures Trading Commission (CFTC) commodity index trader positions, positions do not lead returns, 2004–2009.[a]

$$R_t = \alpha_t + \sum_{i=1}^{m} \gamma_i R_{t-i} + \beta \sum_{j=1}^{n} \frac{\Delta Position_{t-j}}{n} + \varepsilon_t .$$

Market	*m*, *n*	*β*	*t*-statistic	*p*-value
Panel A: Positions measure in net long contracts				
CBOT corn	1,1	−0.000	−0.819	0.413
CBOT soybeans	1,8	0.001	2.246	0.025
CBOT wheat	1,1	−0.000	−0.201	0.841
KCBT wheat	1,1	−0.000	−0.132	0.895
Panel B: Positions measured in percentage of long positions				
CBOT corn	1,1	−0.524	−1.630	0.103
CBOT soybeans	1,1	0.294	1.368	0.171
CBOT wheat	1,1	−0.127	−0.838	0.402
KCBT wheat	1,1	−0.169	−0.871	0.384

[a]CBOT, Chicago Board of Trade; KCBOT, Kansas City Board of Trade. Data for 2009 ends on September 1 2009.

CBOT wheat, and KCBT wheat futures contracts. The data are unique in that they provide the first detailed evidence on commodity index activity in 2004–2005, prior to the much-publicized run-up in commodity prices. The data and empirical results lead to a number of conclusions.

First, there was a fairly dramatic and massive build-up in commodity index fund positions in the US grain futures markets examined. For instance, the number of contracts held by index funds in the CBOT wheat contract increased nearly fourfold from 2004 to 2006. Second, the build-up in commodity index contracts and the peak level of index holdings pre-dates the 2007–2008 increase in commodity prices for which they are blamed. This observation casts serious doubt on the hypothesis that commodity index speculation drove the 2007–2008 commodity price increase. Third, formal econometric tests fail to find a statistical link between commodity index positions and returns in grain futures markets. Both Granger causality tests and long-horizon regressions generally fail to reject the null hypothesis that commodity index positions have no impact on futures prices.

It should be noted that modeling market returns with the traditional time-series approaches used in this study can be criticized for a lack of statistical power due to the considerable volatility in the independent variable (returns). These time-series econometric models may not have sufficient statistical power to reject the null hypothesis of no speculative impact over the relatively short time period studied. Nonetheless, the analysis provides the most rigorous direct test for linkages between index positions and commodity futures returns available to date. Coupled with the data trends, the econometric results are simply not consistent with the bubble theory that has been widely touted.

The presented results are consistent with the majority of academic evidence pertaining to speculation and price behavior (e.g. Bryant *et al.*, 2006; Gorton *et al.*, 2007; Sanders *et al.*, 2009). By the nature of the hypothesis test, empirical studies cannot preclude a speculative impact on commodity prices; it can only fail to reject the null hypothesis of no speculative impact. Unfortunately, those legislators and public-policy commentators who can most easily shape the outcome of this debate do not share the same null hypothesis. Indeed, they have a well-defined enemy: speculators in general and index funds in particular. The outcome of this policy debate has wide-ranging implications not only for the US futures industry but also for futures exchanges outside of the USA. Regulatory miscues could drive financial commodity investments into non-US futures markets or into the world's physical markets. The absence of index funds in US futures markets may reduce liquidity and potentially degrade commodity market performance in terms of price discovery, efficiency, and risk-sharing capacity.

Acknowledgments

The authors are indebted to the staff of the Permanent Subcommittee on Investigations of the US Senate for providing the 2004–2005 index trader position data.

Notes

[1] Original citation: Sanders, D.R. and Irwin, S.H. (2011) New evidence on the impact of index funds in US grain futures markets. *Canadian Journal of Agricultural Economics* 59, 519–532. Reprinted by permission of John Wiley and Sons and the Canadian Agricultural Economics Society.
[2] 'Long-only' refers to the purchase only of futures contracts (as opposed to 'short' sales). For additional futures-related definitions please see the CFTC's glossary of terms (CFTC, 2010).

References

Acworth, W. (2009a) CFTC examines position limits. *The Magazine of the Futures Industry,* September, pp. 24–26.

Acworth, W. (2009b) The Gensler agenda. *The Magazine of the Futures Industry,* September, pp. 18–23.

Agricultural and Food Policy Center (AFPC) (2008) The effects of ethanol on Texas food and feed. *AFPC Research Report* 08-1, Texas A&M University, Texas. Available at: https://www.afpc.tamu.edu/research/publications/515/RR-08-01.pdf (accessed January 14 2022).

Bryant, H., Bessler, D.A. and Haigh, M.S. (2006) Causality in futures markets. *Journal of Futures Markets* 26, 1039–1057.

Childs, N. and Kiawu, J. (2009) Factors behind the rise in global rice prices in 2008. United States Department of Agriculture (USDA), Economic Research Service, RCS-09D-01. Available at: https://www.ers.usda.gov/webdocs/outlooks/38489/13518_rcs09d01_1_.pdf?v=4387 (accessed January 14 2022).

CFTC (Commodity Futures Trading Commission) (2008) Staff Report on Commodity Swap Dealers and Index Traders with Commission Recommendations. Available at: https://www.cftc.gov/sites/default/files/idc/groups/public/@newsroom/documents/file/cftcstaffreportonswapdealers09.pdf (accessed January 14 2022).

CFTC (Commodity Futures Trading Commission) (2010) A Guide to the Language of the Futures Industry. CFTC Glossary. Available at: http://www.cftc.gov/ConsumerProtection/EducationCenter/CFTCGlossary/glossary_f.html (accessed January 14 2022).

De Long, J.B., Shleifer, A., Summers, L.H. and Waldmann, R.J. (1990) Noise trader risk in financial markets. *Journal of Political Economy* 98, 703–738.

Engelke, L. and Yuen, J.C. (2008) Types of commodity investments. In: Fabozzi, F.J., Fuss, R. and Kaiser, D. (eds) *The Handbook of Commodity Investing*. Wiley, Hoboken, New Jersey, pp. 549–569.

FAO (Food and Agriculture Organization of the United Nations) (2009) Financial Speculation and the Food Crisis. Available at: https://reliefweb.int/report/world/financial-speculation-and-food-crisis (accessed November 17 2022).

Gheit, F. (2008) Testimony before the Subcommittee on Oversight and Investigations of the Committee on Energy and Commerce, US House of Representatives, June 23. Available at: https://www.google.com/books/edition/Speculative_Investment_in_Energy_Markets/mCo6_xKYnQMC?hl=en&gbpv=1&dq=fadel+gheit+testimony&pg=PA50&printsec=frontcover (accessed January 14 2022).

Gilbert, C.L. (2009) Speculative influences on commodity futures prices, 2006–2008. Working Paper, Department of Economics, University of Trento, Trento, Italy. Available at: https://www.cftc.gov/sites/default/files/idc/groups/public/@swaps/documents/file/plstudy_14_cifrem.pdf (accessed January 14 2022).

Gilbert, C.L. (2010) How to understand high food prices. *Journal of Agricultural Economics* 61, 398–425.

Gorton, G.B. and Rouwenhorst, K.G. (2006) Facts and fantasies about commodity futures. *Financial Analysts Journal* 62, 47–68.

Gorton, G.B., Hayashi, F. and Rouwenhorst, K.G. (2007) The fundamentals of commodity futures returns. National Bureau of Economic Research (NBER) Working Paper No. 13249. Available at: https://www.nber.org/papers/w13249 (accessed January 14 2022).

Headey, D.D. (2010) Rethinking the global food crisis: the role of trade shocks. IFPRI Discussion Paper 00958. International Food Policy Research Institute, Washington DC. Available at: http://ebrary.ifpri.org/utils/getfile/collection/p15738coll2/id/831/filename/832.pdf (accessed January 14 2022).

Headey, D. and Fan, S. (2008) Anatomy of a crisis: the causes and consequences of surging food prices. *Agricultural Economics* 39, 375–391.

Henriques, D.B. (2008) Odd crop prices defy economics. *New York Times*, March 28. Available at: http://www.nytimes.com/2008/03/28/business/28commodities.html (accessed January 14 2022).

Irwin, S.H., Sanders, D.R. and Merrin, R.P. (2009) Devil or angel? The role of speculation in the recent commodity price boom (and bust). *Journal of Agricultural and Applied Economics* 41, 393–402.

Jegadeesh, N. (1991) Seasonality in stock price mean reversion: evidence from the U.S. and the U.K. *Journal of Finance* 46, 1427–1444.

Krugman, P. (2008) More on oil and speculation. *New York Times*, May 13. Available at: http://krugman.blogs.nytimes.com/2008/05/13/more-on-oil-and-speculation/ (accessed January 14 2022).

Masters, M.W. (2008) Testimony before the Committee on Homeland Security and Governmental Affairs, United States Senate. May 20. Available at: http://hsgac.senate.gov/public/_files/052008Masters.pdf (accessed January 14 2022).

Masters, M.W. and White, A.K. (2008) The accidental Hunt brothers: how institutional investors are driving up food and energy prices. Available at: https://www.cftc.gov/sites/default/files/idc/groups/public/@swaps/documents/file/plstudy_31_ahb.pdf (accessed January 14 2022).

McCalla, A. (2009) World food prices: causes and consequences. *Canadian Journal of Agricultural Economics* 57, 23–34.

McQueen, G. and Thorley, S. (1994) Bubbles, stock returns and duration dependence. *Journal of Financial and Quantitative Analysis* 29, 379–401.

Pirrong, C. (2008) Restricting speculation will not reduce oil prices. *The Wall Street Journal*, July 11. Available at: http://online.wsj.com/article/SB121573804530744613.html (accessed January 14 2022).

Robles, M., Torero, M. and von Braun, J. (2009) When speculation matters. *IFRPR Issue Brief 57*, International Food Policy Research Institute, Washington, DC. Available at: https://www.ifpri.org/publication/when-speculation-matters (accessed January 14 2022).

Sanders, D.R. and Irwin, S.H. (2008) Futures imperfect. *The New York Times*, July 20. Available at: http://www.nytimes.com/2008/07/20/opinion/20irwinsanders.html?_r=1&oref=slogin (accessed January 14 2022).

Sanders, D.R. and Irwin, S.H. (2010) A speculative bubble in commodity futures prices? Cross-sectional evidence. *Agricultural Economics* 41, 25–32.

Sanders, D.R., Irwin, S.H. and Merrin, R.P. (2008) The adequacy of speculation in agricultural futures markets: too much of a good thing? *Marketing and Outlook Research Report 2008-02*, Department of Agricultural and Consumer Economics, University of Illinois at Urbana-Champaign, Illinois. Available at: https://farmdoc.illinois.edu/publications/the-adequacy-of-speculation-in-agricultural-futures-markets-too-much-of-a-good-thing (accessed January 14 2022).

Sanders, D.R., Irwin, S.H. and Merrin, R.P. (2009) Smart money? The forecasting ability of CFTC large traders. *Journal of Agricultural and Resource Economics* 34, 276–296.

Stoll, H.R. and Whaley, R.E. (2010) Commodity index investing and commodity futures prices. *Journal of Applied Finance* 20, 7–46.

Summers, L.H. (1986) Does the stock market rationally reflect fundamental values? *Journal of Finance* 41, 591–601.

USS/PSI (United States Senate, Permanent Subcommittee on Investigations) (2006) The role of market speculation in rising oil and gas prices: a need to put the cop back on the beat. Majority and Minority Staff Report. Available at: https://www.hsgac.senate.gov/imo/media/doc/SenatePrint10965MarketSpecReportFINAL.pdf?attempt=2 (accessed January 14 2022).

USS/PSI (United States Senate, Permanent Subcommittee on Investigations) (2009) Excessive speculation in the wheat market. Majority and Minority Staff Report. Available at: https://www.hsgac.senate.gov/imo/media/doc/REPORTExcessiveSpecullationintheWheatMarketwoexhibitschartsJune2409.pdf?attempt=2 (accessed January 14 2022).

Working, H. (1960) Speculation on hedging markets. *Food Research Institute Studies* 1, 185–220.

4

The Impact of Index and Swap Funds in Commodity Futures Markets[1]

New Author Foreword

Not only did the commodity speculation debate fail to flame out as we expected after the crash in commodity prices in the second half of 2008, but it actually picked up steam heading into the early 2010s. Civic organizations such as Oxfam jumped into the speculation debate, and they tended to react in a fiercely negative manner, basically arguing that speculation in food commodity markets was immoral. In the middle of this swirling debate, we were approached in the summer of 2009 by an economist from the Organisation for Economic Co-operation and Development (OECD), Linda Fulponi, to produce a report on the controversy. We saw this as an important opportunity to dig deeper into the data and to have the stamp of approval from an important and influential international organization such as the OECD.

We had two objectives for the OECD project. First, we wanted to conduct a thorough review and synthesis of the rapidly expanding literature on the role of speculation in commodity futures markets. It did not take long after the 2007–2008 price spike for economists to direct their attention to this very hot topic, and a new study was seemingly popping up every other week. Second, we wanted to provide new empirical evidence on the market impact of commodity index funds. We were in the process of completing our tests using the invaluable data from the Senate Subcommittee (see Chapter 3, this volume), but this was limited to four agricultural futures markets. We wanted to expand the analysis to other non-agricultural markets using another data series from the US Commodity Futures Trading Commission (CFTC) that had been recently made available in the *Disaggregated Commitments of Traders* (DCOT) report. A new systems approach to testing lead-lag dynamics would also be introduced and applied.

We submitted our final report to the OECD in April 2010. The OCED certainly got much more than it expected! We delivered a 130-page report, with 30 pages devoted to a review of the literature. The statistical results in the report were remarkably uniform. No matter how we conducted the tests, we came up with the same answer: there was no consistent evidence of a correlation between index positions and movements in commodity futures prices. In our view, this was a testament to the remarkable ability of commodity futures markets to absorb the increased participation of commodity index investors with apparently minimal impact on the level of prices.

It was obvious that the report was too long for public consumption, so it came as no surprise when we were told to prepare a condensed version. But as this shortened report worked its way through the internal OECD review process, it was not clear that the report would ever see the light of day. Some people apparently did not like our conclusion that the evidence did not support new regulations to limit the participation of index investors in commodity futures markets. Nonetheless, the report was eventually cleared and finally released in June 2010.

We were totally unprepared for the reaction. We knew that the conclusions went against the grain of conventional wisdom in many places and organizations, but we never imagined the intensity of the reaction. It started a global firestorm of controversy, reaching into the pages of major financial publications. As an example, *The Economist* published an article based on the report shortly after it was released and then a rebuttal

DOI:10.1079/9781800622104.0004

letter signed by Sir Richard Branson, the founder of the Virgin Group and a good buddy of President Obama, Michael Masters, a hedge fund manager who was at the center of the furor, and David Frenk, Executive Director for the ironically named lobbying organization 'Better Markets.' For good measure, *The Economist* referenced the OECD report again a few months later in a hilariously titled lead op-ed and accompanying article.[2] This is just a sampling of the reaction to the report, both pro and con. It was the most controversial publication we authored on commodity speculation by a wide margin. Yet, it is still not entirely clear to us why the report was so controversial, as the analysis was quite straightforward and the results were similar to several previous studies. Like they say, timing is everything.

The OECD report was also critical in the role it played in setting our research agenda for years to come. During the process of writing the report and then thinking through criticisms of it, we came up with research idea after research idea. Nearly all the articles in the chapters that follow can all be traced back to conversations during this period. It almost makes one nostalgic.

Abstract

Based on new data and empirical analysis, the study finds that index funds did not cause a bubble in commodity futures prices. There is no statistically significant relationship indicating that changes in index and swap fund positions have increased market volatility. The evidence presented here is strongest for the agricultural futures markets because the data on index trader positions are measured with reasonable accuracy. The evidence is not as strong in the two energy markets studied here because of considerable uncertainty about the degree to which the available data actually reflect index trader positions in these markets. The empirical evidence presented in this preliminary study does not appear at present to warrant extensive changes in the regulation of index funds participation in agricultural commodity markets; any such changes require careful consideration so as to avoid unintended negative impacts.

Key words: commodity, futures market, index funds, price, volatility

JEL categories: G12, G13, G14, price, volatility

4.1 Introduction

The financial industry has developed new products that allow institutions and individuals to invest in commodities through long-only index funds, over-the-counter (OTC) swap agreements, exchange-traded funds, and other structured products.[3] Regardless of form, these instruments have a common goal – provide investors with buy-side exposure to returns from a particular index of commodity prices. The Standard and Poor's Goldman Sachs Commodity Index™ (S&P GSCI) is one of the most widely tracked indexes and generally considered an industry benchmark. It is computed as a production-weighted average of the prices from 24 commodity futures markets.

Several influential studies in recent years purport that investors can capture substantial risk premiums and reduce portfolio risk through relatively modest investment in long-only commodity index funds. Combined with the availability of deep and liquid exchange-traded futures contracts, this evidence fueled a dramatic surge in index fund investment. Some describe this surge and its attendant impacts as the 'financialization' of commodity futures markets. Given the size and scope of the index fund boom it should probably not come as a surprise that a worldwide debate has ensued about the role of index funds in commodity markets. The debate has important ramifications from a policy and regulatory perspective as well as practical implications for the efficient pricing of commodity products.

There are a few indisputable facts about commodity futures markets over 2006–2008, the period of most controversy regarding the impact of money inflows from commodity index funds. First, inflows into long-only commodity index funds did increase rather substantially throughout 2006–2008 (see Fig. 4.1). According to the most widely quoted industry source (Barclays), index fund investment increased from $90 billion at the beginning of 2006 to a peak of just under $200 billion at the end of 2007. Second, commodity prices also increased rather dramatically – 71% as measured by the

Commodity Research Bureau index – from January 2006 through June of 2008 (see Fig. 4.2). Third, prices declined almost equally dramatically from June 2008 through early 2009 (see Fig. 4.2). These facts are clear and not in dispute; it is the interpretation of the interaction among these facts that is so controversial.

On one side, some hedge fund managers, commodity end-users, and policy makers assert that speculative buying by index funds on such a wide scale created a 'bubble,' with the result that commodity futures prices far exceeded fundamental values during the boom. This view has led to new regulatory initiatives to limit speculative positions in commodity futures markets. On the other side, a number of economists have expressed skepticism about the bubble argument, citing logical inconsistencies in bubble arguments and contrary facts. These economists argue that commodity markets were driven by fundamental factors that pushed prices higher. The main factors cited as driving the price of crude oil include strong demand from China, India, and other developing nations, a leveling out of crude oil production, a decrease in the responsiveness of consumers to price increases, and US monetary policy. In the grain markets, the diversion of row crops to biofuel production

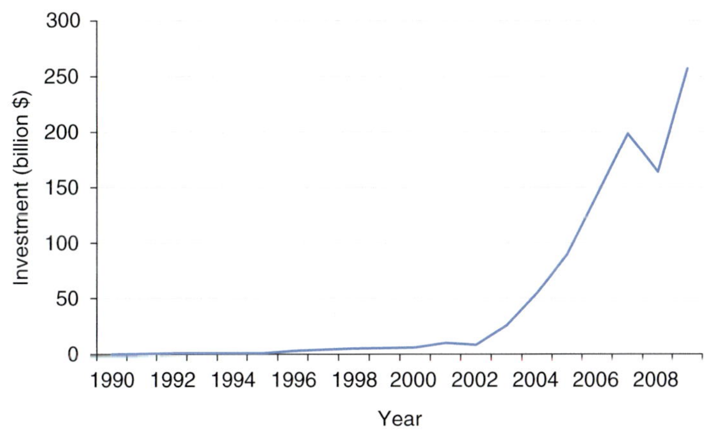

Fig. 4.1. Commodity index fund investment (year end), 1990–2009. (Data from Barclays.)

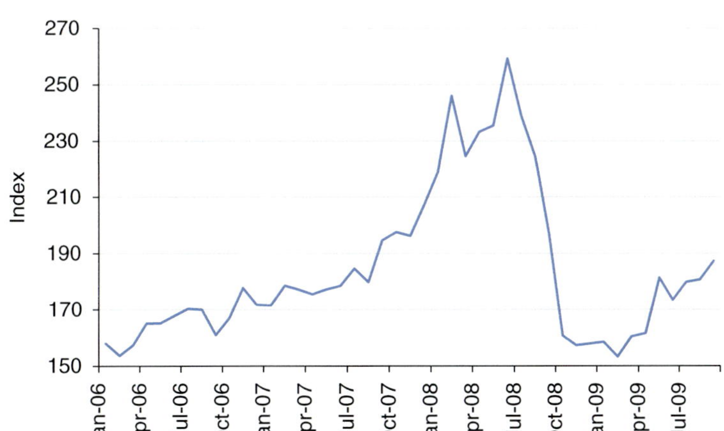

Fig. 4.2. Commodity Research Bureau (CRB) commodity index, January 1990 = 100, January 2006–September 2009.

and weather-related production shortfalls are cited, as well as demand growth from developing nations and US monetary policy.

Even though almost 2 years have passed since the 2008 peak in commodity prices, the controversy surrounding index funds continues unabated. We contend that a detailed and dispassionate synthesis of the arguments and latest research will be of great utility to market observers and policy makers given the strident nature of the debate. Policy makers need to have a full picture of the current state of scientific knowledge on the impact of commodity index funds before imposing costly new regulations. In this article, we provide an overview of the arguments concerning the impact of index funds in commodity futures markets as well as an assessment of the latest research on the subject. We also summarize some new empirical evidence on the market impact of commodity index funds.

4.2 It Was a Bubble

Masters (2008) has interwoven investment and price data to create the most widely cited bubble argument, painting the activity of index funds as akin to the infamous Hunt brothers' cornering of the silver market. He blames the rapid increase in overall commodity prices from 2006 to 2008 on institutional investors' embrace of commodities as an investable asset class. As noted in the introduction, it is clear that considerable dollars flowed into commodity index funds over this period. However, the evidence provided by Masters is limited to anecdotes and the temporal correlation between money flows and prices. Masters and White (2008) recommend specific regulatory steps to address the alleged problems created by index fund investment in commodity futures markets, including the re-establishment of speculative position limits for all speculators in all commodity futures markets and the elimination or severe restriction of index speculation.

A similar position was taken by the US Senate Permanent Subcommittee on Investigations in its examination of the performance of the Chicago Board of Trade's (CBOT) wheat futures contract:

> This Report finds that there is significant and persuasive evidence to conclude that these commodity index traders, in the aggregate, were one of the major causes of "unwarranted changes" – here, increases – in the price of wheat futures contracts relative to the price of wheat in the cash market. The resulting unusual, persistent and large disparities between wheat futures and cash prices impaired the ability of participants in the grain market to use the futures market to price their crops and hedge their price risks over time, and therefore constituted an undue burden on interstate commerce. Accordingly, the Report finds that the activities of commodity index traders, in the aggregate, constituted "excessive speculation" in the wheat market under the Commodity Exchange Act.
>
> (USS/PSI, 2009, p. 2)

Based on these findings, the Subcommittee recommended the: (i) phase out of existing position limit waivers for index traders in wheat; (ii) if necessary, imposition of additional restrictions on index traders, such as a position limit of 5000 contracts per trader; (iii) investigation of index trading in other agricultural markets; and (iv) strengthening of data collection on index trading in non-agricultural markets.

One of the limitations of the bubble argument made by Masters and others is that the link between money inflows from index funds and commodity futures prices is not well developed. This allows critics to assert that bubble proponents make the classical statistical mistake of confusing correlation with causation. In other words, simply observing that large investment has flowed into the long side of commodity futures markets at the same time that prices have risen substantially does not necessarily prove anything without a logical and causal link between the two. One attempt to establish this linkage is found in Petzel's (2009, pp. 8–9) testimony at a CFTC hearing on position limits in energy futures markets:

> Seasoned observers of commodity markets know that as non-commercial participants enter a market, the opposite side is usually taken by a short-term liquidity provider, but the ultimate counterparty is likely to be a commercial. In the case of commodity index buyers, evidence suggests that the sellers are not typically other investors or leveraged speculators. Instead, they are owners of the physical commodity who are willing to sell into the futures market and either deliver at expiration or roll their hedge forward

> if the spread allows them to profit from continued storage. This activity is effectively creating "synthetic" long positions in the commodity for the index investor, matched against real inventories held by the shorts. We have seen high spot prices along with large inventories and strong positive carry relationships as a result of the expanded index activity over the last few years.
>
> (Petzel, 2009, pp. 8–9)

In essence, Petzel argues that unleveraged futures positions of index funds are effectively synthetic long positions in physical commodities, and hence represent new demand. If the magnitude of index fund demand is large enough relative to physically constrained supplies in the short-run, prices and price volatility can increase sharply. The bottom line is that the size of index fund investment is 'too big' for the current size of commodity futures markets.

Hamilton (2009) provides a more formal theoretical treatment of the issues. He begins by noting that the key challenge is reconciling a speculative bubble in crude oil prices with changes in the physical quantities of crude oil. A standard argument is that a price bubble will inevitably lead to a rise in inventories as the quantity supplied at the 'bubble price' exceeds the quantity demanded. Hamilton's theoretical model shows the conditions that must occur for index fund speculation to lead to a bubble in a storable commodity market such as crude oil. First, index fund positions in the futures market must have a positive relationship to the level of futures prices. Otherwise, there is no mechanism for the flow of index fund investment to initiate the bubble that starts in the futures market. Second, the elasticity of demand for the commodity (or the final product, gasoline in the case of crude oil) must be zero or very close to zero. This allows the bubble-related increase in the futures price to be fully passed on to consumers. Third, inventories of the commodity must not increase. These conditions provide an important theoretical framework on which to base empirical tests for the potential of price bubbles in storable commodity futures prices.

4.3 It Was Not a Bubble

A number of economists have expressed skepticism about the bubble argument. These economists cite several contrary facts and argue that commodity markets were driven by fundamental factors that pushed prices higher. Irwin *et al.* (2009) present a useful summary of the counter-arguments made by these economists. Specifically, they note three logical inconsistencies in the arguments made by bubble proponents as well as five instances where the bubble story is not consistent with observed facts. Here, we review these points as well as some additional arguments made by both pro- and anti-bubble proponents in response.

The first possible logical inconsistency within the bubble argument is equating money inflows to commodity futures markets with demand. With equally informed market participants, there is no limit to the number of futures contracts that can be created at a given price level. Index fund buying in this situation is no more 'new demand' than the corresponding selling is 'new supply.' Combined with the observation that commodity futures markets are zero-sum games, this implies that money flows, in and of themselves, do not necessarily impact prices. Prices will only change if new information emerges that causes market participants to revise their estimates of physical supply and/or demand.

What happens when market participants are not equally informed? When this is the case, it is rational for participants to condition demands on both their own information and information about other participants' demands that can be inferred ('inverted') from the futures price. The trades of uninformed participants can impact prices in this more realistic model if informed traders mistakenly believe that trades by uninformed participants reflect valuable information. Hence, it is possible that other traders in commodity futures markets interpreted the large order flow of index funds on the long side of the market as a reflection of valuable private information about commodity price prospects, which would have had the effect of driving prices higher as these traders revised their own demands upward. Of course, this would have required a large number of sophisticated and experienced traders in commodity futures markets to reach a conclusion that index fund investors possessed valuable information that they themselves did not possess.

The second possible logical inconsistency is to argue that index fund investors artificially raised both futures and cash commodity prices when they only participated in futures markets. Futures contracts are financial transactions that only rarely involve the actual delivery of physical commodities. In order to impact the equilibrium price of commodities in the cash market, index investors would have to take delivery and/or buy quantities in the cash market and hold these inventories off the market. Index investors are purely involved in a financial transaction using futures markets; they do not engage in the purchase or hoarding of the cash commodity and any causal linkages between their futures market activity and cash prices is unclear at best. Hence, it is wrong to draw a parallel between index fund positions and past efforts to 'corner' commodity markets, such as the Hunt brothers' effort to manipulate the silver market in 1979–1980.

A third possible logical inconsistency is a blanket categorization of speculators, in particular index funds, as wrongdoers and hedgers as victims of their actions. In reality, the 'bad guy' is not so easily identified since hedgers sometimes speculate and some speculators also hedge. For example, large commercial firms may have valuable information gleaned from their far-flung cash market operations and trade based on that information. The following passage from a recent article on Cargill, Inc. (Davis, 2009) illustrates the point:

> Wearing multiple hats gives Cargill an unusually detailed view of the industries it bets on, as well as the ability to trade on its knowledge in ways few others can match. Cargill freely acknowledges it strives to profit from that information. "When we do a good job of assimilating all those seemingly unrelated facts," says Greg Page, Cargill's chief executive, in a rare interview, "it provides us an opportunity to make money ... without necessarily having to make directional trades, i.e. outguess the weather, outguess individual governments."
>
> (Davis, 2009)

The implication is that the interplay between varied market participants is more complex than a standard textbook description of pure risk-avoiding hedgers and pure risk-seeking speculators. The reality is that market dynamics are ever-changing and it can be difficult to understand the motivations and market implications of trading, especially in real time.

In addition to the logical inconsistencies, there are several ways the bubble story is not consistent with the observed facts. First, as Krugman (2008) asserts, if a bubble raises the market price of a storable commodity above the true equilibrium price, then stocks of that commodity should increase (much like a government-imposed price floor can create a surplus). Stocks were declining, not building, in most commodity markets over 2006–2008, which is inconsistent with the depiction of a price bubble in these markets.

Second, the relationship between prices and inventories for storable commodities is highly convex. Wright (2009) illustrates this point and shows that a given reduction in quantity due to a supply and/or demand shock will have a much larger impact on price when starting with a low quantity (inventories) compared to when starting with a high quantity. He also notes that relatively minor reductions in quantity can result in very large increases in price when the market supply/demand balance is especially tight. Smith (2009) argues that it is plausible that a series of seemingly small supply disruptions in the spring and summer of 2008 could explain the large increase in crude oil prices during this time period in view of the extreme convexity of the pricing function for crude oil in the short run.

Third, theoretical models that show uninformed or noise traders impacting market prices rely on the unpredictable trading patterns of these traders to make arbitrage risky. Because the arbitrage – needed to drive prices to fundamental value – is not riskless, noise traders can drive a wedge between market prices and fundamental values. Importantly, index fund buying is very predictable. That is, index funds widely publish their portfolio (market) weights and rollover periods. Thus, it seems highly unlikely that other large and rational traders would hesitate to trade against an index fund if they were driving prices away from fundamental values.

Fourth, if index fund buying drove commodity prices higher, then markets without index fund investment should not have seen prices advance. Again, the observed facts are inconsistent with this notion. Irwin *et al.* (2009) show that markets without index fund participation (fluid milk and rice futures) and commodities without futures markets (apples and edible beans) also showed price increases over the 2006–2008 period. Stoll and Whaley (2009) report that returns for CBOT wheat, Kansas City Board of Trade (KCBOT) wheat, and Minneapolis Grain Exchange (MGEX) wheat are all highly positively correlated over 2006–2009, yet only CBOT wheat is used heavily by index investors. In a similar fashion, Commodity Exchange (COMEX) gold, COMEX silver, New York Mercantile Exchange (NYMEX) palladium, and NYMEX platinum futures prices are highly correlated over the same time period but only gold and silver are included in popular commodity indexes. Headey and Fan (2008) cite the rapid increases in the prices for 'non-financialized' commodities such as rubber, onion, and iron ore as evidence that rapid price inflation occurred in commodities without futures markets. While certainly instructive, the limits of these kinds of comparisons also need to be kept in mind. Bubble proponents have pointed out that commodity markets selected for the development of futures contracts may be naturally more volatile than those commodities without futures markets.

Fifth, speculation was not excessive when correctly compared to hedging demands. The statistics on long-only index fund trading reported in the media and discussed at hearings tend to view speculation in a vacuum – focusing on absolute position size and activity. Working (1960) argued that speculation must be gauged relative to hedging needs. In particular, speculation can only be considered 'excessive' relative to the level of hedging activity in the market. Utilizing Working's speculative 'T-index,' Sanders *et al.* (2010) demonstrate that the level of speculation in nine agricultural futures markets from 2006 to 2008 (adjusting for index fund positions) was not excessive. Indeed, the levels of speculation in all markets examined were within the realm of historical norms. Across most markets, the rise in index buying was more than offset by commercial (hedger) selling. Buyuksahin and Harris (2009) use daily data from the CFTC's internal large trader database to show that Working's T-index in the crude oil futures market increased in parallel with crude oil prices over 2004–2009 but the peak of the index was still well within historical norms. Till (2009) reports similar results for crude oil, heating oil, and gasoline futures over 2006–2009 using recently available data in the CFTC's DCOT report.

The sixth observable fact revolves around the impact of index funds across markets. *A priori*, there is no reason to expect index funds to have a differential impact across markets given similar position sizes. That is, if index funds can inflate prices, they should have a uniform impact across markets for the same relative position size. It is therefore difficult to rationalize why index fund speculation would impact one market but not another. Further, one would expect markets with the highest concentration of index fund positions to show the largest price increases. Irwin *et al.* (2009) find just the opposite when comparing grain and livestock futures markets. The highest concentration of index fund positions was often in livestock markets, which had the smallest price increases through the spring of 2008. This is difficult to reconcile with the assertion that index buying represents demand.

4.4 Evidence to Date

Not surprisingly, a flurry of studies has been completed recently in an attempt to sort out which side of the debate is correct. Some studies find evidence that commodity index funds have impacted commodity futures prices (Einloth, 2009; Gilbert, 2009; Tang and Xiong, 2010). Results in these studies negate the argument that *no* evidence exists of a relationship between index fund trading and movements in commodity futures prices. However, the evidence is weak because the data and methods used in most of these studies are subject to a number of important criticisms. Hamilton's (2009) study, while not definitive in terms of empirics, is the most important of this group because his theoretical model shows the conditions that must occur for index fund speculation to lead to bubble impacts in a storable commodity market such as crude oil.

A number of studies find little evidence of a relationship between index fund positions and movements in commodity futures prices (Aulerich *et al.*, 2009; Buyuksahin and Harris, 2009; Sanders and Irwin, 2010a, b; Stoll and Whaley, 2009). This constitutes a rejection of the first theoretical requirement for speculative impacts. The most recent evidence in crude oil markets (Kilian and Murphy, 2010) also indicates a rejection of the second theoretical requirement for speculative impacts – a zero or near-zero price elasticity of demand. In sum, the weight of the evidence at this point in time clearly tilts in favor of the argument that index funds did not cause a bubble in commodity futures prices.[4]

There is still a need for further research on the market impact of commodity index funds. The first reason is that direct tests of the relationship between index fund positions and price movements in energy futures markets have been hampered by the lack of publicly available data on positions of index funds in these markets. The second reason is ongoing concerns about the power of time-series statistical tests used in the studies that fail to find evidence of a relationship between index fund positions and movements in commodity futures prices. The time-series tests may lack statistical power to reject the null hypothesis because the dependent variable – the change in futures price – is extremely volatile. In the empirical analysis summarized in the following section, we attempt to address both of these deficiencies.

4.5 New Evidence

Our empirical analysis relies on two related data sets compiled by the US CFTC. The CFTC has long provided the breakdown of each Tuesday's open interest for US markets in the *Commitments of Traders* (COT) report. Open interest for a given market is aggregated across all contract expiration months in the weekly report. The traditional COT categories include: (i) commercials (hedgers); (ii) non-commercials (speculators); and (iii) non-reporting (all traders with position sizes below the reporting level).

Starting in 2007 – in response to complaints by traditional traders about the rapid increase in long-only index money flowing into the markets – the CFTC began releasing the weekly *Supplemental Commitments of Traders* (SCOT) reports, which break out the positions of index traders for 12 agricultural markets. According to the CFTC, the index trader positions reflect both pension funds that would have previously been classified as non-commercials as well as swap dealers who would have previously been classified as commercials hedging OTC transactions involving commodity indices. The commodity index trader data are generally considered the best glimpse of index trading activity in the 12 agricultural markets covered by the report.

While the SCOT data represent an improvement over the traditional COT data, concerns were expressed almost immediately that the data did not extend to other markets, particularly energy and metals futures. In response to requests for more information about the composition of open interest in a broader set of markets, the CFTC began publishing the weekly DCOT report in September 2009 and ultimately provided historical data back to June 2006. The DCOT data are available for the same 12 agricultural markets covered by the SCOT report plus a number of energy and metal futures markets. Like the SCOT report, the positions in the DCOT report represent the combined futures and delta-adjusted options positions aggregated across all contracts for a particular market. Reporting traders are classified into four categories: (i) swap dealers; (ii) managed money; (iii) processors and merchants; and (iv) other reporting traders.

An important question, especially for the energy futures markets, is the degree to which the DCOT swap dealers category represents index fund positions. One can infer from comparisons found in the CFTC's September 2008 report on swap dealer positions (CFTC, 2008) that DCOT swap dealer positions in agricultural futures markets correspond reasonably closely to index trader positions. Since swap dealers operating in agricultural markets conduct a limited amount of non-index long or short swap transactions, there is little error in attributing the net long position of swap dealers in these markets to index funds. However, swap dealers in energy futures markets conduct a substantial amount of non-index swap transactions on both the long and short side of the market, which creates uncertainty about how well the net long position of swap dealers in energy markets represent index fund positions.[5] For example, the CFTC estimates that only 41% of long swap dealer positions in crude oil futures on three dates in 2007 and 2008 were linked to long-only index fund positions (CFTC, 2008). Despite this limitation, swap dealers are used in the present study as the best available proxy for index positions in the energy futures markets.

The SCOT data are available weekly from January 3 2006 through December 29 2009 and the DCOT data are available at the same frequency starting on June 13 2006. To facilitate the comparison of the data sets and results, a common sample starting on June 13 2006 containing 186 weekly observations through December 29 2009 was used in all empirical work.

Index trader positions are collected for the 12 SCOT agricultural markets: (i) CBOT corn; (ii) CBOT soybeans; (iii) CBOT soybean oil; (iv) CBOT wheat; (v) KCBOT wheat; (vi) New York Board of Trade (NYBOT) cotton; (vii) Chicago Mercantile Exchange (CME) live cattle; (viii) CME feeder cattle; (ix) CME lean hogs; (x) NYBOT coffee; (xi) NYBOT sugar; and (xii) NYBOT cocoa. Corresponding DCOT data are collected for these 12 SCOT markets along with the DCOT data for NYMEX crude oil and natural gas. The focus in the DCOT data will be on swap dealer positions because of their potential link to index fund positions.

For the above markets, weekly futures returns (price changes) are calculated using nearby futures contracts, appropriately adjusting for contract roll-overs. In order to test for index trader impact on market variability, two measures of volatility are computed: (i) implied volatility from the options markets; and (ii) realized volatility as measured by Parkinson's (1980) extreme value estimator. It is important to establish whether or not index trader positions impact these market characteristics (returns, implied volatility, and realized volatility). Here, causal linkages are directly tested using Granger causality tests.

A simple graphical analysis of index trader positions and market prices can be misleading. As shown in Fig. 4.3 for CBOT wheat, there are periods of time – such as mid-2007 through late 2008 – where there appears to be a close correspondence between index trader positions and price levels. Conversely, there are periods, such as most of 2009, where any relationship seems remote at best. This type of graphical inspection is commonly presented as establishing an 'obvious' link between index positions and prices. However, it is fraught with statistical complications and begs for a more rigorous test of the linkages, if any.

Granger causality is a standard linear technique for determining whether one time series is useful in forecasting another. In our case the time series of interest are market measures of returns, implied volatility, and realized volatility. The causal variables are measures of trader positions and speculation, including net long positions held by index funds, the percentage of long positions held in each market by index funds, and Working's speculative T-index.

Simply put, Granger's test asks the question: Can past values of trader positions be used to predict either market returns or volatility? This is a much more demanding hurdle than simply looking for a contemporaneous correlation

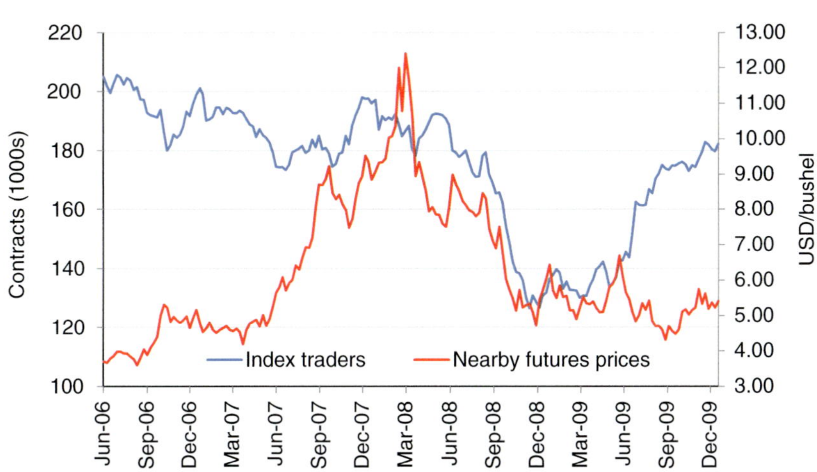

Fig. 4.3. Commodity index trader net long positions in Chicago Board of Trade (CBOT) wheat and nearby futures prices, June 2006–December 2009.

or association between variables. As shown in Fig. 4.4, there is a positive contemporaneous association between changes in net positions held by index traders and price changes (returns) in the CBOT wheat market. The simple correlation coefficient is relatively low at 0.14; but the relationship is statistically significant at the 10% level. However, the magnitude of the impact is quite low since a 3000 contract increase in index traders' long position is associated with just a 0.6% increase in prices during the same week. More importantly, this contemporaneous analysis cannot distinguish between the increase in index traders' positions and other correlated shifts in fundamentals: correlation does not imply causation. Evidence of this point is found in Fig. 4.5, which is the same as Fig. 4.4 except there is a 1 week time lag between the change in index fund positions and the change in the futures price. As clearly shown in Fig. 4.5, increases in net index fund positions are actually followed by small (statistically insignificant) declines in prices the subsequent week. In this example, there is no evidence that changes in index traders' net long positions lead to higher (or lower) market prices.

More formal Granger causality tests are conducted for a number of combinations of causal variables (position measures) and market characteristics. A systems approach is used to test lead-lag dynamics. This improves the power of statistical tests by taking into account the contemporaneous correlation of model residuals across markets. The system test results are summarized in Table 4.1. The formal testing failed to find any reasonably consistent causal links between trader positions and returns. The only statistically significant finding was a negative relationship between positions and market volatility. That is, there is some consistent evidence that increases in index trader positions are followed by lower market volatility. Even these results for market volatility must be interpreted with caution. The possibility still exists that trader positions are correlated with some third variable that is actually causing market volatility to decline.

The data characteristics and empirical results can be summarized with the following general findings and representative results.

1. The overlap between index trader positions (SCOT data set) and those held by swap dealers (DCOT data set) is quite large for the traditional grain and livestock markets. It appears to be a somewhat weaker correspondence for the coffee, sugar, and cocoa markets. It is clear that the swap dealer positions for the energy markets contain many traders other than index funds. Swap dealer positions are at best an imperfect proxy for index fund positions in the energy markets.

This is clearly seen in Table 4.2, which shows the net position (in contracts) held by

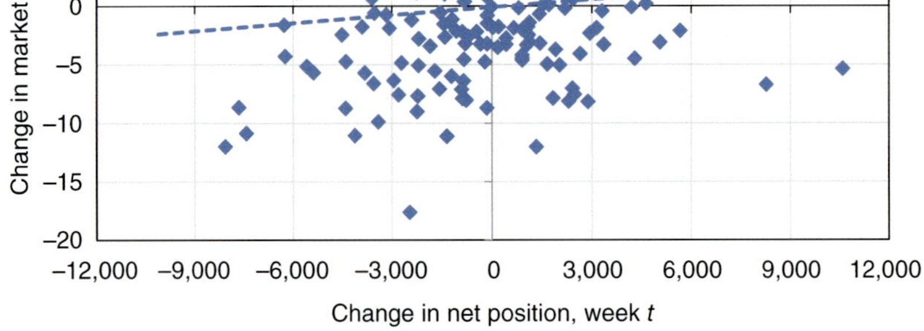

Fig. 4.4. Contemporaneous relationship, Chicago Board of Trade (CBOT) wheat returns (price change) and commodity index trader net long positions, June 2006–December 2009.

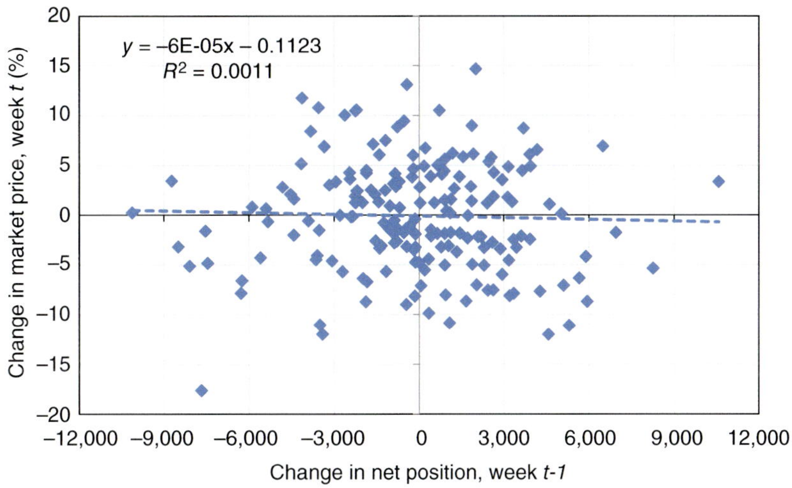

Fig. 4.5. Causal relationship, Chicago Board of Trade (CBOT) wheat returns (price change) and commodity index trader net long positions, June 2006–December 2009.

Table 4.1. Causal relationships estimated for market system, June 2006–December 2009.[a]

	Causal variable		
	Net long position in contracts	Percentage of long positions	Working's speculative T-index[b]
Panel A: Index traders			
Returns	No (negative)	No (positive)	NA
Implied volatility	No (negative)	Yes (negative)	No (positive)
Realized volatility	Yes (negative)	No (positive)	No (positive)
Panel B: Swap dealers			
Returns	No (positive)	No (positive)	NA
Implied volatility	No (negative)	Yes (negative)	NA
Realized volatility	Yes (negative)	No (positive)	NA

[a]A 'Yes' indicates a statistically significant (5% level) causal relationship running from the causal variables (column headings) to the market factors (row headings) for the overall system test. A 'No' indicates that no relationship was found. The direction of the causal relationship is indicated by 'positive' or 'negative' in parenthesis, regardless of whether the impact was statistically significant or not.
[b]NA, not applicable.

index traders (Panel A) and swap dealers (Panel B) over the sample period. In Panel A, the minimum net long position held by index traders is never negative (short); whereas in Panel B, the minimum net long position for sugar, cocoa, crude oil, and natural gas is negative. In these markets, swap dealers clearly hold positions other than those representing long-only index investments.

2. Index fund and swap dealer positions are large. In an absolute sense, the largest average position sizes held in nearly every market is by long index funds or swap dealers. In some markets, such as CBOT wheat, the average position size for these traders is in excess of the speculative position limits. In a relative sense, index and swap dealer positions can also be quite large. Index traders often hold as much as 40% of the long positions in a market and the swap dealer category frequently holds over 30% of the long positions in a given market.

Despite the large average position size, the total size of index funds within a given market is not overwhelming. Table 4.3 shows the

Table 4.2. Summary statistics, net long positions held by index traders and swap dealers (no. of contracts), June 2006–December 2009.[a]

Market[b]	Mean	Maximum	Minimum	Standard deviation
Panel A: Index traders				
Corn	354,043	452,568	223,985	64,877
Soybeans	140,651	198,707	89,731	26,004
Soybean oil	66,011	77,752	36,630	10,192
CBOT wheat	174,677	205,585	126,545	21,769
KCBOT wheat	28,654	46,527	16,293	6,011
Cotton	84,985	122,555	57,841	15,209
Live cattle	110,006	156,752	80,276	20,632
Feeder cattle	7,479	10,889	4,972	1,456
Lean hogs	80,616	127,379	46,004	18,538
Coffee	44,451	67,021	30,572	9,697
Sugar	231,756	392,740	135,745	74,836
Cocoa	18,910	31,883	5,117	5,830
Panel B: Swap dealers				
Corn	313,172	430,100	163,606	77,941
Soybeans	121,557	193,888	73,898	27,892
Soybean oil	61,453	89,502	27,442	16,234
CBOT wheat	142,550	189,217	91,681	25,373
KCBOT wheat	22,073	33,863	9,952	6,906
Cotton	72,092	118,380	42,637	16,797
Live cattle	88,844	128,967	65,368	16,351
Feeder cattle	4,161	6,723	1,730	1,194
Lean hogs	69,149	114,377	36,326	16,858
Coffee	37,179	56,959	21,667	8,718
Sugar	132,099	271,255	–32,149	81,371
Cocoa	8,380	16,474	–5,103	4,763
Crude oil	40,912	106,176	–10,534	27,504
Natural gas	49,018	253,500	–67,553	78,063

[a]Net positions are simply calculated as long positions minus short positions.
[b]CBOT, Chicago Board of Trade; KCBOT, Kansas City Board of Trade.

percentage of the market that is comprised of each trader category in the SCOT (Panel A) and DCOT (Panel B) data.[6] In each market, the largest participant is a category other than index funds or swap dealers. In fact, in the SCOT categories, index traders are the smallest category in four of the 12 markets and the second smallest in the other eight markets. The exception is swap dealers in the crude oil market who account for 37% of the open interest. Again, this inconsistency indicates that the link between swap dealer positions and index traders may be weak in the energy markets.

3. There is no convincing evidence that positions held by index traders or swap dealers impact market returns. Except for a few instances in individual markets, Granger-style causality tests fail to reject the null hypothesis that that trader positions do not lead market returns.

The full results for testing if commodity index traders lead market returns are shown in Table 4.4. In the individual markets, the null hypothesis of no causality can be rejected at the 5% level (with 95% confidence) as shown by the *p*-values for the null hypothesis that $\beta_i = 0, \forall j$. Importantly, however, the directional impact for corn is negative while it is positive for cotton. This makes very little sense in the context of the current debate. Not surprisingly, the system-wide impact, which takes into account the opposing directional findings across markets, is negative (-0.4010) and indistinguishable from zero.

4. Larger long positions by index traders and swap dealers lead to lower market volatility in a Granger sense. There is a consistent tendency across a number of position and volatility measures to reject the null hypothesis that index trader positions do not lead market volatility. The

Table 4.3. Percentage of total open interest held by *Supplemental Commitments of Traders* (SCOT) and *Disaggregated Commitments of Traders* (DCOT) categories, June 2006 to December 2009.

Panel A: SCOT categories

Market	Non-commercial (%)	Commercial (%)	Index (%)	Non-reporting (%)
Corn	39	35	13	14
Soybeans	40	33	14	14
Soybean oil	35	44	12	8
CBOT wheat	41	26	23	10
KCBOT wheat	28	39	12	20
Cotton	39	38	17	6
Live cattle	38	28	20	14
Feeder cattle	38	17	14	31
Lean hogs	39	25	21	14
Coffee	43	39	13	5
Sugar	31	44	17	8
Cocoa	33	54	7	6

Panel B: DCOT categories

Market	Managed money (%)	Producers and merchants (%)	Swap dealers (%)	Other reporting (%)	Non-reporting (%)
Corn	16	32	13	25	14
Soybeans	19	31	13	23	14
Soybean oil	17	42	14	18	8
CBOT wheat	22	23	22	22	10
KCBOT wheat	19	38	10	13	20
Cotton	16	35	18	25	6
Live cattle	25	27	18	16	14
Feeder cattle	23	17	9	20	31
Lean hogs	23	25	19	19	14
Coffee	20	37	14	25	5
Sugar	16	39	19	18	8
Cocoa	26	48	12	9	6
Crude oil	18	18	37	23	3
Natural gas	43	12	28	11	5

direction of the impact is routinely negative. While index positions lead to lower volatility in a statistical sense, it is possible that trader positions coincide with some other fundamental variable that is actually causing the lower market volatility. Still, this result is contrary to popular notions about index traders increasing market volatility.

These general conclusions apply to both the volatility implied in the options markets and the realized volatility. As a representative example, consider the Granger causality test of the null hypothesis that DCOT swap dealers' net positions do not lead realized market volatility. The system estimation results are presented in Table 4.5. The null hypothesis is rejected at the 5% level in soybeans and cocoa. In both of these markets, the directional impact is negative: increases in net long positions held by swap dealers predict lower market volatility in the subsequent week. More convincing than the individual market results, the system results show that the aggregate directional impact is statistically negative (-36.1) with nearly 99% confidence (1 - 0.0131).

5. Excessive speculation – as measured by Working's T-index – is associated with greater subsequent variability in a few markets. These results conflict with negative relationships found between index trader positions and market volatility.

Table 4.4. Granger causality test results for commodity index trader net positions do not lead returns, June 2006–December 2009.[a, b]

$$R_{t,k} = \alpha_k + \sum_{i=1}^{m} \gamma_{i,k} R_{t-i,k} + \sum_{j=1}^{n} \beta_{j,k} \Delta NET_{t-j,k} + \varepsilon_{t,k}$$

for each market, k, and time, t.

Market	m, n	p-value $\beta_j = 0, \forall j$	Estimate $\Sigma\, \beta_j$	p-value $\Sigma\, \beta_j = 0$
Corn	1,1	0.0002	−0.1210	
Soybeans	1,1	0.4206	−0.0444	
Soybean oil	1,1	0.2922	0.0874	
CBOT wheat	1,1	0.3629	0.0319	
KCBOT wheat	1,1	0.1261	−0.1460	
Cotton	1,1	0.0018	0.3590	
Live cattle	2,2	0.1812	0.0008	0.9861
Feeder cattle	2,1	0.1300	−0.3730	
Lean hogs	1,1	0.2078	−0.1320	
Coffee	1,1	0.3348	−0.1730	
Sugar	1,1	0.2647	−0.0520	
Cocoa	1,1	0.4591	0.1610	
		p-value $\beta_{j,k} = 0, \forall j,k$	Estimate $\Sigma\Sigma\, \beta_{j,k}$	p-value $\Sigma\Sigma\, \beta_{j,k} = 0$
System		0.0001	−0.4010	0.3836

[a] $\Sigma\, \beta_j$ values are taken to the 10^5 power.
[b] The models are estimated across the K markets as a seemingly unrelated regressions (SUR) system. Wald tests could not reject the following cross-market coefficient restrictions: $\alpha_1 = \alpha_2 = ... = \alpha_K$; $\gamma_{1,1} = \gamma_{1,2} = ... = \gamma_{1,K}$; and $\gamma_{2,1} = \gamma_{2,2} = ... = \gamma_{2,K}$ for all K markets. These restrictions are imposed on the system and the common coefficients are estimated as a single pooled parameter across all K markets.

The contrasting results suggest that excessive speculation is broader than just index fund activity and may be better measured with Working's T-index, which measures excessive speculation relative to hedging demands.

Table 4.6 shows the summary statistics for Working's T-index adjusted for index trader positions. For example, the average T-index for corn is 1.15 – indicating speculation in the corn market is 15% greater than that needed to meet hedging needs. Historically, this would have been considered a potentially inadequate amount of speculation to efficiently meet hedging demands. Notably, some of the markets with high T-values (livestock and CBOT wheat) are also those markets with a relatively high portion of index traders (see Table 4.3, Panel A). Still, even in these markets, the maximums are not beyond those recorded by prior researchers, the average values are near historic norms, and the minimums could be considered inadequate.

Working's T-index is silent on the direction of speculation (long versus short). Instead, the amount of speculation is gauged relative to what is needed to balance hedging positions. Because it is directionless, Working's T-index is only tested as a causal variable for market volatility. Table 4.7 shows the results for testing if the T-index Granger causes realized market volatility. Causality is found in four markets at the 95% confidence level. In all four markets, the directional impact is positive – higher levels of excessive speculation, as measured by Working's T, are followed by greater realized market volatility. For example, if the speculative index in lean hogs increases by 0.10, then actual volatility the following week increases by 1.18%. These individual market results are notable in comparison to the negative directional impacts found when simply measuring speculation with net index fund positions (Table 4.5). Still, the impact is not pervasive across markets as no system impact is found at even a modest confidence level.

In sum, our results tilt the weight of the evidence even further in favor of the argument that

Table 4.5. Granger causality test results for swap dealer net positions do not lead realized volatility, June 2006–December 2009.[a, b]

$$RV_{t,k} = \alpha_k + \sum_{i=1}^{m} \gamma_{i,k} RV_{t-i,k} + \sum_{j=1}^{n} \beta_{j,k} \Delta NET_{t-j,k} + \varepsilon_{t,k}$$

for each market, *k*, and time, *t*.

Market	m, n	p-value $\beta_j = 0, \forall j$	Estimate $\Sigma \beta_j$	p-value $\Sigma \beta_j = 0$
Corn	2,1	0.8258	0.2000	
Soybeans	4,1	0.0242	–3.3700	
Soybean oil	2,1	0.5347	–0.9500	
CBOT wheat	2,1	0.6975	0.4370	
KCBOT wheat	3,1	0.1308	–5.5000	
Cotton	3,1	0.9358	0.2340	
Live cattle	3,1	0.0600	–2.4600	
Feeder cattle	3,1	0.5317	–5.8200	
Lean hogs	3,1	0.1531	3.7900	
Coffee	1,2	0.1568	–11.8200	0.0581
Sugar	3,1	0.8018	–0.3200	
Cocoa	4,1	0.0420	–12.0300	
Crude oil	3,1	0.0889	1.0500	
Natural gas	1,1	0.5975	0.4610	
		p-value $\beta_{j,k} = 0, \forall j,k$	Estimate $\Sigma\Sigma \beta_{j,k}$	p-value $\Sigma\Sigma \beta_{j,k} = 0$
System		0.0408	–36.1	0.0131

[a] $\Sigma \beta_j$ values are taken to the 10^5 power.
[b] The models are estimated across the *K* markets as an SUR system. Wald tests could not reject the following cross-market coefficient restrictions: $\gamma_{3,1} = \gamma_{3,2} = \ldots = \gamma_{23,K}$ for all *K* markets. These restrictions are imposed on the system and the common coefficients are estimated as a single pooled parameter across all *K* markets.

Table 4.6. Summary statistics, Working's speculative T-index, adjusted for index trader positions, June 2006–December 2009.[a]

Market	Mean	Maximum	Minimum	Standard deviation
Corn	1.15	1.34	1.07	0.06
Soybeans	1.17	1.53	1.09	0.09
Soybean oil	1.12	1.36	1.04	0.07
CBOT wheat	1.44	1.87	1.19	0.16
KCBOT wheat	1.18	1.34	1.08	0.06
Cotton	1.16	1.48	1.03	0.11
Live cattle	1.33	1.50	1.15	0.07
Feeder cattle	1.86	3.28	1.32	0.38
Lean hogs	1.43	2.01	1.17	0.19
Coffee	1.17	1.41	1.04	0.08
Sugar	1.15	1.26	1.06	0.04
Cocoa	1.14	1.28	1.06	0.05

[a] Working's speculative 'T'-index is easily calculated using the traditional COT trader categories:
T = 1 + SS / (HL + HS) if (HS ≥ HL)
OR
TL = 1 + SL / (HL + HS) if (HL > HS)
where open interest held by speculators (non-commercials) and hedgers (commercials) is denoted as follows: SS, speculation, short; SL, speculation, long; HL, hedging, long; and HS, hedging, short.

Table 4.7. Granger causality test results for T-index does not lead realized volatility, June 2006–December 2009.[a]

$$RV_{t,k} = \alpha_k + \sum_{i=1}^{m} \gamma_{i,k} RV_{t-i,k} + \sum_{j=1}^{n} \beta_{j,k} TIndex_{t-j,k} + \varepsilon_{t,k}$$

for each market, *k*, and time, *t*.

Market	*m,n*	*p*-value $\beta_j = 0, \forall j$	Estimate $\Sigma\,\beta_j$	*p*-value $\Sigma\,\beta_j = 0$
Corn	1,1	0.0470	24.8261	
Soybeans	4,1	0.6982	–2.5196	
Soybean oil	2,1	0.7590	2.3205	
CBOT wheat	2,1	0.5745	–1.7284	
KCBOT wheat	3,1	0.7993	–1.8937	
Cotton	3,1	0.4823	–4.7687	
Live cattle	3,1	0.3602	3.2854	
Feeder cattle	3,1	0.0208	1.8090	
Lean hogs	3,1	0.0003	11.7991	
Coffee	1,1	0.6234	–4.0321	
Sugar	4,1	0.2101	–30.5000	
Cocoa	4,1	0.0308	34.0968	
		p-value $\beta_{j,k} = 0, \forall j,k$	Estimate $\Sigma\Sigma\,\beta_{j,k}$	*p*-value $\Sigma\Sigma\,\beta_{j,k} = 0$
System		0.0028	32.6945	0.3844

[a]The models are estimated across the *K* markets as an SUR system. Wald tests could not reject the following cross-market coefficient restrictions: $\gamma_{2,1} = \gamma_{2,2} = ... = \gamma_{2,K}$; $\gamma_{3,1} = \gamma_{3,2} = ... = \gamma_{3,K}$ for all *K* markets. These restrictions are imposed on the system and the common coefficients are estimated as a single pooled parameter across all *K* markets.

index funds did *not* cause a bubble in commodity futures prices.[7] The evidence in our study is strongest for the agricultural futures markets because the data on index trader positions are measured with reasonable accuracy. The evidence is not as strong in the two energy markets studied because of considerable uncertainty about the degree to which the available data actually reflect index trader positions in these markets. Perhaps the most surprising result is the consistent tendency for increasing index fund positions to be associated with *declining* volatility. This result is contrary to popular notions about the market impact of index funds but is not so surprising in light of the traditional problem in commodity futures markets of the *inadequacy* of speculation (see Sanders *et al.*, 2010).

4.6 Conclusions

The policy implications of the available evidence on the market impact of commodity index funds are straightforward: current regulatory proposals to limit speculation – especially on the part of index funds – are not justified and likely will do more harm than good. In particular, limiting the participation of index fund investors would rob commodity futures markets of an important source of liquidity and risk-absorption capacity at a time when both are in high demand. More ominously, tighter position limits on speculation in commodity futures markets combined with the removal of hedge exemptions could force commodity index funds into cash markets, where truly chaotic results could follow. The net result is that moves to tighten regulations on index funds are likely to make commodity futures markets less-efficient mechanisms for transferring risk from parties who do not want to bear it to those that do, creating added costs that ultimately are passed back to producers in the form of lower prices and to consumers as higher prices.

These conclusions do not necessarily imply that commodity futures markets have functioned flawlessly during the last several years. In particular, the lack of consistently acceptable convergence performance for CBOT corn, soybean, and wheat contracts since late 2005 has been widely discussed (e.g. Henriques, 2008). The failure of cash and futures prices to converge at contract expiration has existed for extended and varied periods. Performance has been consistently weakest in wheat, with delivery location basis at times exceeding $1/bushel, a level of disconnect between cash and futures not previously experienced in grain futures markets. The possible role of index funds in contributing to convergence problems has also been widely discussed (USS/PSI, 2009). Further research is needed to better understand the impact of index fund trading on this aspect of commodity market performance.

Notes

[1] Original citation: Irwin, S.H. and Sanders, D.R. (2010) The impact of index and swap funds in commodity futures markets: preliminary results. OECD Food, Agriculture and Fisheries Working Papers No. 27. OECD Publishing, Paris. Available at: https://doi.org/10.1787/5kmd40wl1t5f-en.

[2] See the following articles:

Buttonwood. (2010) Clearing the usual suspects: investors may not have caused commodity price rises. *The Economist*, June 24. Available at: https://www.economist.com/finance-and-economics/2010/06/24/clearing-the-usual-suspects (accessed January 14 2022).

Branson, Sir R., Masters, M. and Frenk, D. (2010) Letter: swaps, spots and bubbles. *The Economist*, July 29. Available at: https://www.economist.com/letters/2010/07/29/on-speculation-free-trade-indian-infrastructure-wheat-rust-cyber-security-mental-health-car-insurance-general-mcchrystal (accessed January 14 2022).

The Economist (2010) Dr Evil, or drivel? The charge-sheet against commodity speculators is flimsy. *The Economist*, November 11. Available at: https://www.economist.com/finance-and-economics/2010/11/11/dr-evil-or-drivel (accessed January 14 2022).

The Economist (2010) Know your onions: commodity speculators do more good than harm. *The Economist*, November 11. Available at: https://www.economist.com/leaders/2010/11/11/know-your-onions (accessed January 14 2022).

[3] In the remainder of this report, the term 'commodity index fund' or 'index fund' is used generically to refer to all of the varied long-only commodity investment instruments.

[4] The full version of this technical report contains detailed reviews of the studies cited in this section. See Irwin and Sanders (2010).

[5] This was precisely the reason that the CFTC excluded energy futures markets from the SCOT report.

[6] The denominator in these calculations is the sum of total long and short open interest, or two times either the long or short total open interest.

[7] The full version of this technical report contains a detailed presentation of all statistical test results. See Irwin and Sanders (2010).

References

Aulerich, N.M., Irwin, S.H. and Garcia, P. (2009) The price impact of index funds in commodity futures markets: evidence from the CFTC's daily large trader reporting system. In: *Proceedings of the NCCC-134 Conference on Applied Commodity Price Analysis, Forecasting, and Market Risk Management, St Louis, Missouri, April 20–21*. Available at: https://legacy.farmdoc.illinois.edu/nccc134/conf_2009/pdf/confp19-09.pdf (accessed January 14 2022).

Buyuksahin, B. and Harris, J.H. (2009) The role of speculators in the crude oil futures markets. Working Paper, U.S. Commodity Futures Trading Commission. Available at: https://www.cftc.gov/sites/default/files/idc/groups/public/@swaps/documents/file/plstudy_19_cftc.pdf (accessed January 14 2022).

CFTC (Commodity Futures Trading Commission) (2008) *Staff Report on Commodity Swap Dealers and Index Traders with Commission Recommendations*. Available at: https://www.cftc.gov/sites/default/

files/idc/groups/public/@newsroom/documents/file/cftcstaffreportonswapdealers09.pdf (accessed January 14 2022).

Davis, A. (2009) Cargill's inside view helps it buck downturn. *The Wall Street Journal*, January 14. Available at: http://online.wsj.com/article/SB123189501407679581.html?mod=rss_whats_news_us (accessed January 14 2022).

Einloth, J. (2009) Speculation and recent volatility in the price of oil. FDIC Center for Financial Research Working Paper No. 2009-08. Federal Deposit Insurance Corporation. Available at: https://www.fdic.gov/bank/analytical/cfr/2009/wp2009/2009-08.pdf (accessed January 14 2022).

Gilbert, C.L. (2009) Speculative influences on commodity futures prices, 2006–2008. Working Paper, Department of Economics, University of Trento, Trento, Italy. Available at: https://www.cftc.gov/sites/default/files/idc/groups/public/@swaps/documents/file/plstudy_14_cifrem.pdf (accessed January 14 2022).

Hamilton, J.D. (2009) Causes and consequences of the oil shock of 2007–08. *Brookings Papers on Economic Activity,* spring, 215–261.

Headey, D. and Fan, S. (2008) Anatomy of a crisis: the causes and consequences of surging food prices. *Agricultural Economics* 39, 375–391.

Henriques, D.B. (2008) Odd crop prices defy economics. *The New York Times*, March 28. Available at: http://www.nytimes.com/2008/03/28/business/28commodities.html (accessed January 14 2022).

Irwin, S.H. and Sanders, D.R. (2010) *The Impact of Index and Swap Funds in Commodity Futures Markets*. Technical report prepared for the Organisation for Economic Co-operation and Development (OECD). OECD Food, Agriculture and Fisheries Papers No. 27. OECD Publishing, Paris.

Irwin, S.H., Sanders, D.R. and Merrin, R.P. (2009) Devil or angel? The role of speculation in the recent commodity price boom (and bust). *Journal of Agricultural and Applied Economics* 41, 393–402.

Kilian, L. and Murphy, D. (2010) The role of inventories and speculative trading in the global market for crude oil. Working Paper, Department of Economics, University of Michigan, Ann Arbor, Michigan, May. Available at: https://papers.ssrn.com/sol3/papers.cfm?abstract_id=1578103 (accessed January 14 2022).

Krugman, P. (2008) More on oil and speculation. *The New York Times*, May 13. Available at: http://krugman.blogs.nytimes.com/2008/05/13/more-on-oil-and-speculation/ (accessed January 14 2022).

Masters, M.W. (2008) Testimony before the Committee on Homeland Security and Governmental Affairs, United States Senate. May 20. Available at: http://hsgac.senate.gov/public/_files/052008Masters.pdf (accessed January 14 2022).

Masters, M.W. and White, A.K. (2008) The accidental Hunt brothers: how institutional investors are driving up food and energy prices. Available at: https://www.cftc.gov/sites/default/files/idc/groups/public/@swaps/documents/file/plstudy_31_ahb.pdf (accessed January 14 2022).

Parkinson, M. (1980) The extreme value method for estimating the variance of the rate of return. *Journal of Business* 53, 61–65.

Petzel, T.E. (2009) Summary comments on the testimony before the CFTC, July 28. Available at: https://www.cftc.gov/sites/default/files/idc/groups/public/@newsroom/documents/file/hearing072809_petzel1.pdf (accessed January 14 2022).

Sanders, D.R. and Irwin, S.H. (2010a) A speculative bubble in commodity futures prices? Cross-sectional evidence. *Agricultural Economics* 41, 25–32.

Sanders, D.R. and Irwin, S.H. (2010b) Bubbles, froth, and facts: The impact of index funds on commodity futures prices. Working Paper, Department of Agricultural and Consumer Economics, University of Illinois at Urbana-Champaign, Illinois.

Sanders, D.R., Irwin, S.H. and Merrin, R.P. (2010) The adequacy of speculation in agricultural futures markets: too much of a good thing? *Applied Economic Perspectives and Policy* 32, 77–94.

Smith, J.L. (2009) World oil: market or mayhem? *Journal of Economic Perspectives* 23, 145–164.

Stoll, H.R. and Whaley, R.E. (2009) Commodity index investing and commodity futures prices. Working Paper, Owen Graduate School of Management, Vanderbilt University, Nashville, Tennessee. Available at: https://www.cftc.gov/sites/default/files/idc/groups/public/@swaps/documents/file/plstudy_45_hsrw.pdf (accessed January 14 2022).

Tang, K. and Xiong, W. (2010) Index investment and the financialization of commodities. Working Paper, Department of Economics, Princeton University, Princeton, New Jersey. Available at: https://papers.ssrn.com/sol3/papers.cfm?abstract_id=1683135 (accessed January 14 2022).

Till, H. (2009) Has there been excessive speculation in the U.S. oil futures markets? What can we (carefully) conclude from new CFTC data? EDHEC-Risk Institute position paper. Available at: https://papers.ssrn.com/sol3/papers.cfm?abstract_id=2608027 (accessed January 14 2022).

USS/PSI (United States Senate, Permanent Subcommittee on Investigations) (2009) *Excessive Speculation in the Wheat Market*. Majority and Minority Staff Report. Available at: https://www.hsgac.senate.gov/imo/media/doc/REPORTExcessiveSpecullationintheWheatMarketwoexhibitschartsJune2409.pdf?attempt=2 (accessed January 14 2022).

Working, H. (1960) Speculation on hedging markets. *Food Research Institute Studies* 1, 185–220.

Wright, B. (2009) International grain reserves and other instruments to address volatility in grain markets. Policy Research Working Paper WPS5028. The World Bank. Available at: https://documents.worldbank.org/en/publication/documents-reports/documentdetail/375561468336329144/international-grain-reserves-and-other-instruments-to-address-volatility-in-grain-markets (accessed January 14 2022).

5

Testing the Masters Hypothesis in Commodity Futures Markets[1]

New Author Foreword

Critics of our OECD report focused on several data and methodology issues. The main data issue was the lack of accurate data on index positions in energy futures markets, particularly West Texas Intermediate (WTI) crude oil which was at the center of the entire speculation controversy. We were aware of this limitation and highlighted it in the OECD report. The principal methodology issue was a supposed lack of statistical power of the Granger causality time-series tests we used in the analysis. Given the intensity of the reaction to the report, it did not take us long to realize that we needed to respond to the criticisms in a constructive manner.

Another important motivation for this research was a paper by Kenneth Singleton that first appeared in working paper form in May 2010. This was after we had submitted the final version of our OECD report so we could not cite it or include it in our literature review. Singleton's paper (ultimately published in 2014) would end up having an outsized impact on the debate about the impact of index funds on commodity futures prices, in no small part because he was a prominent financial economist from Stanford. He was an intellectual heavyweight and his work had instant credibility. But from the start, we were skeptical of Singleton's findings, which suggested commodity index investment had a large and economically significant impact on WTI crude oil futures prices.

We knew that the first step in our new research was to obtain better data on commodity index positions. Fortunately, the US Commodity Futures Trading Commission (CFTC) had recently begun publishing the *Index Investment Data* report, or IID for short. This data series addressed something called the 'netting' problem with respect to measuring index positions in commodity futures markets. The bulk of commodity index investments are offered to investors by swap dealers, and these dealers cover their risk by hedging in the futures market. This is how most commodity index positions are ultimately placed in the market, which is a relatively minor technical issue except when swap dealers have a 'book' of business on both the long and the short side of the market. When this is the case, dealers only hedge the net of the long and short positions in commodity futures markets, thereby masking the true size and variation of long commodity index positions. This is mainly a problem in metal and energy markets, which include the critical futures market for crude oil. The IID report solved this problem by requiring swap dealers to report their gross long index positions ('pre-netting') to the CFTC for all major markets.

We began the article with a detailed and careful explanation of the different index position series available from the CFTC, as there seemed to be a good bit of confusion about what each one measured and how they related to one another. This allowed us to develop the important argument that the IID data was the most accurate available across all types of commodity futures markets, and especially for crude oil. This turned out to be critical for responding to Singleton's work. He showed a huge spike in index trading during 2008 that seemed to almost perfectly coincide with the spike in crude oil futures prices. By comparison, the IID data showed that index positions in crude oil futures went down, not up, during the crucial period in 2008. Something seemed seriously amiss with Singleton's data.

DOI:10.1079/9781800622104.0005

We discovered that Singleton used a method of projecting index positions in crude oil futures that seemed logical on the surface. In particular, he used the relatively accurate index position data from the CFTC in agricultural markets to project positions in crude oil futures. The IID data allowed us to show that this algorithm produced inaccurate estimates of positions in crude oil, and this explained why Singleton's finding of an economically large impact of index positions on crude oil futures prices was faulty. However, we did not yet know precisely why the algorithm produced incorrect index positions. That would have to wait for a later article (see Chapter 10, this volume).

We also made a major methodological improvement in the article by using cross-sectional statistical tests instead of time-series tests. The problem with time-series tests is that there is always the possibility that you have not properly controlled for other important factors that impact commodity prices through time besides index investment. Cross-sectional tests address this problem by examining variation in index positions across commodities at a point in time, thereby eliminating the need to specify time-series conditioning variables. After using better data and more powerful statistical tests, the results were pretty much the same as before – no consistent relationship between index positions and commodity futures price movements.

When we were writing the early drafts of the article, we knew that it needed a catchy title given the rapidly expanding literature on the subject. We believed that we had produced the best empirical evidence to date, and we wanted to make sure the paper was not lost in the shuffle. We decided to be bold and title it 'Testing the Masters Hypothesis in Commodity Futures Markets.' The name in the title refers to Michael W. Masters, a hedge fund manager and leading proponent of the view that index funds were a major driver of the spike in commodity futures prices. Naming the hypothesis in this way turned out to be a masterstroke. First, it focused the debate about speculation in a very specific manner on the idea that index funds had caused large and long-lasting bubbles in commodity futures prices. Second, it personalized the controversy in a way that made it easy to remember and reference.

Looking back, it is impressive that we completed the research and writing for the first version of the article so quickly, at least by academic standards. We were ready to submit to a journal in October 2010, just 6 months after the release of the OECD report. Thinking we had a real winner on our hands, we submitted the paper to a couple of top finance journals, but without any luck. Rereading the reviews today, it is still not clear why these journals rejected the paper: pretty weak sauce. Publishing in top-level finance and economic journals is very clubby and we definitely were not part of the club.

Fortunately, a good field journal, *Energy Economics*, accepted the paper without much revision in late 2011 and it appeared in published form in 2012. It is not an overstatement to say that the article was an instant hit and quickly started garnering an impressive number of citations. The article has become one of our most widely cited and is well on its way to 400 citations. While we were very gratified by the positive reception, we also have to admit that the success of the article had as much to do with the catchy title as the substance of the work. We would still argue today that the cross-sectional approach using IID positions is the most powerful test of the market impact of index funds. But the cross-sectional approach has never really caught on in the literature, for reasons that are not obvious to us. Differences of opinion make a horse-race, right?

Abstract

The Masters Hypothesis is the claim that long-only index investment was a major driver of the 2007–2008 spike in commodity futures prices and energy futures prices in particular. Index position data compiled by the CFTC are carefully compared. In the energy markets, index position estimates based on agricultural markets are shown to contain considerable error relative to the CFTC's *Index Investment Data* (IID). Fama-MacBeth tests using the CFTC's quarterly IID find very little evidence that index positions influence returns or volatility in 19 commodity futures markets. Granger causality tests and long-horizon regression tests also show no causal links between daily returns or volatility in the crude oil and natural gas futures markets and the positions for two large energy exchange-traded index funds. Overall, the empirical results of this study offer no support for the Masters Hypothesis.

Key words: commodity, futures market, index funds, Michael Masters, price

JEL categories: D84, G13, G14, Q13, Q41

5.1 Introduction

Commodity futures prices spiked in 2007–2008, led by an increase in crude oil futures prices to a new (nominal) all-time high of $145/barrel. As the spike developed, concerns emerged that the increase was being driven by inflows into new commodity index investments.[2] These financial investments are packaged in a variety of forms but share a common goal – provide the investor with long-only exposure to returns from an index of commodity prices.[3] For example, the Standard and Poor's Goldman Sachs Commodity Index™ (S&P GSCI) is one of the most widely tracked commodity indexes and generally considered an industry benchmark. It is computed as a production-weighted average of the prices from 24 commodity futures markets with a relatively heavy weighting towards energy markets.

Hedge fund manager Michael W. Masters is a leading proponent of the view that commodity index investment was a major driver of the spike in commodity futures prices. He has testified numerous times before the US Congress and Commodity Futures Trading Commission (CFTC) with variations of the following argument:

> Institutional Investors, with nearly $30 trillion in assets under management, have decided en masse to embrace commodities futures as an investable asset class. In the last five years, they have poured hundreds of billions of dollars into the commodities futures markets, a large fraction of which has gone into energy futures. While individually these Investors are trying to do the right thing for their portfolios (and stakeholders), they are unaware that collectively they are having a massive impact on the futures markets that makes the Hunt brothers pale in comparison. In the last 4½, years assets allocated to commodity index replication trading strategies have grown from $13 billion in 2003 to $317 billion in July 2008. At the same time, the prices for the 25 commodities that make up these indices have risen by an average of over 200%. Today's commodities futures markets are excessively speculative, and the speculative position limits designed to protect the markets have been raised, or in some cases, eliminated. Congress must act to re-establish hard and fast position limits across all markets.
>
> (Masters and White, 2008, p. 1)

In essence, Masters argues that massive buy-side 'demand' from index funds created a bubble in commodity prices, with the result that prices, and crude oil prices in particular, far exceeded fundamental values at the peak. We use the term 'Masters Hypothesis' as a short-hand label for this argument.

As highlighted in the above quote, the regulatory response to concerns about the impact of index investments in commodity futures markets centers on speculative position limits. Limits on speculative positions in US agricultural futures markets have been set by the CFTC and its precursors for decades.[4] The CFTC proposed to extend this regulatory regime to four energy futures markets in early 2010. The fact that the CFTC received over 8000 responses during the public comment period for the proposed rule-making (Acworth and Morrison, 2010), the second highest number of responses in its 36-year history, highlights the economic importance of the policy debate surrounding commodity index investments. Most recently, the 2010 Dodd-Frank Wall Street Reform and Consumer Protection Act granted the CFTC broad authority to set aggregate speculative position limits on futures and swap positions in all non-exempt 'physical commodity markets' in the USA.

From a theoretical perspective, the impact of index funds in commodity futures markets hinges on the predictability of their position changes. If position changes are perfectly predictable, index funds will not have an impact in a rational expectations equilibrium because other market participants will anticipate their activity, trade against them, and thereby negate any potential impact (De Long *et al.*, 1990). If index fund position changes are less than perfectly predictable, a market impact is possible. However, unpredictability of index fund activity is a necessary but not sufficient condition for a market impact. For example, assume index fund position changes are unpredictable but related to changing fundamentals (speculation in the traditional sense), then the changes would be positively correlated with contemporaneous or subsequent changes in commodity futures prices, but this would be a reflection of valuable information on fundamentals rather than a 'flow' impact.

One might expect actual index position changes to be highly predictable since most funds track well-known commodity indexes and publish their mechanical procedures for 'rolling' to new contract months. However, this ignores the fact that positions also change due to investment flows into and out of index funds and these may be quite large. A more plausible scenario is one where index fund position changes are largely unpredictable and unrelated to market fundamentals, since portfolio diversification is a main driver of investment in index funds (Stoll and Whaley, 2010) and this has the effect of making index positions 'insensitive to the supply and demand fundamentals' (Masters and White, 2008, p. 29). Prices could be impacted for several reasons under this scenario:

1. The commodity futures market may not be sufficiently liquid to absorb the large order flow of index funds. This implies prices are temporarily pushed away from fundamental value. Since the impact is temporary, contemporaneous index fund position and price changes are positively correlated and current position changes and subsequent price changes are negatively correlated. This is the classic problem of illiquidity arising from the asynchronous arrival of traders to the marketplace (Grossman and Miller, 1988).
2. Index investors are in effect noise traders who make arbitrage risky. This opens the possibility of index investors 'creating their own space' if their positions are large enough (De Long *et al.*, 1990). Once again a positive contemporaneous correlation is implied between index position changes and price changes; however, subsequent price changes may be positively or negatively correlated with the current position change, depending on whether prices are pushed above or below fundamentals.
3. Other traders in commodity futures markets have difficulty distinguishing signals from noise. The large order flow of index funds on the long side of the market may be seen as a reflection of valuable private information about commodity price prospects, which has the effect of driving the futures price higher as other traders subsequently revise their own demands upward (Grossman, 1986). This increase in the expected future cash price leads to an increase in inventories and also raises current cash prices (Hamilton, 2009). Index position changes are positively correlated with contemporaneous price changes and, possibly, subsequent price changes if the reaction of other traders to index order flow is not instantaneous.

However plausible from a logical standpoint, it is nonetheless an empirical question whether these types of impacts are discernable in actual market observations.

Several recent studies test whether there is a statistical link between market positions of index funds and commodity futures price movements.[5] Gilbert (2009) reports evidence of a significant relationship between index fund trading activity and returns in three commodity futures markets: (i) crude oil; (ii) aluminum; and (iii) copper. He estimates the maximum impact of index funds in these markets to be a price increase of 15%. In subsequent work, Gilbert (2010) finds evidence of significant relationship between index fund trading and food price changes. Singleton (2011) estimates a regression model of crude oil futures prices and finds that index investment flows are an important determinant of price changes along with several other conditioning variables. His estimates indicate that a 1 million contract increase in index fund positions in WTI crude oil over the previous 13-week period results in a 0.272% increase in nearby crude oil futures prices in the next week. Both Gilbert (2009, 2010) and Singleton (2011) rely on measures of index positions in energy markets imputed from positions held in agricultural commodities.

Alternatively, Brunetti and Buyuksahin (2009) conduct a battery of Granger causality tests and do not find a statistical link between swap dealers' positions (a proxy for commodity index fund positions) and subsequent returns or volatility in the crude oil, natural gas, and corn futures markets. Stoll and Whaley (2010) also use a variety of tests, including Granger causality tests, and find no evidence that the position of commodity index traders impacts prices in agricultural futures markets. Sanders and Irwin (2010, 2011a, b) report similar results for agricultural and energy futures markets. Buyuksahin and Harris (2011) do not find a statistical link between swap dealers' positions and changes in crude oil futures prices. Buyuksahin and Robe (2011) show that index fund activity (again, as measured by swap dealer positions) is not associated with the increasing correlation between commodity and stock returns.

Irwin and Sanders (2011) survey this literature and conclude that the weight of the available empirical evidence tilts against the Masters Hypothesis. However, proponents of the Masters Hypothesis sharply criticize the data and methods used in the studies that fail to detect a market impact of index fund investment.[6] The first criticism is due to reliance on the CFTC's *Commitments of Traders* (COT) database of the positions held by reporting traders. In this database the futures positions of swap dealers represent the net of long commodity index positions with other long and short over-the-counter (OTC) positions and this may mask the true level of index investment in commodities markets; that is, to the degree that long and short OTC positions are internally netted by swap dealers in a particular commodity market, the futures positions reported to the CFTC by swap dealers may not accurately reflect the total effective 'demand' of index investors. The second criticism is that the impact of index investment is estimated for daily or weekly time horizons; shorter than the time horizon implicit in the Masters Hypothesis. While the horizon is not stated explicitly, the basic idea is that a 'wave' of index investment pressures commodity futures prices over much longer time intervals such as a month or a quarter. This is similar to the argument of Summers (1986) and others regarding tests for a slowly accumulating 'fads' component in stock returns. The third criticism is that nearly all of the studies employ time-series Granger causality tests, which may lack the statistical power necessary to reject the null hypothesis of non-causality because the dependent variable – the change in commodity futures prices – is extremely volatile.

The present study addresses criticisms of previous work by using new data that better matches the basic tenets of the Masters Hypothesis and applying more powerful statistical tests. The CFTC's quarterly *Index Investment Data* (IID) report provides the best measure of actual commodity index investment and our study is the first to use this data in empirical tests. The IID provides not only a better measure of index investment – because positions are measured before internal netting by swap dealers – but it also covers the complete spectrum of agricultural, energy, metal, and soft commodity futures markets. A cross-sectional Fama-MacBeth regression test is applied to the IID positions at a quarterly horizon. Ibragimov and Muller (2010a, b) show that the Fama-MacBeth test has good power properties for the sample sizes considered here. Quarterly horizons should capture slowly accumulating price pressure if it is present in the data. Consistent with theory, both lagged and contemporaneous effects are considered as well.

We supplement the cross-sectional tests using a large sample of daily positions held by the US Oil and Natural Gas Funds, two of the largest exchange-traded index funds (ETFs), in the crude oil and natural gas futures markets, respectively. In order to address power concerns, both Granger causality and long-horizon regression tests (Valkanov, 2003) are applied to the daily time-series of ETF positions.

Fama-MacBeth tests using the quarterly IID positions reveal very little evidence that index positions influence returns or volatility in 19 commodity futures markets. The cross-sectional IID results are robust to whether lagged or contemporaneous effects are considered and the addition of the nearby-deferred futures spread as a conditioning variable. Granger causality tests show no causal links between daily returns or volatility in crude oil and natural gas futures markets and the positions of the two large ETFs. Long-horizon regression tests likewise fail to reject the null hypothesis of no market impact for the ETFs. Overall, the empirical results of this study fail to support the Masters Hypothesis.

5.2 Measures of Commodity Index Fund Investment

A key issue in testing the Masters Hypothesis is the measurement of index fund investment in commodity futures markets. Private vendors have collected data for over a decade; however, only the aggregate dollar investment is available for all markets combined. As noted in the introduction, the CFTC has developed several data series on commodity index fund investment and/or positions by individual futures markets. Since these data are used in nearly all previous studies on the market impact of commodity index investment, as well as the present study, it is important to understand how the data are collected and categorized as well as any limitations.

5.2.1 *Commitments of Traders* (COT) reports

The core of the CFTC's market surveillance program is the Large Trader Reporting System (LTRS) established under the Commodity Exchange Act (CEA). The CEA authorizes the CFTC to collect market data and position information from market participants who have position levels above the reporting level for a specific futures market (CFTC, 2010c). The CFTC is prohibited from disclosing the positions held by individuals, but they have long categorized traders within the system as those traders that are largely hedgers (commercials) and those that are largely speculating (non-commercials). The classification of traders is heavily based on the information provided by traders in the CFTC's Form 40, where traders must describe the nature of their futures transactions (including the associated physical market activities, if any).

The CFTC releases the combined futures and delta-adjusted option positions aggregated by trader category each Friday in the *Futures-and-Options-Combined Commitments of Traders* (COT) report. Open interest reflects Tuesday's closing positions for a given market and is aggregated across all contract expiration months. Non-commercial open interest is divided into long, short, and spreading; whereas, commercial and non-reporting open interest is simply divided into long or short. The following relation explains how the market's total open interest (TOI) is disaggregated:

$$\underbrace{\left[NCL+NCS+2(NCSP)\right]+[(CL+CS)]}_{\text{Reporting}}+\underbrace{\left[NRL+NRS\right]}_{\text{Non-reporting}}=2(TOI) \quad (5.1)$$

where NCL, NCS, and NCSP are non-commercial long, short, and spreading positions, respectively. CL (CS) represents commercial long (short) positions, and NRL (NRS) are long (short) positions held by non-reporting traders. Reporting and non-reporting positions must sum to the market's total open interest (TOI), and the number of longs must equal the number of short positions.

There have been ongoing complaints that the legacy COT trader designations may be inaccurate (e.g. Ederington and Lee, 2002). As one example, speculators may have an incentive to self-classify their activity as commercial hedging to circumvent speculative position limits in some markets. But the CFTC implements a fairly rigorous process – including statements of cash positions in the underlying commodity – to ensure that commercial traders have an underlying risk associated with futures positions. However, in recent years, industry participants began to suspect that these data were contaminated because the underlying risk for many reporting commercials was not a position in the physical commodity (CFTC, 2006a, b). Rather, the reporting commercials were banks and other swap dealers hedging risk associated with OTC derivative positions.

In response to these concerns the CFTC added two variations to the legacy COT reports. The first is the *Disaggregated Commitments of Traders* (DCOT) report, which, as the title suggests, simply further disaggregates the legacy COT commercial and non-commercial trader categories. The second report, *Supplemental Commitments of Traders* (SCOT) adds a new category specifically to capture commodity index traders.

5.2.2 *Disaggregated Commitments of Traders* (DCOT) report

The CFTC began publishing the DCOT report in September 2009 and ultimately provided historical data back to June 2006 (CFTC, 2009). The DCOT data are available for most commodity futures markets. Constructed in a manner analogous to the legacy reports, DCOT reports break down combined futures and options positions as follows:

$$[SDL + SDS + 2(SDSP)] + [MML + MMS + 2(MMSP)] + [PML + PMS] + [ORL + ORS + 2(ORSP)] + [NRL + NRS] = 2(TOI) \quad (5.2)$$

where reporting non-commercial traders are disaggregated into managed money (MM) and other reportables (OR). Commercial traders from the legacy COT reports are segregated into processors and merchants (PM) as well as swap dealers (SD). Positions are divided into long (L), short (S), and spreading (SP) as indicated by the corresponding suffixes. For example, the SDL, SDS, and SDSP are the swap dealers' long, short, and spreading positions, respectively.

Taken from the legacy commercial category, swap dealers (SD) are those traders who deal primarily in swaps and hedge those transactions in the futures market. There is considerable uncertainty whether swap dealer positions represent an underlying speculative or hedging position. A large portion of swap dealers' trading represents commodity index investors and swap dealer positions are often used as a proxy for their activity (e.g. Brunetti and Buyuksahin, 2009; Buyuksahin and Harris, 2011; Sanders and Irwin, 2011b). Processors and merchants (PM) include traditional commercial users – processors and producers of the commodity who are actively engaged in the physical markets and are using the futures to hedge associated price risks.

Disaggregated from the legacy non-commercial category, managed money (MM) represents positions held by commodity trading advisors, commodity pool operators and hedge funds that manage and conduct futures trading on behalf of clients. The traders included in the managed money category would largely represent the more traditional class of speculative traders. Other reportable (OR) are non-commercial traders who are large enough to report but do not fit into one of the other categories. These might include large individual speculative traders or market-makers as well as firms managing their own assets. For swap dealers, managed money and other reportables, offsetting long and short positions in the same market but with different contract months are reported as spreading positions. Index investment positions in the DCOT report can be a part of the swap dealers, managed money, and other reportable categories.

5.2.3 *Supplemental Commitments of Traders* (SCOT) report

Starting in 2007, the CFTC began releasing the SCOT reports (sometimes referred to as the commodity index traders or CIT report), which specifically breaks out the positions of index traders for 12 agricultural markets. Index traders are identified by reviewing the CFTC Form 40 and through confidential interviews with traders known to be index traders or who exhibit trading patterns consistent with indexing (CFTC, 2010a). It is important to note that the CFTC does not distinguish index and non-index positions in this process. So if a trader is identified as an index trader, then all of their positions are counted as index positions. In addition, the index trader positions reflect both institutional investment funds that would have previously been classified as non-commercials in the legacy COT reports as well as swap dealers who would have previously been classified as commercials hedging OTC transactions. Sanders *et al.* (2010) show that approximately 85% of index trader positions in the 12 SCOT markets are in fact drawn from the long commercial category with the other 15% from the long non-commercial category. This implies that the bulk of index positions in the 12 SCOT markets are initially established in the OTC market and the underlying position is then transmitted to the futures market by swap dealers hedging OTC exposure.

The supplemental data are released in conjunction with the legacy COT report showing combined futures and options positions. The index trader positions are simply removed from their prior categories and presented as a new category of reporting traders. The supplemental data include the long and short positions held by commercials (less index traders), non-commercials (less index traders), index traders, and non-reporting traders aggregated across all contracts for a particular market:

$$[(NCL - CITL) + (NCS - CITS) + 2(NCSP)] + [(CL - CITL) + (CS - CITS)] + [CITL + CITS] + [NRL + NRS] = 2(TOI) \tag{5.3}$$

The above relation is analogous to that for the traditional COT report shown in Eqn 5.1, except commercial and non-commercial positions are adjusted for the commodity index trader long (CITL) and index trader short (CITS) positions.

There is a relatively straightforward disaggregation of the legacy COT categories into the DCOT categories and then back into the SCOT classifications as shown in Fig. 5.1. In this illustration, the source of errors in compiling index trader positions from the DCOT data becomes clear. First, and most obvious, the use of the swap dealer positions in the DCOT as a proxy for index trader positions will clearly exclude those index positions found in the managed money and the other reportables categories. Second, and less obvious, to the degree that positions are netted within firms – especially among swap dealers – reportable futures positions will understate true levels of index investment.

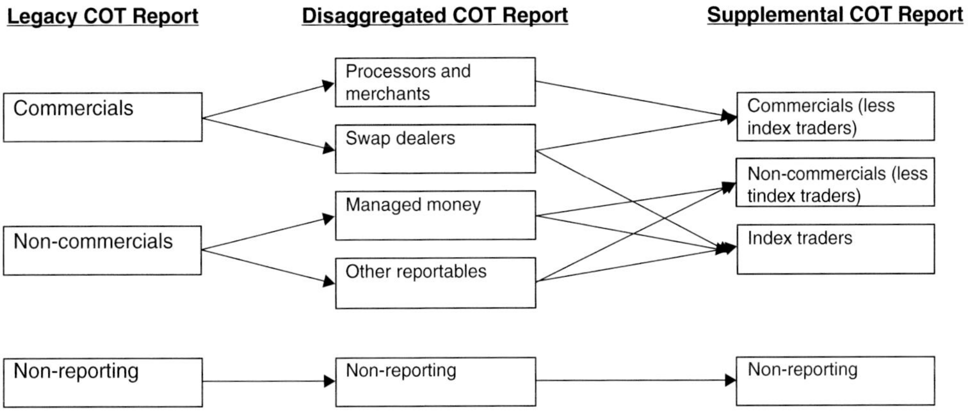

Fig. 5.1. Relationship between legacy, disaggregated, and supplemental *Commitments of Traders* (COT) reports.

One can infer from comparisons found in the CFTC's September 2008 report on swap dealer positions (CFTC, 2008) that DCOT swap dealer positions in agricultural futures markets correspond reasonably well to index trader positions. Since swap dealers operating in agricultural markets conduct a limited amount of non-index long or short swap transactions, there is limited error in attributing the net long futures position of swap dealers in these markets to index funds. In contrast, swap dealers in energy futures markets conduct a substantial amount of non-index swap transactions on both the long and the short side of the market, which creates uncertainty about how well the net long position of swap dealers in energy markets represent index fund positions. For example, the CFTC estimates that only 41% of long swap dealer positions in crude oil futures markets on three dates in 2007 and 2008 were linked to long-only index fund positions (CFTC, 2008).

In sum, while the netting effect seems to have a relatively modest impact on the measurement of index investing in agricultural futures markets, it likely creates considerable measurement error for the energy and metals futures markets. Realizing this, the CFTC utilized their regulatory power to first issue a 'special call' in June 2008 to assess the total investment activities (on- and off-exchange) of commodity index funds. The index investment data collected through the CFTC's special call provides the most accurate measure of total commodity index investment available to date.

5.2.4 *Index Investment Data* (IID) report

The 'special call' of all swap dealers and index funds known to be significant users of US futures markets has allowed the CFTC to collect the total notional value of these firms' commodity index business and the equivalent number of futures contracts (CFTC, 2010b). The original call in June 2008 gathered data from 43 entities engaging in index activities in commodity markets. These entities included index funds, swap dealers, pension funds, hedge funds, mutual funds, exchange-traded funds (ETFs), and exchange-traded notes (ETNs). Since the call included the financial institutions known to be the largest swap dealers in the world and all entities granted exemptions from federal position limits or 'no action' letters, the coverage of the special call is the most comprehensive to date in terms of commodity index activity (CFTC, 2008, 2010b).

Firms subject to the special call report the total notional value of commodity index positions, whether the positions are for a firm's own account or on behalf of a client. Notional value is also reported separately for US and non-US markets. Crucially, each firm's entire 'book of business' in futures and OTC markets that is related to commodity index investment is reported, not just the netted amount that may be managed ultimately in the futures markets. The data reported as part of the special call is cross-checked by comparing it to the positions in the CFTC's larger trader reporting system and by engaging the firms in extensive discussions.

As pointed out by the CFTC (2010b), because the special call corrects for netting, the positions reported in the *Index Investment Data* (IID) report are more precise than the SCOT and DCOT in representing index investment. Moreover, because the netting effect that plagues the COT data is corrected, the IID is more comprehensive in terms of markets covered. The IID is available for the 12 agricultural markets in the SCOT report plus seven major energy and metals markets. The IID report does have some limitations. First, small entities or entities unknown to the CFTC may be omitted from the special call. Second, trading records are not independently examined by the CFTC. Third, 'index' activity is not specifically defined so there could be some inconsistency in the reported data across firms. Despite these limitations, the comprehensive nature of the data set makes the IID the best candidate for evaluating potential market impacts from commodity index investments. In the words of the CFTC, 'The index investment data represents the Commission's best effort to provide a one-day snapshot of the positions of swap dealers and index funds' (CFTC, 2010b).

5.3 Data and Descriptive Statistics

5.3.1 Comparison of IID, DCOT, and SCOT data

The IID is collected at the end of each quarter from December 31 2007 through March 31 2011 (14 quarter-end observations). The markets covered by the IID are listed in Table 5.1 along with the notional values as of March 31 2010. Table 5.1 also includes the percentage allocated across markets, and comparable weightings for three popular commodity indices: (i) S&P GSCI; (ii) Dow Jones-UBS Commodity Index™ (DJ-UBS); and (iii) the Reuters/Jefferies Commodity Research Bureau Index™ (RJ-CRB). A total of $161.2 billion was invested in commodity index investments as of March 31 2010. The IID show that 78% of index investments are in US futures markets with the other 22% in non-US markets. Not surprisingly, the energy markets – crude oil in particular – have the greatest weighting at 22.6%, while minor agricultural markets – such as feeder cattle – have allocations less than 1%. As reported in the subsequent columns, implied weightings in the IID are generally within the range of those used by the commercial indices. Although not detailed in the IID reports, the non-US markets also likely mirror commercial index investments in the London-based Brent crude oil and base metal markets (e.g. aluminum and nickel).

Figure 5.2 shows total index fund investment in US commodity futures markets from the IID reports between December 31 2007 and March 31 2011. Investment increased notably during the first half of 2008, reaching a peak of $163.3 billion at the end of June 2008. A steep slide followed during the second half of 2008, with a trough of $68.3 billion at the end of the year. Investment subsequently recovered to about the same level as that observed in December 2007 and then increased to a new high of $195 billion by the end of March 2011. It is important to remember that the data shown in Fig. 5.2 are total notional values and as such are influenced by both commodity price movements and investment inflows and outflows. We also note that while the entire time period of the recent commodity price spike is not covered by the IID, the first three-quarters of 2008 are included, which is the period of the most rapid rise of commodity futures prices and greatest concern from a public policy standpoint.

It was noted earlier that the theoretical impact of index funds in commodity futures markets hinges on the predictability of their position changes. If index fund position changes are predictable, other market participants will anticipate their activity, trade against them, and negate any potential impact (De Long *et al.*, 1990). The predictability in overall index positions – as measured by the IID – is tested in a simple time-series framework. Specifically, an $AR(1)$ model is estimated after pooling quarterly percentage growth rates across all 19 markets. First-order autocorrelation estimates are small (0.060 for contract growth rates and 0.221 for notional value growth rates) and statistically insignificant. Other conditioning variables may potentially be useful to predict index fund position changes, which would permit other traders to anticipate their trades and thereby negate any market impact. However, from a pure time-series statistical perspective there is negligible predictability in

Table 5.1. Notional value of net long index fund positions based on *Index Investment Data* (IID) and comparisons to popular commodity indices, March 31 2010.[a]

Commodity futures market[b]	IID notional value (billion $)	IID portfolio weight (%)	S&P GSCI portfolio weight (%)	DJ-UBS portfolio weight (%)	RJ-CRB porfolio weight (%)
Corn	6.9	4.3	3.1	6.6	6.0
Soybeans	8.0	5.0	2.2	7.9	6.0
Soybean oil	2.2	1.4	0.0	3.0	0.0
Wheat, CBOT	5.5	3.4	2.8	4.4	1.0
Wheat, KCBT	0.7	0.4	0.6	0.0	0.0
Cotton	3.0	1.9	1.2	2.4	5.0
Live cattle	5.1	3.2	2.7	4.0	6.0
Feeder cattle	0.5	0.3	0.5	0.0	0.0
Lean hogs	2.9	1.8	1.8	2.8	1.0
Coffee	2.7	1.7	0.7	2.5	5.0
Sugar	3.8	2.4	1.5	1.6	5.0
Cocoa	0.7	0.4	0.4	0.0	5.0
WTI crude oil	36.4	22.6	36.4	15.7	23.0
RBOB unleaded gas	7.8	4.8	4.7	4.1	5.0
Heating oil	6.7	4.2	4.7	3.9	5.0
Natural gas	12.5	7.8	4.0	8.4	6.0
Gold	10.3	6.4	3.1	9.8	6.0
Silver	3.1	1.9	0.4	3.4	1.0
Copper	5.5	3.4	3.0	7.7	6.0
Other US markets	1.4	0.9	0.0	0.0	1.0
US markets total	125.8	78.0	74.2	88.0	93.0
Non-US markets total	35.4	22.0	25.8	12.0	7.0
All markets total	161.2	100.0	100.0	100.0	100.0
Sector					
Grains	23.3	14.5	8.7	21.8	13.0
Livestock	10.2	6.3	3.8	6.5	20.0
Softs	8.5	5.3	5.0	6.8	7.0
Metals	18.9	11.7	7.1	20.9	13.0
Energy	63.4	39.4	49.7	32.0	39.0
Other US markets	1.4	0.9	0.0	0.0	1.0
Non-US markets	35.4	22.0	25.8	12.0	7.0

[a]DJ-UBS, Dow Jones-UBS Commodity Index; RJ-CRB, Reuters/Jefferies Commodity Research Bureau Index; S&P GSCI, Standard and Poor's Goldman Sachs Commodity Index.
[b]CBOT, Chicago Board of Trade; KCBOT, Kansas City Board of Trade; RBOB, reformulated blendstock for oxygenate blending; WTI, West Texas Intermediate (here and in subsequent tables).

quarterly index fund position changes. As noted earlier, this unpredictability is a necessary (but not sufficient) condition for market impact in theoretical models.

For comparison purposes, the SCOT data and DCOT data are also collected for the same dates or the dates nearest to the IID quarter-end dates. Four of the 14 release dates of the traditional COT reports coincide precisely with quarter-end dates for the IID. The other COT dates are all within 2 days of the IID compilation dates. Table 5.2 shows the position sizes (in contracts) for index investors as reported in the IID, SCOT, and DCOT reports for a date when the reports coincide, June 30 2009. For the DCOT reports, it is assumed that the best measure of index positions is that held by swap dealers. The data show that the discrepancy between the IID positions and other COT-based position measures can be large. Focusing on the 12 SCOT agricultural markets, the reported index trader positions are on average overstated by 3% with an average absolute error of 10% versus the IID positions. This is possibly due to the fact that the SCOT counts

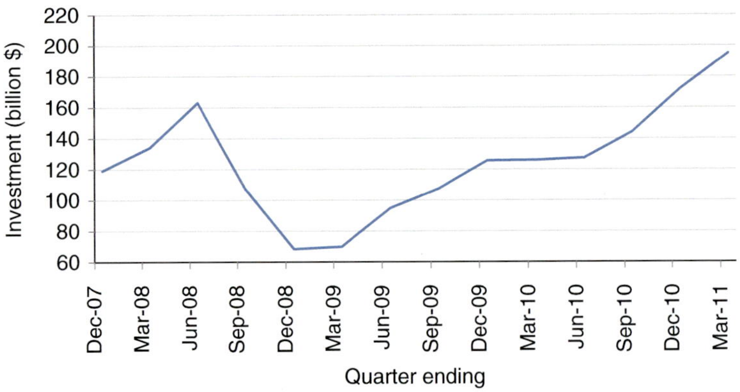

Fig. 5.2. Total net long index investment in US commodity futures markets, quarterly, December 2007–March 2011. (*Index Investment Data* (IID).)

Table 5.2. Comparison of alternative measures of index fund net long positions on June 30 2009.[a]

Commodity futures market	IID (thousand contracts)	SCOT index traders (thousand contracts)	Difference from IID (%)	DCOT swap dealers (thousand contracts)	Difference from IID (%)
Corn	298	305	2	219	–27
Soybeans	122	137	12	109	–11
Soybean oil	62	61	–2	54	–13
Wheat, CBOT	171	152	–11	120	–30
Wheat, KCBT	24	27	15	14	–43
Cotton	53	67	26	51	–4
Live cattle	97	94	–4	74	–23
Feeder cattle	7	7	4	4	–41
Lean hogs	62	59	–5	52	–16
Coffee	36	42	17	34	–6
Sugar	237	207	–13	27	–88
Cocoa	18	17	–7	10	–43
WTI crude oil	430			50	–88
RBOB gas	66			40	–39
Heating oil	59			55	–6
Natural gas	233			12	–95
Gold	73			–43	–160
Silver	27			5	–83
Copper	63			48	–23

[a] This table compares three measures of the net long position (total long minus short positions) of index funds in commodity futures markets on June 30 2009. The data are collected by the US Commodity Futures Trading Commission (CFTC) and reported in the *Index Investment Data* (IID), *Supplemental Commitments of Traders* (SCOT), and *Disaggregated Commitments of Traders* (DCOT) reports.

all of an index trader's futures positions as index positions, even if they represent some other activity. The other dates where the COT and IID reports coincide show similar comparisons.

In the same 12 agricultural markets, the DCOT swap dealer positions are consistently low estimates of index positions. This is partially due to the fact that some index positions are held by non-commercials in the managed money and other reportables categories. It is also due to the netting of positions that occurs within a swap dealer's book. Although the netting effect is expected to be relatively small in the agricultural markets, the average absolute error is still 29% for the 12 agricultural markets. The largest errors are concentrated in sugar and cocoa where there is a greater tendency for offsetting non-index OTC transactions.

In the seven non-SCOT energy and metal markets shown in Table 5.2, the average absolute difference between swap dealer positions and the IID increases markedly to 71%. Indeed, on this date, the swap dealer position in the gold market is short, the opposite of the IID's long position. Likewise, on other report dates, net short swap dealer positions are recorded for silver, sugar, natural gas, and crude oil, whereas the IID only shows net long positions for these markets. This is clear evidence of the netting effect impacting reported DCOT swap dealer positions, where the internal crossing of positions effectively masks the total underlying level of index investment.

The differences between the index trader and swap dealer positions and the IID may be irrelevant if the measures are highly correlated. Correlation coefficients (Pearson) of changes in net positions are presented in Table 5.3. Consider first the correlation between SCOT positions and IID positions. Statistically significant correlations are observed in 11 of the 12 SCOT markets (10% level for a two-tailed *t*-test) with an average correlation coefficient at 0.81. For the same 12 markets, DCOT swap dealer positions are significantly correlated with IID positions in only eight markets. The correlation between IID net positions and swap dealer positions breaks down even more in the energy and metals markets, where one of the correlations is negative (gold) and none is statistically different from zero at the 10% level.

It is clear that the single best measure of total commodity index positions is that available in the IID report. The DCOT report's swap dealer net positions are relatively poor proxies for total index positions as they often show negative net long positions that are not statistically correlated with IID positions in the energy and metals markets. The SCOT report provides useful measures of index positions in agricultural futures markets.

5.3.2 Comparison of mapping algorithm to IID

Masters (2008) and others impute index positions for individual commodity futures markets

Table 5.3. Correlation between *Index Investment Data* (IID) net long positions and alternative measures of index positions.[a]

Commodity futures market	IID and SCOT index traders	IID and DCOT swap dealers
Corn	0.86	0.80
Soybeans	0.81	0.88
Soybean oil	0.76	0.66
Wheat, CBOT	0.76	0.67
Wheat, KCBT	0.45	0.47
Cotton	0.83	0.72
Live cattle	0.95	0.91
Feeder cattle	0.73	0.57
Lean hogs	0.95	0.93
Coffee	0.82	0.71
Sugar	0.93	0.59
Cocoa	0.85	0.55
WTI crude oil		0.13
RBOB unleaded gas		0.42
Heating oil		0.09
Natural gas		0.33
Gold		–0.36
Silver		0.10
Copper		0.59

[a]Simple Pearson correlation coefficients are calculated between the changes in net trader positions reported in the *Index Investment Data* (IID), *Supplemental Commitments of Traders* (SCOT), and *Disaggregated Commitments of Traders* (DCOT) reports. The data are based on end-of-quarter positions between March 31 2008 and March 31 2011.

(e.g. energy and metals) that are not included in the SCOT report. This is accomplished by taking a SCOT market that is unique to a particular index and using the notional value of index trader positions in that market (as reported in the SCOT) to estimate total investment in a particular index. Then, any non-SCOT market is simply assigned the notional value according to its weight in the index.

As an example, consider the calculation using SCOT data from December 31 2007 and following the procedure presented in the Appendix of Masters (2008), soybean oil futures had a reported index position of 77,752 contracts with a notional value of $2279 million. Soybean oil is not included in the S&P GSCI but is a component of the Dow Jones-AIG (DJ-AIG) Commodity Index, where it has a weight of 2.85%. Scaling this position up accordingly (2279/0.0285) yields a total notional investment of $79,965 million in the DJ-AIG index. WTI crude oil receives a weight of 12.72% in the index; therefore, the notional value of the crude oil investment stemming from DJ-AIG investments is $10,172 million or 106,000 contracts at $95.98/barrel. For the S&P GSCI an analogous process is followed using feeder cattle and Kansas City Board of Trade (KCBT) wheat futures, both of which are unique to the S&P GSCI. The result is a range of possible WTI holdings from 286,000 (imputed via feeder cattle) to 468,000 (imputed via KCBT wheat) contracts. Next, the WTI crude oil contracts attributed to S&P GSCI and DJ-AIG investments are summed together to get a total index position in WTI crude oil contracts ranging from 392,000 to 574,000 contracts with an average of 483,000.

This mapping algorithm rests on a number of key considerations that may not be an accurate reflection of reality (see Appendix 2 in Buyuksahin and Robe (2011)). First, it assumes that total investment is allocated to individual markets according to one of the widely tracked commodity price indexes. Second, it assumes that fund managers strictly adhere to this weighting even in relatively minor markets such as feeder cattle. Third, as shown above, it can generate a wide range of estimates depending on the market on which the estimates are based. Finally, as pointed out by Singleton (2011), the measurement error using this mapping procedure will be amplified by the process of scaling minor market positions (such as soybean oil) up to major market positions (such as crude oil). For instance, in the example above, feeder cattle had a weight of 0.47% in the S&P GSCI, which created a scaling factor of 213 for imputing total investment. So any error in the SCOT feeder cattle position for index traders is increased many times over in the Masters (2008) scaling procedure.

Indeed, the data generated suggest that the level of error in these mapping procedures is likely to be quite large. In the example presented in Masters (2008) errors are of a relatively large magnitude even for SCOT markets. For example, the mapping algorithm underestimates cocoa index positions by 35% and corn by 25%. If the errors for markets within the SCOT report are that large, then it is likely that the errors for non-SCOT markets, like energy and metals, are even larger.

Even if the mapping algorithm is not accurate in an absolute sense, it may still provide useful estimates for statistical testing if there is a reasonably close correlation between the index positions held in the reported SCOT market (e.g. soybean oil) and those markets not reported (e.g. crude oil). Table 5.4 shows the simple (Pearson) correlation between growth rates in IID positions in those markets included in the SCOT and those that are not. Of the 84 correlations calculated, only 18 are statistically significant at the 10% level (two-tailed *t*-test). Most alarming, the correlations between the IID positions in the 12 SCOT markets and WTI crude oil are negative, which suggests large errors may result by using any mapping algorithm to infer WTI crude oil index positions from those held in agricultural markets.

To further demonstrate the potential errors associated with a mapping algorithm, the method proposed by Masters (2008) is applied to positions reported in the SCOT report at the end of each quarter.[7] The dates utilized are those that coincide as closely as possible (usually within 1 or 2 days) with the quarter-end dates for the IID reports. The average estimated positions (1000s contracts) for WTI crude oil futures are shown in Fig. 5.3. Under the assumption that the IID provides the best estimates of actual index positions, the positions inferred using the Masters algorithm have a mean absolute percentage error of 52%. Moreover, the positions do

Table 5.4. Cross-market correlations of growth rates in index investment net long positions.[a]

	Non-SCOT market						
	WTI	RBOB	Heating	Natural			
SCOT market	crude oil	gasoline	oil	gas	Gold	Silver	Copper
Corn	−0.40	0.19	0.60	0.63	0.81	0.77	0.42
Soybeans	−0.36	0.27	0.82	0.52	0.69	0.78	0.42
Soybean oil	−0.37	0.21	0.88	0.61	0.66	0.59	0.23
Wheat, CBOT	−0.41	0.30	0.81	0.57	0.77	0.63	0.29
Wheat, KCBT	−0.31	0.02	0.54	0.23	0.09	0.20	0.04
Cotton	−0.23	0.23	0.44	0.35	0.54	0.69	0.33
Live cattle	−0.50	0.19	0.43	0.40	0.87	0.51	0.19
Feeder cattle	−0.10	0.43	0.08	0.19	0.17	0.21	0.34
Lean hogs	−0.52	0.23	0.29	0.34	0.76	0.62	0.32
Coffee	−0.38	0.31	0.19	0.22	0.48	0.77	0.43
Sugar	−0.07	0.38	0.13	0.33	0.48	0.32	0.59
Cocoa	−0.37	0.08	0.28	0.46	0.42	0.21	0.18
Average	−0.34	0.24	0.46	0.41	0.56	0.53	0.31

[a]Simple Pearson correlation coefficients calculated between the log-relative growth rates in net trader positions reported in the *Index Investment Data* (IID) reports. SCOT denotes the *Supplemental Commitments of Traders* report. The data are based on end-of-quarter positions between March 31 2008 and March 31 2011.

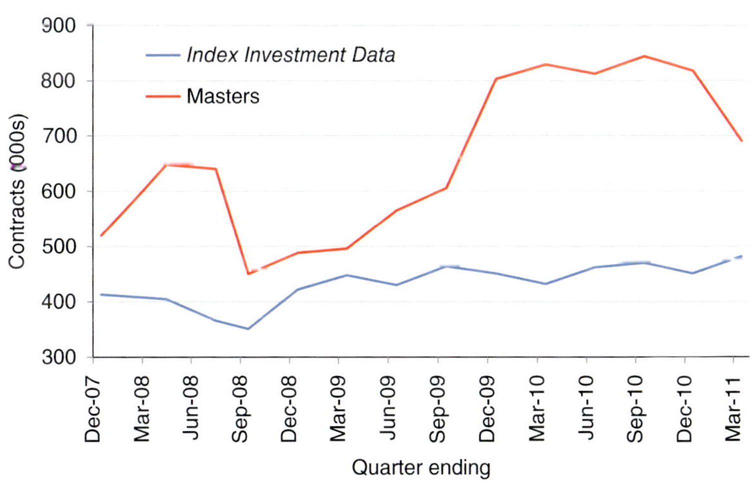

Fig. 5.3. Comparison of quarterly West Texas Intermediate (WTI) crude oil net long index positions based on *Index Investment Data* (IID) and Masters algorithm estimates, December 2007–March 2011.

not necessarily move together and the growth rates have a statistically insignificant correlation coefficient of 0.03. In the first half of 2008, when energy prices were moving rapidly higher, the IID showed WTI crude oil index positions declined while the Masters estimates increased. From 2009 forward, the gap between the IID and the imputed WTI crude oil positions increased dramatically.[8] A close examination of the SCOT data shows that the combined notional value held by index funds in feeder cattle, KCBT wheat, and soybean oil (the markets used to impute WTI crude oil positions) increased by 241% through 2009 and 2010. Over those same 2 years, the notional value of IID positions held in WTI crude oil rose by half as much, 107%. Clearly, any mapping algorithm that scales up positions in the three SCOT markets will grossly overestimate index positions held in WTI crude oil futures.

In sum, mapping algorithms based on index positions reported in the SCOT report generate highly suspect data. The imputed data not only contain large absolute errors but also a lack of correlation with the best available estimate of actual positions – those reported in the IID report. Consequently, empirical research (e.g. Gilbert, 2009; Stoll and Whaley, 2010; Singleton, 2011) that uses imputed data at best suffers from severe error in measurement and at worst leads to highly misleading statistical estimates and inference. The IID report is the only accurate measure of total index positions in the energy and metal markets and it is used in the cross-sectional regression tests of market impact that follow.

5.4 Cross-sectional Regression Tests

We use a cross-sectional regression framework to test the market impact of IID commodity index positions. This approach is adopted for two reasons. First, cross-sectional regression tests may be more powerful than traditional time-series tests because the variation in index fund positions across markets at a point in time may be more informative than the variation in fund positions across time for a given market (Sanders and Irwin, 2010). This is precisely the motivation for the wide use of such tests when estimating the factors impacting equity returns (e.g. Fama and French, 1992). Second, the IID positions are available only on a quarterly basis since December 2007, providing a limited number of time-series observations for any given market.[9] The IID does include broad coverage of 19 commodity futures markets, lending itself to cross-sectional tests of market impact.

A potential concern when using the cross-sectional approach is whether there is sufficient variation in the index investment growth rates across commodities to estimate market impacts with a reasonable degree of precision. The earlier analysis of SCOT versus non-SCOT cross-market correlations (see Table 5.4) showed that, despite the widespread use of similar commodity price indexes, growth rates do not move in lockstep. Further evidence is presented in Fig. 5.4, which shows the quarterly growth rates in index investment in terms of net long contracts for the 19 IID markets. In each quarter ending between March 31 2008 and March 31 2011, there are usually markets with positive growth in excess of 50% and some with declines in the 20–30% range. The average cross-sectional standard deviation (by quarter across markets), 11.2%, is not markedly smaller than the average time-series standard deviation (by market across quarters),

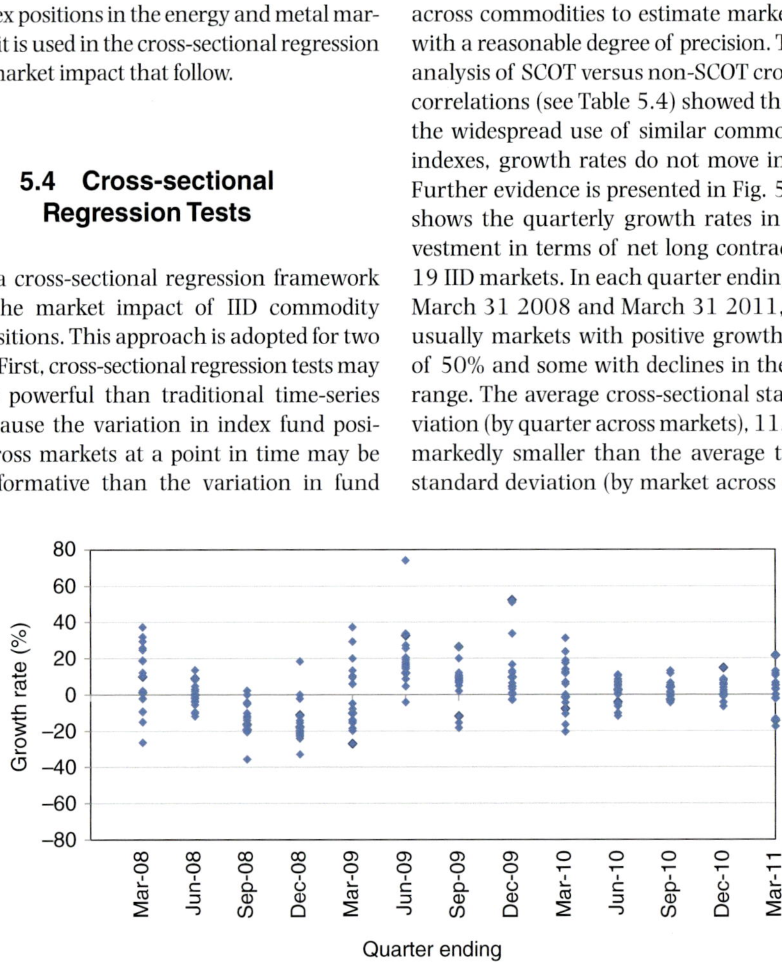

Note: Each marker represents the percentage growth rate of index investment in a commodity futures market for each quarter.

Fig. 5.4. Growth rate of quarterly *Index Investment Data* (IID) positions in 19 US commodity futures markets, March 2008–March 2011.

14.7%, and the average pair-wise correlation of growth rates across the 19 markets is only 0.36. The evidence presented here and earlier in Table 5.4 suggests there is ample cross-sectional variation in growth rates for a valid cross-sectional regression analysis.

The relationship between commodity index positions and subsequent commodity market returns can be expressed in the following regression model:

$$RQ_{i,t} = \alpha + \beta \Delta IND_{i,t-j} + e_{i,t} \quad i = 1,\ldots,N \qquad t = 1,\ldots,T \quad j = 0 \text{ or } 1 \tag{5.4}$$

where $RQ_{i,t}$ is the return in commodity futures market i and quarter t and $\Delta IND_{i,t-j}$ is the growth rate of commodity index fund investment in market i and quarter $t - j$.[10] The null hypothesis of no impact on returns is that the slope coefficient, β, in Eqn 5.4 equals zero. An alternative bubble-type hypothesis is that $\beta > 0$, such that an increase in fund positions in market i portends relatively large subsequent returns in that market. In the first set of regressions, the data are organized such that the linkages run from the positions at time $t - 1$ ($j = 1$) to the subsequent returns at time t. In the second set of regressions, the data are organized such that the linkages are contemporaneous ($j = 0$) at time t.

The quarterly return for market i in Eqn 5.4 is calculated as $RQ_{i,t} = ln\left(p^1_{i,t} / p^1_{i,t-1}\right) \cdot 100$, where $p^1_{i,t}$ is the commodity futures price of the nearest-to-expiration contract (but not entering the expiration month) on the last business day of each quarter. In order to avoid distortions associated with contract roll-overs $p^1_{i,t-1}$ is always calculated using futures prices for the same nearest-to-expiration contract as $p^1_{i,t}$. Returns are calculated for the 19 US commodity futures markets listed in Table 5.1 for each quarter ending between March 31 2008 and March 31 2011. Futures prices are obtained from Barchart, Inc. (available at: https://www.barchart.com/ (accessed January 14 2022)).

The growth rate of commodity index investment for the same 19 markets and quarters is computed two ways for the IID.[11] The first is $\Delta IND^{**}_{i,t} = ln\left(INDN_{i,t} / INDN_{i,t-1}\right) \cdot 100,$, where $INDL_{i,t}$ is the net long position (number of long contracts minus short contracts) of commodity index funds in market i at the end of quarter t and the second is $\Delta IND^{**}_{i,t} = ln\left(INDN_{i,t} / INDN_{i,t-1}\right) \cdot 100,$ where $INDN_{i,t}$ is the notional value (net long position times the nearby futures price) of index funds in market i at the end of quarter t.

Volatility impacts are also tested using the following regression model:

$$VQ_{i,t} = \alpha + \beta \Delta IND_{i,t-j} + e_{i,t} \quad i = 1,\ldots,K \qquad t = 1,\ldots,T \quad j = 0 \text{ or } 1 \tag{5.5}$$

where $VQ_{i,t}$ is the volatility of futures returns in commodity futures market i and quarter t. Two measures of volatility are considered in each commodity futures market. The first is the market's realized volatility, estimated for each quarter by the annualized standard deviation of the daily log-relative returns for the nearby futures contract, and the second is implied volatility, computed via Black's (1976) model as average of implied volatilities for the two nearest-to-the-money put and call options for the commodity futures contract expiring nearest to the end of each quarter but not during the quarter. The implied volatilities are obtained from Barchart, Inc. (available at: https://www.barchart.com/ (accessed January 14 2022)).

Fama and MacBeth (1973) propose a method of estimating β in Eqns 5.4 and 5.5 that is still widely used in the literature (see Campbell *et al.*, 1997). With the Fama-MacBeth regression procedure, Eqns 5.4 and 5.5 are estimated via ordinary least squares (OLS) regression for each time period $t = 1, 2, 3, \ldots, T$ across the $i = 1, 2, 3, \ldots, N$ markets. The average of the estimated slopes (β) is calculated for the T regressions and the associated standard error is $\sigma_{\bar{\beta}} / T^{1/2}$. The basic estimation strategy is to exploit the information in the cross-section of markets about the relationship between index investment and returns or volatility and then treat each cross-section as an independent sample. In our case, there are $T = 12$ quarters (accounting for calculation of growth rates and a one-period lag) that end between June 30 2008 and March 31 2011 and $N = 19$ markets for a total of 228 observations to be used in cross-sectional estimation.

Ibragimov and Muller (2010a, p. 454) provide a formal justification for the Fama-MacBeth test and show that as long as 'coefficient estimators are approximately normal (or scale mixtures of normals) and independent, the Fama-MacBeth method results in valid inference even for a short panel that is heterogeneous over time.' Monte Carlo simulations indicate that the maximal loss in power is only 6% for a 5% test when the number of groups (quarters in our case) is 16, and 13% when the number is of groups is eight. The crucial assumption when applying the Fama-MacBeth test is independence in the time dimension. Not surprisingly, the average first-order autocorrelation of quarterly time-series returns across our sample of 19 commodity futures markets is only 0.22, which indicates that independence in the time dimension is a plausible assumption. The average cross-sectional correlation of returns across the 12 quarters, 0.49, is over twice as high, but independence in this dimension is not required. Finally, supplementary simulations reported in Ibragimov and Muller (2010b) that roughly match the characteristics of our sample – eight groups, low correlation of returns through time for a given market, and moderately high correlation of returns across markets for a given period – show that the Fama-MacBeth estimator performs well in terms of size and power relative to alternative estimators.[12]

The Fama-MacBeth estimation results are presented in Table 5.5. Panel A shows results for the cases where the independent variable is the growth rate of net long commodity index fund positions in the prior period ($j = 1$). The average slope coefficient estimates are negative, whether the dependent variable is returns, realized volatility, or implied volatility. The coefficient on implied volatility is statistically significant at the 5% level (p-value = 0.0179). The estimated coefficient indicates that a one-percentage-point increase in the IID growth rate decreases subsequent implied volatility (annualized) about 0.17 percentage points. Panel B of Table 5.5 shows the cross-sectional regressions estimated using the lagged growth in notional value as the explanatory variable. All but one of the average estimated slope coefficients is again negative and none is statistically different from zero. The contemporaneous results for the growth rate of net long commodity index fund positions ($j = 0$) are shown in Panel C of Table 5.5. Notional values cannot be used in the contemporaneous model estimation because concurrent notional values are arithmetically related to concurrent returns through the period-ending price. The return model has negative average estimated slope coefficient when contemporaneous changes in index investment are considered. The estimated slope coefficient suggests that a 1% increase in the notional value of index investments coincides with a statistically significant (but relatively small) 0.38% decrease in market returns. The average estimated slopes for volatility are positive but not statistically different from zero. Overall, six of the nine average slope coefficients in Table 5.5 are negative – the opposite of what is predicted under the Masters Hypothesis – and there is little evidence of statistical significance for the average slope coefficients.

Table 5.5. Quarterly Fama-MacBeth cross-sectional regression tests for index fund impacts in commodity futures markets.

Market variable	Intercept (α)	Slope (β)	p-value ($\beta = 0$)
Panel A: Position variable: one-period lagged growth in contracts			
Returns	−1.77	−0.0329	0.8441
Realized volatility	−0.70	−0.1865	0.1282
Implied volatility	−0.84	−0.1739	0.0179
Panel B: Position variable: one-period lagged growth in notional value			
Returns	−0.74	0.1340	0.1311
Realized volatility	−1.90	−0.0633	0.6531
Implied volatility	−0.27	−0.0151	0.9020
Panel C: Position variable: contemporaneous growth in contracts			
Returns	−1.99	−0.3807	0.0394
Realized volatility	−1.01	0.2271	0.1948
Implied volatility	−0.55	0.0378	0.7652

[a]The data are available for $T = 12$ quarters (accounting for calculation of growth rates and a one-period lag) that end between June 30 2008 and March 31 2011 and $N = 19$ markets for a total of 228 observations. The p-value refers to a two-tailed t-test of the null hypothesis that the slope equals zero.

Additional detail on the cross-sectional estimation results is found in Fig. 5.5, which shows the individual quarterly slope estimates for returns and the growth rate in IID net long positions. The plot reveals the wide variation of slope estimates around zero from quarter to quarter, consistent with the lack of statistical significance reported in Table 5.5. There is also no evidence of a trend in slope estimates, which eliminates the possibility of a relatively large and positive relationship early in the sample but declining in subsequent periods as commodity futures markets adjust to the entry of index investment. If there is a surprising result it is the consistency of signs for contemporaneous IID growth rates – nine of the 12 quarterly point estimates for the slope are negative. Finally, similar plots for growth rates in notional value and volatility (not shown) also show no discernable trends.

As with any regression model the inclusion of important 'conditioning variables' may improve the reliability of the cross-sectional regression estimates. In the equity markets, fundamental pricing factors – such as book-to-market ratios or price-to-earnings ratios – can be measured in a consistent manner across companies and are often used as conditioning variables in tests of the factors influencing equity returns (e.g. Fama and French, 1992). In commodity markets it is more difficult to identify analogous conditioning variables that can be quantified in a consistent manner across markets. However, Gorton *et al.* (2007) show that the spread between nearby and deferred futures contracts is a good proxy for relative supply and demand (or inventories) across commodity markets. If the deferred futures price is higher than the nearby futures price (contango), then inventories are relatively abundant. Conversely, if the nearby futures price is higher than the deferred price (backwardation), then the spread signals that inventories are scarce.

To match up with the cross-sectional data, nearby and deferred futures prices are collected at the end of each quarter. The spread for market i in quarter t is calculated as $S_{i,t} = ln\left(p_{i,t}^{1} / p_{i,t}^{2}\right) \cdot 100$, where, as earlier, $p_{i,t}^{1}$ is the commodity futures price of the nearest-to-expiration contract on the last business day of each quarter, and $p_{i,t}^{2}$ is the price of the next-nearest-to-expiration futures contract on the last business day of each quarter. A negative value for $S_{i,t}$ indicates a market in contango and a positive value indicates a market in backwardation. If the market fundamentals – as captured by the spread – have an impact on the cross-section of returns it is expected that those markets that are backwardated (signaling relatively tight inventories) will have greater returns (Gorton *et al.*, 2007). That is, the slope coefficient will be positive.

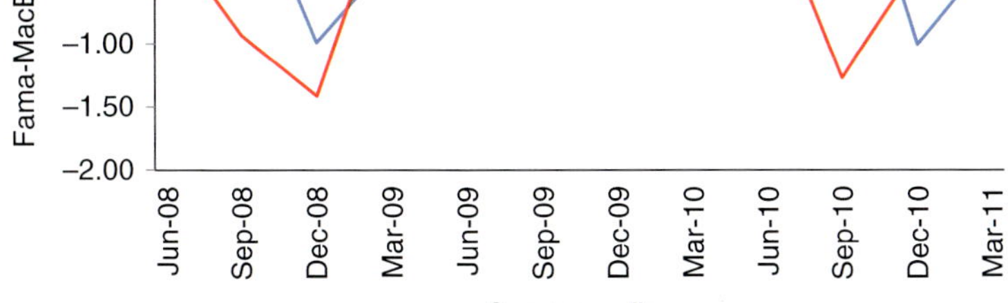

Note: The blue (red) line shows the Fama-MacBeth cross-sectional slope estimate between returns and the one-period lagged (contemporaneous) growth rate of net long index fund positions in 19 US commodity futures markets for quarters ending between June 30 2008 and March 31 2010.

Fig. 5.5. Fama-MacBeth slope estimates for returns and growth rate of Index Investment Data (IID) positions.

Equations 5.4 and 5.5 are re-estimated and the futures market spread is included as a second explanatory variable. Note that in a contemporaneous specification of Eqn 5.4 the quarter ending price of the nearby futures, $t/\sqrt{T}$, would be on both sides of the regression equation, which would build in a predictive component to the specification. To avoid this problem, the model is only specified using lagged spread values. The estimation results for the equations using lagged explanatory variables are shown in Table 5.6. Consistent with expectations, there is a tendency for the slope coefficient on the spread variable to be positive, indicating that a market in greater contango (relatively higher inventories) has lower returns and less volatility. However, none of the estimated slope coefficients on either index positions or the spread is statistically different from zero, confirming the results in Table 5.5. Because differences in storage costs and other factors may affect the absolute size of futures spreads across markets, the models were also estimated using lagged changes in the spreads to capture relative tightening or loosening of perceived fundamentals. The results were not materially different from those presented here.

In sum, the cross-sectional regression tests provide very little evidence that increases in commodity index investment (contracts or notional value) are associated with contemporaneous or subsequent commodity futures returns or volatility. Only two of nine average estimated slope coefficients are statistically significant and both of these significant cases have negative signs that contradict the claim that index investment increases returns or volatility. Similarly, cross-sectional regressions that included the nearby-deferred futures spread as a conditioning variable did not produce statistically significant results. The findings cast serious doubt on the Masters Hypothesis because: (i) IID growth rates are the best available measurement of the increase in the total effective 'demand' of index investors; and (ii) the Fama-MacBeth test has good power properties even for a short panel that is heterogeneous over time (Ibragimov and Muller, 2010a, b).

5.5 Time-series Tests

While the IID report provides the most accurate measure of aggregate index investment in commodity futures markets, especially in energy and metals, the relatively small number of available time-series observations precludes the estimation of market-specific estimates of index impacts. This limitation is especially relevant given that energy futures markets are at the center of the present policy debate. As identified earlier, time-series tests using DCOT swap dealer positions (e.g. Buyuksahin and Harris, 2011) in the energy futures markets do not capture total index investment. An alternative approach is to utilize the actual position data reported by

Table 5.6. Quarterly Fama-MacBeth cross-sectional regression tests for index fund impacts in commodity futures markets conditioned on nearby-deferred spreads.[a]

Market variable	Intercept (α)	Positions slope (β_1)	p-value ($\beta_1 = 0$)	Spread slope (β_2)	p-value ($\beta_2 = 0$)
Panel A: Position variable: one-period lagged growth in contracts					
Returns	−2.28	0.0049	0.9727	−0.0559	0.8965
Realized volatility	−0.51	−0.0793	0.5949	0.5366	0.3339
Implied volatility	0.83	−0.1013	0.3112	0.2774	0.5655
Panel B: Position variable: one-period lagged growth in notional value					
Returns	−0.57	0.1141	0.1837	0.0652	0.8766
Realized volatility	−0.71	−0.0477	0.6759	0.5125	0.2823
Implied volatility	−0.72	−0.0156	0.8931	0.4247	0.3588

[a]The data are available for $T = 12$ quarters (accounting for calculation of growth rates and a one-period lag) that end between June 30 2008 and March 31 2011 and $N = 19$ markets for a total of 228 observations. The p-value refers to a two-tailed t-test of the null hypothesis that the slope equals zero.

energy-related exchange-traded funds (ETFs) as the basis for time-series tests in energy futures markets.

The US Oil Fund, LP (USO) and the US Natural Gas Fund, LP (UNG) are ETFs designed to track the percentage change in the price of light sweet crude oil in Cushing, Oklahoma and natural gas at Henry Hub, Louisiana, respectively. Not coincidently, these are the cash markets that underlie the New York Mercantile Exchange's (NYMEX) crude oil and natural gas futures contracts. USO and UNG gain this exposure by directly purchasing futures contracts, forward contracts, and swap contracts (USOF, 2010; USNGF, 2010). Each fund publishes their daily position (futures contract equivalents) and pending transactions. Futures positions are generally held in the nearby futures contracts with clearly publicized dates for rolling from the expiring futures contract to the next expiration month.

Futures contracts held directly by the USO or UNG are categorized as 'managed money' in the DCOT report. However, any swap positions held by the funds may or may not show up in the swap dealer category depending on whether the positions are netted internally by the particular swap dealer. The published positions held by these funds clearly provide the most exact measure of their market activity. Figures 5.6 and 5.7 demonstrate the size of these funds relative to total index positions in WTI crude oil and natural gas futures as represented by the IID report. USO accounted for nearly 15% of index open interest in crude oil during late 2008 and early 2009. Since then, it has declined to about 4–5% of the index positions in the crude oil market. UNG's share of natural gas index positions rose to over 40% in early 2009 and then declined to near 7% in the most recent quarter. The correlation between the growth in positions for USO and UNG and the corresponding IID net positions is 0.55 and 0.58, respectively.

We use the growth in positions held by the USO and UNG ETFs in direct tests for index investment impacts in the energy futures markets. Daily data for USO are available from July 5 2006 through May 24 2011 for a total of 1231 observations. Daily data for USG are available from July 2 2007 through May 24 2011 for a total of 983 observations. The data are especially useful because: (i) a relatively large number of daily observations are available; (ii) these ETFs only trade at the daily close, so the change in the daily closing positions accurately represents

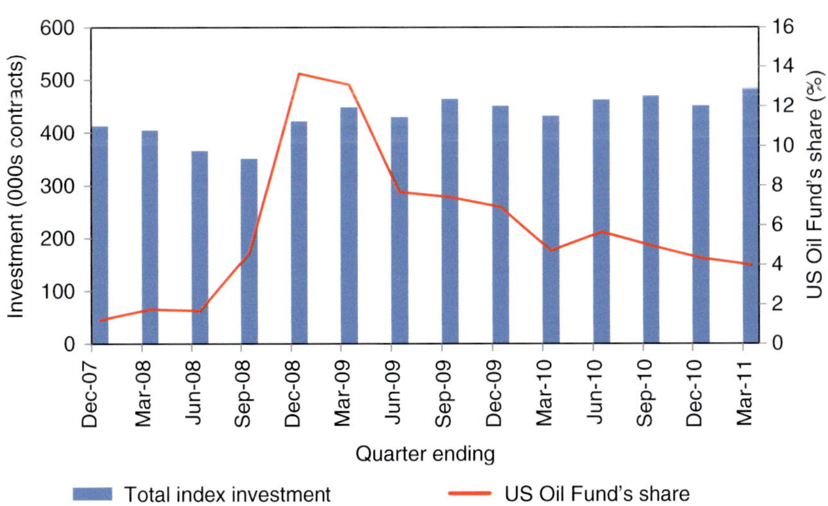

Note: Bars (left-scale) show the net long index fund position (thousand contracts) in WTI crude oil as of the end of each quarter between December 31 2007 and March 31 2010. The line (right-scale) shows the net long position of the US Oil Fund exchange-traded fund (ETF) as a percentage of the total index open interest in WTI crude oil futures on the same dates.

Fig. 5.6. US Oil Fund's share of index investment in the WTI crude oil futures market.

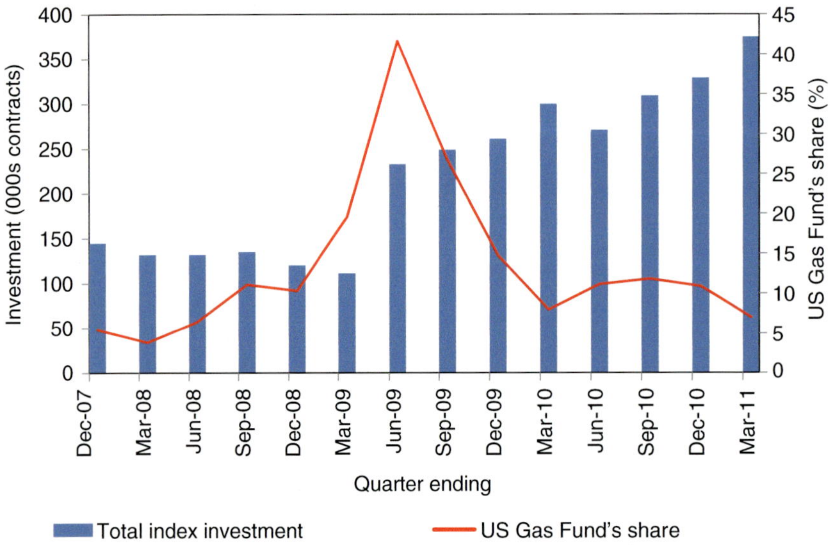

Notes: Bars (left-scale) show the net long index fund position (thousand contracts) in natural gas futures as of the end of each quarter between December 31 2007 and March 31 2010.
The line (right-scale) shows the net long position of the US Natural Gas Fund exchange-traded fund (ETF) as a percentage of the total index open interest in natural gas futures on the same dates.

Fig. 5.7. US Natural Gas Fund's share of index investment in the natural gas futures market.

all of the fund's buying or selling for a given day;[13] and (iii) the sample period encompasses the entire 2007–2008 spike in energy futures prices. While it is certain that trading activity of any two ETFs will not completely reflect overall commodity index investments in energy futures markets, the USO and USG ETFs actively trade and represent an important part of index investments in these markets. The availability of their positions over a reasonably long time also affords a unique opportunity to identify temporal index investment effects in energy markets where DCOT observations are of less value and IID observations are available for a more limited time span.

The predictability of ETF position changes is examined before conducting empirical tests. The daily growth in positions for USO and USG were modeled in a simple autoregressive framework. The Schwartz criterion was used to select the order of the autoregressive model with a maximum lag structure of 20 days. The USO model is specified as an AR(6) with the first-order autocorrelation coefficient insignificantly different from zero. Lags 2 through 6 are statistically significant but of both positive and negative signs. The model only explains 3% of the variation in positions. The USG model is specified as an AR(9) with statistically significant autocorrelation coefficients at lags of 1, 2, and 5 days. While the estimated coefficients at these lags are statistically different from zero, they are small in magnitude and the autoregressive model explains only 4% of the daily changes in USG positions. Similar results are found when modeling growth rates for notional value. Overall, the AR models provide very little evidence of persistence in the market activity of the USO and USG ETFs.

5.5.1 Granger causality tests

The first step of the empirical analysis is to test the causal relationship between futures returns and positions of a given ETF in a Granger-type framework:

$$RD_t = \alpha + \sum_{i=1}^{m} \gamma_i RD_{t-i} + \sum_{j=0}^{n} \beta_j \Delta ETF_{t-j} + \varepsilon_t \; t = 1,\ldots,T \tag{5.6}$$

where RD_t is the return in crude oil or natural gas commodity futures on day t and ΔETF_{t-i} is the growth rate of USO or UNG positions on day $t - j$. In a similar fashion to the Fama-MacBeth regressions, returns are measured as the daily returns of the relevant nearby futures contract and growth rates are measured as the log-relative changes in contracts or notional value. To test for conventional Granger causality between ETF trading activity and returns, Eqn 5.6 is restricted to $j = 1, \ldots, n$. When the contemporaneous association between ETF trading activity and returns is also considered, Eqn 5.6 is respecified allowing $j = 0, \ldots, n$. The regression model is also specified with the volatility of futures returns as the dependent variable:

$$VD_t = \alpha + \sum_{i=1}^{m} \gamma_i VD_{t-i} + \sum_{j=0}^{n} \beta_j \Delta ETF_{t-j} + \varepsilon_t \; t = 1, \ldots, T \tag{5.7}$$

where VD_t is either the growth in realized volatility or implied volatility of crude oil or natural gas futures for day t. Realized volatility is estimated using Parkinson's (1980) estimator.[14] In all of the above cases the null hypothesis of no causality is $\beta_j = 0$ for all j.

The lag structure (m, n) in Eqns 5.6 and 5.7 is determined by a search procedure over $m = 10$ and $n = 10$. The alternative model specifications are estimated using OLS and the model that minimizes the Schwartz criterion is selected to avoid over-parmeterization (Enders, 1995, p. 88). The OLS residuals are tested for serial correlation (Breusch-Godfrey Lagrange multiplier test) and heteroskedasticity. In the event of serial correlation, additional lags of the dependent variable are added until the null of no serial correlation cannot be rejected. If the errors are heteroskedastic, then the model is re-estimated using White's heteroskedastic-consistent covariance estimator.[15]

The estimation results and hypothesis tests are presented in Table 5.7 for various combinations of market and position variables. For instance, Panel A shows the results for USO (crude oil) and USG (natural gas) using lagged values of the growth in contracts. With nearly all position measures, the selection process chooses the simplest model with a single lag (or contemporaneous value) of each independent variable. Therefore, the null hypothesis of interest is simply $\beta_1 = 0$. This null is rejected in only four of the 18 cases presented in Table 5.7 at the 5% level. There is some evidence of a linkage between USG positions and implied volatility in the natural gas market (Panel A), which implies that a one percentage point growth in contracts leads to just over a tenth of a percentage point increase in implied volatility. There is also a significant positive contemporaneous relationship between USG and implied volatility (Panel C). While statistically significant, the magnitude of these relationships is of debatable economic importance.

A significant (p-value = 0.0531) and negative contemporaneous relationship is found between USO contract positions and crude oil futures returns (Panel C). However, the estimated coefficient implies that a one percentage point increase in the growth rate of USO contracts is associated with only about a 0.02% decline in crude oil futures prices. Likewise, a negative and significant contemporaneous relationship is found for UNG in the natural gas futures markets. The negative impact on returns is larger in this case, just under a tenth of a percentage point.

5.5.2 Long-horizon regression tests

The Granger causality tests are designed to detect the relationship, if any, between daily positions and returns. As noted earlier, such tests may have low power to detect relationships over longer horizons (e.g. Summers, 1986). Index trader positions may flow in 'waves' that build slowly – pushing prices higher – and then fade slowly. In this scenario, horizons longer than a day may be necessary to capture the predictive component of index trader positions. So we implement the long-horizon regression model as specified by Valkanov (2003):

$$\sum_{i=0}^{k-1} RD_{t+i} = \alpha + \beta \sum_{i=0}^{k-1} \Delta ETF_{t+i-1} + \varepsilon_{t+i} \quad t = 1, \ldots, T \tag{5.8}$$

and

$$\sum_{i=0}^{k-1} VD_{t+i} = \alpha + \beta \sum_{i=0}^{k-1} \Delta ETF_{t+i-1} + \varepsilon_{t+i} \quad t = 1, \ldots, T \tag{5.9}$$

where all variables are defined as before. In essence, Eqns 5.8 and 5.9 are OLS regressions of a

Table 5.7. Daily Granger causality regression tests for exchange-traded index fund (ETF) impacts in the crude oil and natural gas futures markets.[a]

	Crude oil futures			Natural gas futures		
Market variable	Lag specification (*m*,*n*)	Position slope (β_j)	*p*-value ($\beta_j = 0$)	Lag specification (*m*,*n*)	Position slope (β_j)	*p*-value ($\beta_j = 0$)
Panel A: Position variable: lagged growth in contracts						
Returns	1,1	0.0039	0.6570	2,1	0.0025	0.9050
Realized volatility	10,1	0.0651	0.6091	9,1	–0.3670	0.2116
Implied volatility	3,1	–0.0088	0.7780	5,1	0.1185	0.0009
Panel B: Position variable: lagged growth in notional value						
Returns	1,1	0.0036	0.6839	2,1	0.0183	0.4188
Realized volatility	10,1	–0.0573	0.6428	9,1	0.3648	0.1366
Implied volatility	3,1	–0.0098	0.7419	3,1	0.0027	0.9543
Panel C: Position variable: contemporaneous growth in contracts						
Returns	1,0	–0.0230	0.0531	2,0	–0.0949	0.0001
Realized volatility	10,0	–0.0039	0.9758	9,0	0.3206	0.2833
Implied volatility	3,0	–0.0277	0.4834	5,2	0.1203	0.0256

[a] Daily data for the US Oil Fund is available from July 5 2006 through May 24, 2011 for a total of 1231 observations. Daily data for the US Natural Gas Fund is available from July 2 2007 through May 24 2011 for a total of 983 observations. Note that the test of the null hypothesis that the slope equals zero in the case of natural gas futures and implied volatility in Panel C is based on the sum of the two estimated β coefficients. All other tests are based on a single β coefficient.

k-period moving sum of the dependent variable at time t against a k-period moving sum of the independent variable in the previous period, time $t-1$. If the estimated β is positive (negative), then it indicates a fads-style model where prices tend to increase (decrease) slowly over a relatively long period after widespread index buying (selling). The fads stylization captured in Eqns 5.8 and 5.9 – with a positive β – is consistent with the Masters Hypothesis of an index investment-driven bubble in commodity futures prices.

The underlying dependent variable (returns or volatility) and independent variable (growth in positions or notional value) in Eqns 5.8 and 5.9 are both stationary, so the sums are also stationary. Valkonov (2003) demonstrates that the OLS slope estimator in this specification is consistent and converges at a high rate of T. The specification in Eqns 5.8 and 5.9 clearly creates an overlapping horizon problem for inference. Valkanov shows that Newey-West t-statistics do not converge to well-defined distributions and suggests using the rescaled t-statistic, $t/\sqrt{T}$, along with simulated critical values for inference. Valkanov also demonstrates that the rescaled t-statistic generally is the most powerful among several alternative long-horizon test statistics.

The long-horizon regressions represented by models in Eqns 5.8 and 5.9 are estimated using daily data at alternative horizons of 5, 20, and 60 trading days, which roughly correspond to weekly, monthly, and quarterly time horizons. The estimated OLS β coefficients are shown in Table 5.8 along with the rescaled t-statistic. Critical values for the rescaled t-statistic (–0.563, 0.595) are taken from Valkanov's (2003) Table 4 for Case 2 and $c = -5.0$, $\delta = 0.00$, $T = 750$, and tail values representing the 10% significance level. These represent a conservative case that, if anything, favors a rejection of the null hypothesis that the slope equals zero. The estimated slope coefficients for return regressions are presented in Panels A and B of Table 5.8 for the growth in contracts and growth in notional value, respectively. In no case is the null hypothesis for USO or USG rejected using Valkanov's critical values. Panels C through F show the estimation results for volatility models. Again, there is not a single case where the slope coefficient is statistically significant. There is simply no evidence that the change in positions by these ETFs impact crude oil or natural gas futures prices over longer horizons.

In sum, time-series tests using positions held by the USO and UNG ETFs provide no support for the Masters Hypothesis. With daily data and Granger-style tests, we find no causal linkages between fund positions and subsequent market returns. On a contemporaneous basis, there is some evidence of a negative association between fund buying and market returns; just the opposite of what would be expected under the Masters Hypothesis. Long-horizon regressions provide a powerful time-series test of bubble and fad-style models, and the null hypothesis of no impact is not rejected in a single case.

5.6 Conclusions

Hedge fund manager Michael W. Masters is a leading proponent of the view that commodity index investment was a major driver of the 2007–2008 spike in commodity prices. In essence, Masters argues that massive buy-side pressure from index funds created a bubble in commodity prices, with the result that prices, and crude oil prices, in particular, far exceeded fundamental values at the peak. Empirical tests of the Masters Hypothesis in commodity futures markets require data that match the basic tenets of the hypothesis. In particular, swap dealer netting of index positions with other OTC transactions results in futures positions – as reported in the US CFTC's COT reports – that may not accurately reflect the total effective 'demand' of index investors. The netting effect is most severe in energy futures markets, which are at the heart of the current public policy debate about speculative impacts. The best measure of actual commodity index investment is found in the CFTC's quarterly IID report, which is collected independently of the CFTC's large trader reporting system.

The IID shows that the statistical relationships between index positions in different markets can be weak. Most pointedly, the net long IID position in crude oil is negatively correlated with those of agricultural commodities from 2007 to 2011. WTI crude oil futures positions created using the Masters (2008) algorithm do

Table 5.8. Long-horizon regression tests for ETF impacts in the crude oil and natural gas futures markets.[a]

	Crude oil		Natural gas	
Horizon (k)	Slope (β_k)	Rescaled t-statistic ($t/\sqrt{T}$)	Slope (β_k)	Rescaled t-statistic ($t/\sqrt{T}$)
Panel A: Dependent variable: returns, independent variable: growth in contracts				
Weekly (k = 5)	−0.0452	−0.0770	−7.2859	−0.0770
Monthly (k = 20)	−0.1803	−0.1442	−5.6635	−0.0421
Quarterly (k = 60)	−0.3386	−0.1563	−9.2808	−0.0515
Panel B: Dependent variable: returns, independent variable: growth in notional value				
Weekly (k = 5)	0.0221	0.0434	19.4459	0.1611
Monthly (k = 20)	−0.0543	−0.0272	15.6069	0.0915
Quarterly (k = 60)	−0.2143	−0.0587	6.8231	0.0536
Panel C: Dependent variable: realized volatility, independent variable: growth in contracts				
Weekly (k = 5)	0.0462	0.0175	−0.1593	−0.0294
Monthly (k = 20)	0.1612	0.0648	0.1335	0.0493
Quarterly (k = 60)	0.2278	0.0790	0.1545	0.0801
Panel D: Dependent variable: realized volatility, independent variable: growth in notional value				
Weekly (k = 5)	−0.0209	−0.0081	−0.0101	−0.0022
Monthly (k = 20)	0.0358	0.0127	0.0826	0.0252
Quarterly (k = 60)	0.1306	0.0514	0.0986	0.0559
Panel E: Dependent variable: implied volatility, independent variable: growth in contracts				
Weekly (k = 5)	0.0250	0.0255	0.1277	0.0834
Monthly (k = 20)	0.1355	0.0809	0.1859	0.1440
Quarterly (k = 60)	0.1750	0.0701	0.1940	0.1212
Panel F: Dependent variable: implied volatility, independent variable: growth in notional value				
Weekly (k = 5)	−0.0064	−0.0069	0.0602	0.0447
Monthly (k = 20)	0.0587	0.0282	0.0929	0.0461
Quarterly (k = 60)	0.0923	0.0404	0.1216	0.0630

[a]Daily data for the US Oil Fund is available from July 5 2006 through May 24 2011 for a total of 1231 observations. Daily data for the US Natural Gas Fund is available from July 2 2007 through May 24 2011 for a total of 983 observations. Critical values for the rescaled t-statistic (−0.563, 0.595) are taken from Valkanov's (2003) Table 4 for Case 2 and $c = -5.0$, $\delta = 0.00$, $T = 750$, and tail values representing the 10% significance level.

not correspond closely with those reported in the IID. The average absolute error is 52% and the correlation is near zero at the quarterly horizon. Empirical studies that use the Masters algorithm (Stoll and Whaley, 2010; Singleton, 2011) or other mapping methods (Gilbert, 2009) to generate index positions in WTI crude oil futures are likely seriously flawed as a result. The IID appears to be the only reliable measure of index investments in the energy and metals futures markets.

Cross-sectional Fama-MacBeth regression tests using the IID find very little evidence that index positions influence returns or volatility in 19 commodity futures markets. The results are robust to whether lagged or contemporaneous effects are considered and the addition of the nearby-deferred futures spread as a conditioning variable. The findings are the strongest evidence against the Masters Hypothesis to date because: (i) IID growth rates are the best available measurement of the increase in the total effective 'demand' of index investors; and (ii) the Fama-MacBeth test has good power properties for the sample sizes considered in this study (Ibragimov and Muller, 2010a, b).

The cross-sectional tests are supplemented with time-series tests based on extensive samples of daily position data for two large energy ETFs – the US Oil Fund and the US Natural Gas Fund. In order to address power concerns, both Granger causality and long-horizon regression tests (Valkanov, 2003) are applied to the daily time-series of ETF positions. Granger causality tests show no causal links between daily returns or volatility in crude oil and natural gas futures markets and the positions of the ETFs. Long-horizon regression tests likewise fail to reject the null hypothesis of no market impact for the two ETFs.

The failure of the empirical results to support the Masters Hypothesis has important implications for the pricing performance of commodity futures markets. It implies that the markets were sufficiently liquid to absorb the large order flow of index funds in recent years, at least over daily or longer time horizons. Traders in commodity futures markets also did not confuse index fund position changes with signals about market fundamentals; rather, they reacted rationally to the market activity of index funds. The evidence strongly suggests that index funds – while a sizable participant – did *not* in fact harm price discovery in commodity futures markets. From this vantage point, recent regulatory plans to impose speculative position limits on index fund investors in all US commodity futures markets appear to be ill-conceived.

Acknowledgments

The authors are indebted to John Hyland, Chief Investment Officer, United States Commodity Funds, LLC, for providing the ETF data used in this study. Hongxia Jiao provided superb assistance in collecting the futures and options data. Wade Brorsen, Phil Garcia, Jim Hamilton, and Aaron Smith provided numerous helpful comments on an earlier draft of the paper.

Notes

[1] Original citation: Irwin, S.H. and Sanders, D.R. (2012) Testing the Masters Hypothesis in commodity futures markets. *Energy Economics* 34, 256–269. Reprinted by permission of Elsevier B.V.

[2] See USS/PSI (2009), Irwin *et al*. (2009), and Pirrong (2010) for detailed discussions of the controversy surrounding the role of index investment in the 2007–2008 commodity price spike.

[3] Investors may enter directly into over-the-counter (OTC) contracts with swap dealers to gain the desired exposure to returns from a particular index of commodity prices. Some firms also offer investment funds whose returns are tied to a commodity index. Exchange-traded funds (ETFs) and structured notes (ETNs) also have been developed that track commodity indexes. In the remainder of this paper, the term 'commodity index fund' or 'index fund' is used generically to refer to long-only commodity investments. See Engelke and Yuen (2008) for further details on commodity index investments.

[4] A detailed history of position limit regulations in US commodity futures markets can be found at: http://www.cftc.gov/PressRoom/SpeechesTestimony/berkovitzstatement072809.html (accessed January 14 2022).

[5] Other studies investigate the impact of speculation in the recent commodity price spike without directly testing for statistical linkages between index fund positions and price movements (e.g. Einloth, 2009; Kilian and Murphy, 2010; Phillips and Yu, 2010; Tang and Xiong, 2010). Conclusions are mixed with regard to the impact of speculation or whether a price bubble occurred.

[6] For an example, see the letter 'Swaps, Spots, and Bubbles' by Sir Richard Branson, Michael Masters, and David Frenk published in the July 29 2010 issue of *The Economist* magazine (available at: http://www.economist.com/node/16690679 (accessed January 14 2022)).

[7] The Masters (2008) procedure was followed as closely as possible. The weights for the S&P GSCI markets were those reported at the end of each quarter (1–2 days within the SCOT report dates). The DJ-UBS weights were not available on those specific dates, so the target weights reported for each year were utilized. While the raw data used by Masters (2008) were not available, we compared our estimates where possible to estimates presented graphically in Masters (2008, 2009), Masters and White (2008), and Singleton (2011). Our estimates appear to track their data closely through 2008, but diverge by roughly 100,000 contracts beginning in mid-2009 with our estimates being lower. The source of this difference is not known. Additional research is needed to determine the source of the divergence.

[8] One possibility is the rising popularity of single commodity or sector-specific index funds.

[9] A number of pooled or system estimation methods – such as a seemingly unrelated regressions – could potentially be used with this data. At this time, the application of those methods is limited due to the fact that the number of time-series observations (14) is less than the cross-sectional observations (19). As a longer history of the data set becomes available, these estimation methods may prove to be a useful means to analyze these data.

[10] It is important to use an explanatory variable that can be normalized across markets. Since contract and market size varies widely across the 19 markets, we follow Stoll and Whaley (2010) and use the growth rate in commodity index investment. This assumes pricing pressure is related to the flow of commodity index investment rather than the absolute size of the investment.
[11] Futures equivalent positions in the IID report are inferred from total investment for each market. We follow previous researchers (e.g. Stoll and Whaley, 2010) and relate these aggregate positions to nearby futures returns because it is well known that index positions generally are rolled from one nearby futures contract to the next.
[12] Peterson (2009) also shows that the Fama-MacBeth slope estimator is unbiased and has correctly sized confidence intervals in the absence of serial correlation for a given market.
[13] Based on email communication with John Hyland, Chief Investment Officer, United States Commodity Funds, LLC, in 2011.
[14] Akay *et al.* (2010) and others show that volatility estimation using extreme values is highly efficient and generally free from bias caused by microstructure noise.
[15] Using the Augmented Dickey Fuller (ADF) test, the null of a unit root could not be rejected (10% level) for the (log) level of USO contracts but it was for USG positions in terms of contracts. The null is rejected (10% level) for USG but not for USO using the (log) level of notional values. The null is not rejected (10% level) in any case using the log-relative change in levels. Based on this evidence, the use of growth rates is assumed to be most appropriate for achieving stationarity and balanced regressions. In addition, there is a strong initial trend in the size of the funds as they 'ramped up' after the initial offering. The first 3 months of data for each fund are deleted in order to eliminate growth rate outliers during this period.

References

Acworth, W. and Morrison, J. (2010) Massive response to CFTC position limit proposal. *The Magazine of the Futures Industry*, June/July, 38–41.

Akay, O., Griffiths, M.D. and Winters, D.B. (2010) On the robustness of range-based volatility estimators. *Journal of Financial Research* 33, 179–199.

Black, F. (1976) The pricing of commodity contracts. *Journal of Financial Economics* 3, 167–169.

Brunetti, C. and Buyuksahin, B. (2009) Is speculation destabilizing? Working paper, US Commodity Futures Trading Commission. Available at: https://papers.ssrn.com/sol3/papers.cfm?abstract_id=1393524 (accessed January 14 2022).

Buyuksahin, B. and Harris, J.H. (2011) Do speculators drive crude oil futures prices? *Energy Journal* 32, 167–202.

Buyuksahin, B. and Robe, M.A. (2011) Does 'paper oil' matter? Energy markets' financialization and equity-commodity co-movements. Working paper, US Commodity Futures Trading Commission. Available at: https://papers.ssrn.com/sol3/papers.cfm?abstract_id=1855264 (accessed January 14 2022).

Campbell, J.Y., Lo, A.W. and MacKinlay, A.C. (1997) *The Econometrics of Financial Markets*. Princeton University Press, Princeton, New Jersey.

CFTC (Commodity Futures Trading Commission) (2006a) Comprehensive review of the commitments of traders reporting program. Federal Register, 71, FR 35627. Available at: https://www.federalregister.gov/documents/2006/06/21/E6-9722/comprehensive-review-of-the-commitments-of-traders-reporting-program (accessed January 14 2022).

CFTC (Commodity Futures Trading Commission) (2006b) Commodity Futures Trading Commission actions in response to the 'comprehensive review of the commitments of traders reporting program (June 21, 2006). Available at: https://www.cftc.gov/sites/default/files/idc/groups/public/@commitmentsoftraders/documents/file/noticeonsupplementalcotrept.pdf (accessed January 14 2022).

CFTC (Commodity Futures Trading Commission) (2008) *Staff Report on Commodity Swap Dealers and Index Traders with Commission Recommendations*. Available at: https://www.cftc.gov/sites/default/files/idc/groups/public/@newsroom/documents/file/cftcstaffreportonswapdealers09.pdf (accessed January 14 2022).

CFTC (Commodity Futures Trading Commission) (2009) *Disaggregated Commitments of Traders*: explanatory notes. Available at: https://www.cftc.gov/sites/default/files/idc/groups/public/@commitmentsoftraders/documents/file/disaggregatedcotexplanatorynot.pdf#:~:text=Explanatory%20Notes%20The%20

Commodity%20Futures%20Trading%20Commission%20%28Commission%29,2009%2C%20 the%20remaining%20physical%20commodity%20markets%20were%20included.1 (accessed January 14 2022).

CFTC (Commodity Futures Trading Commission) (2010a) *Commitments of Traders* (COT) Reports Descriptions. Available at: https://www.cftc.gov/MarketReports/CommitmentsofTraders/index.htm (accessed January 14 2022).

CFTC (Commodity Futures Trading Commission) (2010b) Index Investment Data: Explanatory Notes. Available at: https://www.cftc.gov/MarketReports/IndexInvestmentData/ExplanatoryNotes/index.htm#:~:-text=The%20index%20investment%20data%20is%20an%20outgrowth%20of,number%20of%20futures%20contracts%20for%20each%20U.S.%20market (accessed January 14 2022).

CFTC (Commodity Futures Trading Commission) (2010c) Large Trader Reporting Program. Available at: https://www.cftc.gov/IndustryOversight/MarketSurveillance/LargeTraderReportingProgram/index.htm#:~:text=Large%20Trader%20Data%20Under%20the%20Commission%E2%80%99s%20 LTRS%2C%20clearing,specific%20reporting%20levels%20as%20set%20by%20the%20Commission (accessed January 14 2022).

De Long, J.B., Shleifer, A. Summers, L.H. and Waldmann, R.J. (1990) Noise trader risk in financial markets. *Journal of Political Economy* 98, 703–738.

Ederington, L. and Lee, J.H. (2002) Who trades futures and how: evidence from the heating oil market. *Journal of Business* 75, 353–373.

Einloth, J. (2009) *Speculation and recent volatility in the price of oil*. FDIC Center for Financial Research Working Paper No. 2009-08. Federal Deposit Insurance Corporation. Available at: https://www.fdic.gov/bank/analytical/cfr/2009/wp2009/2009-08.pdf (accessed January 14 2022).

Enders, W. (1995) *Applied Econometric Time Series*. Wiley, New York.

Engelke, L. and Yuen, J.C. (2008) Types of commodity investments. In: Fabozzi, F.J., Fuss, R. and Kaiser, D. (eds) *The Handbook of Commodity Investing*. Wiley, Hoboken, New Jersey, pp. 549–569.

Fama, E.F. and French, K.R. (1992) The cross-section of expected stock returns. *Journal of Finance* 47, 427–465.

Fama, E.F. and MacBeth, J.D. (1973) Risk, return, and equilibrium: empirical tests. *Journal of Political Economy* 81, 607–636.

Gilbert, C.L. (2009) Speculative influences on commodity futures prices, 2006–2008. Working Paper, Department of Economics, University of Trento, Trento, Italy. Available at: https://www.cftc.gov/sites/default/files/idc/groups/public/@swaps/documents/file/plstudy_14_cifrem.pdf (accessed January 14 2022).

Gilbert, C.L. (2010) How to understand high food prices. *Journal of Agricultural Economics* 61, 398–425.

Gorton, G.B., Hayashi, F. and Rouwenhorst, K.G. (2007) The fundamentals of commodity futures returns. National Bureau of Economic Research (NBER) Working Paper No. 13249. Available at: https://www.nber.org/papers/w13249 (accessed January 14 2022).

Grossman, S.J. (1986) An analysis of 'insider trading' on futures markets. *Journal of Business* 59, S129–S146.

Grossman, S.J. and Miller, M.H. (1988) Liquidity and market structure. *Journal of Finance* 43, 617–633.

Hamilton, J.D. (2009) Causes and consequences of the oil shock of 2007–08. *Brookings Papers on Economic Activity,* spring, 215–261.

Ibragimov, R. and Muller, U.K. (2010a) *t*-statistic based correlation and heterogeneity robust inference. *Journal of Business and Economics Statistics* 28, 453–468.

Ibragimov, R. and Muller, U.K. (2010b) Supplementary material to *t*-statistic based correlation and heterogeneity robust inference. Available at: https://www.princeton.edu/~umueller/tstat_sup.pdf (accessed January 14 2022).

Irwin, S.H. and Sanders, D.R. (2011) Index funds, financialization, and commodity futures markets. *Applied Economic Perspectives and Policy* 33, 1–31.

Irwin, S.H., Sanders, D.R. and Merrin, R.P. (2009) Devil or angel? The role of speculation in the recent commodity price boom (and bust). *Journal of Agricultural and Applied Economics* 41, 393–402.

Kilian, L. and Murphy, D. (2010) The role of inventories and speculative trading in the global market for crude oil. Working Paper, Department of Economics, University of Michigan, Ann Arbor, Michigan. May. Available at: https://papers.ssrn.com/sol3/papers.cfm?abstract_id=1578103 (accessed January 14 2022).

Masters, M.W. (2008) Testimony before the Committee on Homeland Security and Governmental Affairs, United States Senate. May 20. Available at: http://hsgac.senate.gov/public/_files/052008Masters.pdf (accessed January 14 2022).

Masters, M.W. (2009) Testimony before the Commodity Futures Trading Commission. August 5. Available at: https://www.cftc.gov/sites/default/files/idc/groups/public/@newsroom/documents/file/hearing080509_masters.pdf (accessed January 14 2022).

Masters, M.W. and White, A.K. (2008) The accidental Hunt brothers: how institutional investors are driving up food and energy prices. Available at: https://www.cftc.gov/sites/default/files/idc/groups/public/@swaps/documents/file/plstudy_31_ahb.pdf (accessed January 14 2022).

Parkinson, M. (1980) The extreme value method for estimating the variance of the rate of return. *Journal of Business* 53, 61–65.

Peterson, M.A. (2009) Estimating standard errors in finance panel data sets: comparing approaches. *Review of Financial Studies* 22, 435–480.

Phillips, P.C.B. and Yu, J. (2010) Dating the timeline of financial bubbles during the subprime crisis. Cowles Foundation Discussion Paper No. 1770, Yale University. Available at: https://cowles.yale.edu/sites/default/files/files/pub/d17/d1770.pdf (accessed January 14 2022).

Pirrong, C. (2010) No theory? No evidence? No problem! *Regulation* 33, 38–44.

Sanders, D.R. and Irwin, S.H. (2010) A speculative bubble in commodity futures prices? Cross-sectional evidence. *Agricultural Economics* 41, 25–32.

Sanders, D.R. and Irwin, S.H. (2011a) New evidence on the impact of index funds in U.S. grain futures markets. *Canadian Journal of Agricultural Economics* 59, 519–532.

Sanders, D.R. and Irwin, S.H. (2011b) The impact of index funds in commodity futures markets: a systems approach. *Journal of Alternative Investments* 14, 40–49.

Sanders, D.R., Irwin, S.H. and Merrin, R.P. (2010) The adequacy of speculation in agricultural futures markets: too much of a good thing? *Applied Economic Perspectives and Policy* 32, 77–94.

Singleton, K.J. (2011) Investor flows and the 2008 boom/bust in oil prices. Working Paper No. 3116, Graduate School of Business, Stanford University, Stanford, California. Available at: https://www.gsb.stanford.edu/faculty-research/working-papers/investor-flows-2008-boombust-oil-prices (accessed January 14 2022).

Singleton, K.J. (2014) Investor flows and the 2008 boom/bust in oil prices. *Management Science* 60, 300–318.

Stoll, H.R. and Whaley, R.E. (2010) Commodity index investing and commodity futures prices. *Journal of Applied Finance* 20, 7–46.

Summers, L.H. (1986) Does the stock market rationally reflect fundamental values? *Journal of Finance* 41, 591–601.

Tang, K. and Xiong, W. (2010) Index investment and the financialization of commodities. Working Paper, Department of Economics, Princeton University, Princeton, New Jersey. Available at: https://papers.ssrn.com/sol3/papers.cfm?abstract_id=1683135 (accessed January 14 2022).

USNGF (United States Natural Gas Fund, LP) (2010) Fund objectives and key features. Available at: https://www.uscfinvestments.com/ung (accessed January 14 2022).

USOF (United States Oil Fund, LP) (2010) Fund objectives and key features. Available at: https://www.uscfinvestments.com/uso (accessed January 14 2022).

USS/PSI (United States Senate, Permanent Subcommittee on Investigations) (2009) Excessive speculation in the wheat market. Majority and Minority Staff Report. Available at: https://www.hsgac.senate.gov/imo/media/doc/REPORTExcessiveSpecullationintheWheatMarketwoexhibitschartsJune2409.pdf?attempt=2 (accessed January 14 2022).

Valkanov, R. (2003) Long-horizon regressions: theoretical results and applications. *Journal of Financial Economics* 68, 201–232.

6

Financialization and Structural Change in Commodity Futures Markets[1]

New Author Foreword

While working on the OECD report it became clear to us that there was a great deal of confusion about the nature of 'financialization' and the types of market impacts associated with it. What exactly was financialization? Could the market impacts due to financialization be disentangled from those associated with other structural changes occurring at the same time? What was happening to the composition of market participants in commodity futures markets while the process of financialization evolved? The timing was perfect for writing an article that explored these questions.

Once again, we were provided the opportunity to write the needed article by our old friend Hector Zapata of Louisiana State. He contacted us in May 2011 about being part of an invited paper session for the Annual Meeting of the Southern Agricultural Economics Association to be held in Birmingham, Alabama during February 2012. Given the success of our previous invited paper for the Southern Association Annual Meeting (see Chapter 2, this volume), we were very enthusiastic about participating.

When we began writing the article, we came up with what we think is some of our most memorable phrasing. In the first paragraph of the paper we wrote:

> A trader from the latter part of the 19th century magically transported to the trading pits of the waning years of the 20th century might have been surprised by the size of the commodity futures markets but not by the way trading was conducted or the main types of participants.
>
> (Irwin and Sanders, 2012a, p. 371)

We used this colorful imagery to emphasize the stability of the basic nature of commodity futures markets for so long, and then contrast this with the enormous changes that had occurred in the previous decade.

Around this time, there was some movement towards a broader definition of financialization that included hedge funds and other managed money traders. We argued that this only confused matters and deflected attention away from the core question of the impact of large-scale buying of financial index investors in commodity futures markets. This is the original definition of financialization provided by Domanski and Heath (2007), and we did not see any compelling reason to change it. While there will never be universal acceptance of any definition, this now seems to be the most widely accepted use of the term in the field of financial economics.

A major complication in any analysis of the impact of financialization in commodity futures markets is that other historically large and important structural changes were taking place at roughly the same time. The first and most obvious was the switch from open outcry pit trading to electronic trading, which basically ran in parallel with financialization. We did not realize it at the time, but the data we collected to document the transition to electronic trading in agricultural futures markets had not been presented anywhere else. In our view, every good paper needs a 'money chart,' and our charts showing how the share of electronic trading exploded during 2006–2007 certainly fit the bill.

 Speculation by Commodity Index Funds: The Impact on Food and Energy Prices (S.H. Irwin and D.R. Sanders)
DOI:10.1079/9781800622104.0006

The second structural change was the revolution in market access that arose from the combination of incredible changes in communication tools (e.g. cell phones, the Web, social media), new financial instruments like exchange-traded funds (ETFs) and index funds, and, of course, electronic trading. These changes dramatically lowered the cost of trading and the ease with which one could participate in the commodity futures markets.

We then argued that the structural changes could have irrational and harmful impacts or rational and beneficial impacts. The latter had hardly been considered with all of the brouhaha over the Masters Hypothesis. Our review of the literature available at the time suggested that the rational and beneficial impacts more than likely outweighed any irrational and harmful impacts.

Even though this paper is not one of our best known, it has still been cited more than 250 times since it was published, according to Google Scholar. With the passage of time, we are convinced that the main point of the paper has become even more important; that is, it can be very difficult to disentangle market impacts due to financialization and other structural changes occurring at the same time. In more technical terms, simple approaches to identifying financialization impacts can be easily contaminated by other structural changes. It is not unusual to find tests of financialization impacts in the literature that are simple 'before and after' tests using 2004 as the breakpoint (we are guilty of doing this too). This is not to say that such tests are a waste of time but rather, that they need to be interpreted with a great deal of care. It is also why further research is needed that pays very careful attention to identification issues.

Abstract

The first decade of the 21st century has perhaps witnessed more structural change in commodity futures markets than all previous decades combined. Not only have trading volumes and open interest increased markedly, but this time period also saw historic changes in both trading and participants. The available literature indicates that the irrational and harmful impacts of the structural changes in commodity futures markets over the last decade have been minimal. In particular, there is little evidence that passive index investment caused a massive bubble in commodity futures prices. There is intriguing evidence of several other rational and beneficial impacts of the structural changes over the last decade. In particular, the expanding market participation may have decreased risk premiums and hence, the cost of hedging, reduced price volatility, and better integrated commodity markets with financial markets.

Keywords: index funds, commodity, futures markets, prices, speculation, bubble

JEL categories: D84, G12, G13, G14, Q13, Q41

6.1 Introduction

Since their modern inception in Chicago during the 1850s, dozens of commodity futures contracts have been launched and countless alterations have been made to the relatively few contracts that have survived over time. Nonetheless, the basic structure of the markets has been remarkably stable over time – a trader from the latter part of the 19th century magically transported to the trading pits of the waning years of the 20th century might have been surprised by the size of the commodity futures markets but not by the way trading was conducted or the main types of participants.[2] This stability was not fated to last however.

The first decade of the 21st century has perhaps witnessed more structural change in commodity futures markets than all previous decades combined. Not only have trading volumes and open interest increased markedly, but this period also saw historic changes in both trading and participants. Commodity futures markets transitioned from a primarily telephone/open outcry trading platform to a computer/electronic order matching platform. As a result, market access was greatly expanded, and trading costs declined. Perhaps not coincidently, the same period saw new 'financial' participants enter the commodity futures arena. Investments that track a commodity index became an accepted alternative investment for institutions and pension funds. The increasing importance of these non-traditional participants has been labeled the 'financialization' of commodity futures markets (Domanski and Heath, 2007). Finally, exchange-traded funds (ETFs) were introduced that tracked commodity indices or even single futures markets. These changes undoubtedly contributed to the increase in the volume of trade on commodity futures markets.

The structural shifts seen in commodity futures markets during the last decade can impact the marketplace along both rational and irrational avenues. Rational market impacts – such as improved market liquidity and potentially reduced risk premiums – stem from broader market participation and more active trade. Irrational impacts – such as a commodity price bubble – would stem from the markets' inability to adjust to these changes. In this article, we first review recent trends in open interest and volume for important agricultural futures markets. Next, we examine the forces of structural change within the commodity futures markets driving the trends in market participation over the last decade. Finally, we provide an overview of the literature on rational and irrational market impacts. The emphasis in this part of the paper is on the potential irrational impacts of financialization, since this has received the most attention in terms of public policy and academic research.

6.2 Trends in Open Interest and Volume

There is little doubt that 'something happened' from 2003 to 2008 in the commodity futures markets. For example, combined futures and option (delta-adjusted) open interest in Chicago Board of Trade (CBOT) soybeans over 1995–2002 was relatively stable at an average of 223,000 contracts (Fig. 6.1).[3] In 2003, open interest started to build and moved up to a peak of 878,000 contracts by February 2008. Open interest declined during the financial crises of late 2008 and 2009 but then moved higher again in 2011, with a February peak in excess of 1 million futures and options contracts. A similar story can be told for the other commodity futures such as wheat (Fig. 6.1), live cattle and lean hogs (Fig. 6.2). Across these markets, open interest still appears to be on a solid upward path.

Not surprisingly, trading volumes over the same period increased rather dramatically. From 2000 to 2003, soybean futures had a monthly average trading volume of just under 1.2 million contracts (Table 6.1 and Fig. 6.3). Over the next 6 years, trading volume nearly tripled with a record 4.4 million futures contracts changing hands in February 2011. Similar proportional increases in trading volume were seen in wheat (Fig. 6.3). However, in the livestock futures markets, the increases in futures trading volume are even more dramatic (Fig. 6.4). Lean hog futures contracts have seen a nearly fivefold increase in monthly volume, and live cattle futures volume increased from an average of 338,000 contracts in 2000–2003 to 925,000 in 2009–2011 (Table 6.1).

Despite some of the marked increases in the absolute levels of volume and open interest, the

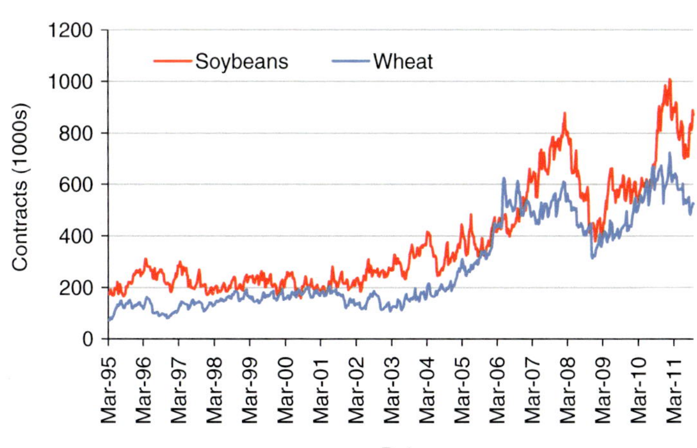

Fig. 6.1. Soybean and wheat combined futures and options open interest, 1995–2011.

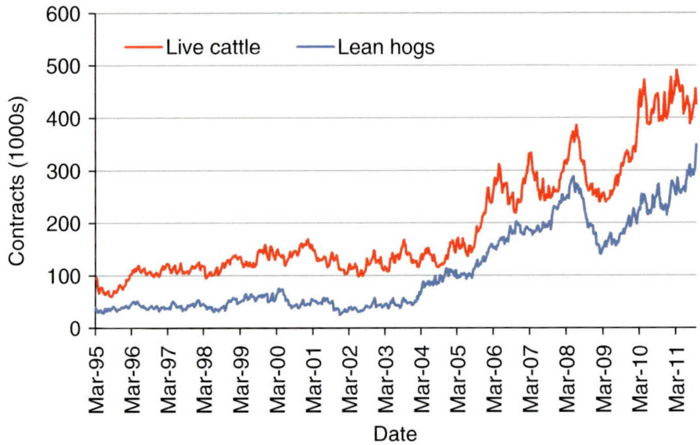

Fig. 6.2. Live cattle and lean hogs combined futures and options open interest, 1995–2011.

Table 6.1. Summary market statistics, 2000–2011.[a]

Time period	Monthly futures volume (1000s)	Futures volume on electronic (%)	Futures and options open interest (1000s)	Futures volume-to-open interest (ratio)	Options share of open interest (%)	Open interest in deferred contracts (%)
Panel A: Corn						
2000–2003	1483	1	571	3.42	23.8	6
2004–2008	3405	37	1396	3.30	25.9	14
2009–2011	5316	92	1703	4.51	31.5	14
Panel B: Soybeans						
2000–2003	1184	1	240	6.35	22.1	3
2004–2008	2075	35	490	5.81	25.8	7
2009–2011	3070	92	675	6.35	27.3	8
Panel C: Wheat						
2000–2003	564	1	158	4.71	22.7	2
2004–2008	1198	36	390	3.88	18.6	7
2009–2011	1797	96	517	4.37	21.9	10
Panel D: Live cattle						
2000–2003	338	0	131	3.01	14.0	11
2004–2008	615	7	228	3.11	11.6	10
2009–2011	925	62	370	3.15	18.3	11
Panel E: Lean hogs						
2000–2003	171	0	46	4.18	8.5	10
2004–2008	493	11	159	3.38	6.9	10
2009–2011	684	68	223	3.67	13.6	12

[a]Volume and open interest are measured in thousands of contracts.

proportion of trading activity to market size remained relatively constant. One measure of activity or turnover is the trading volume-to-open interest ratio. Bessembinder and Seguin (1993) have shown that greater trading volume can be associated with greater price volatility; but market price volatility can be mitigated by large open interest. A market trending toward higher volume-to-open interest might be more susceptible to volatility shocks.

The data for commodity futures suggest remarkably steady volume-to-open interest ratios

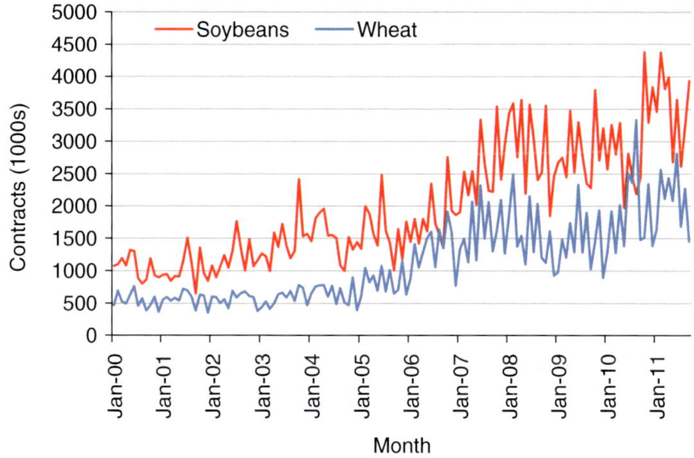

Fig. 6.3. Soybeans and wheat monthly futures trading volume, 2000–2011.

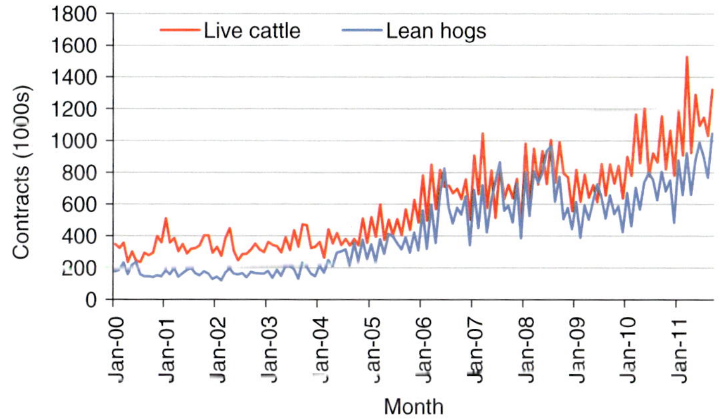

Fig. 6.4. Live cattle and lean hogs monthly futures trading volume, 2000–2011.

for the commodity futures markets studied (see Table 6.1). Among the grain markets (Fig. 6.5), soybean futures averaged a ratio of 6.35 from 2000 to 2003 and 6.35 from 2009 to 2011. Likewise, corn and wheat have ratios that are relatively stable with averages of 3.62 and 4.28 over the entire sample. In the livestock markets, the volume-to-open interest ratios are even more stable. Live cattle and live hog ratios have been consistently between 3.0 and 4.0 with live cattle averaging 3.08 and lean hogs 3.72 over the entire 2000–2011 period (Fig. 6.6 and Table 6.1). These data suggest that despite the large increases in market participation, trading volume and open interest have remained in fairly constant proportions. Overall, this may suggest no general change in the markets' ability to absorb price shocks (Bessembinder and Seguin, 1993).

The large increases in futures and options open interest could be driven by an expanded use of futures, options, or both. While there is no theoretical reason to anticipate that this would create a market impact, it could provide clues as to the nature of the increase in activity. To gauge the relative positions in the options market, the options open interest (delta-adjusted) is simply expressed as a percentage of the total open interest in each market. The data are plotted to observe any changes over the last decade and a summary is provided in Table 6.1.

As shown in Fig. 6.7, the percentage of open interest that is held in the options markets has

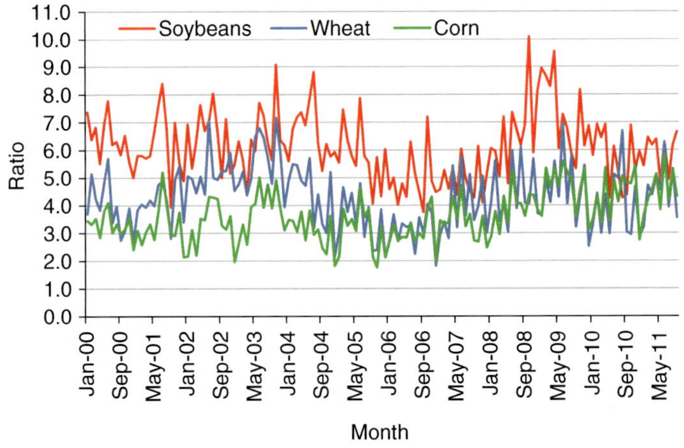

Fig. 6.5. Futures volume-to-open interest ratios, grains, 2000–2011.

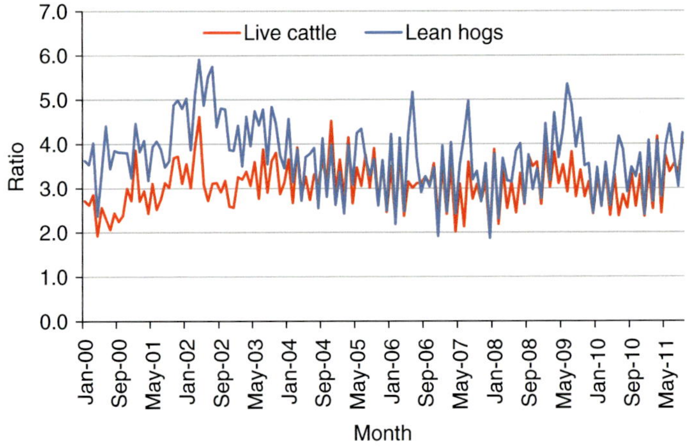

Fig. 6.6. Futures volume-to-open interest ratios, livestock, 2000–2011.

been relatively stable over the 2000–2011 sample with corn, wheat, and soybeans averaging 26%, 20%, and 25%, respectively. There has been a gradual increase in corn options' share of total open interest, increasing from 23.8% in 2000–2003 to 31.5% in 2009–2011 (Table 6.1). Conversely, the share of open interest held in wheat options actually declined from 22.7% in 2000–2003 to a low of 18.6% in 2004–2008.

While there has been some trend towards higher relative option participation in the 2009–2011 period in the grain markets, it has been a gradual increase and not uniform across markets. In contrast, the livestock markets show a clear upward move in option participation. In particular, lean hog options accounted for 5% of total open interest in 2005–2007 and by 2011 options totaled over 15% of the market (Fig. 6.8). While less dramatic, live cattle options have also increased their share of the open interest from the lows in 2004–2008 of under 11.6% to over 18% in 2009–2011 (Table 6.1 and Fig. 6.8). The timing of the increases (2008–2011) tends to coincide with the increase in trading volume and open interest that occurred in 2009–2011. The reason for the increase in the share of open interest held in options markets for livestock is not clear. It may be related to improved market-making services or trading platforms or other exchange-related issues.

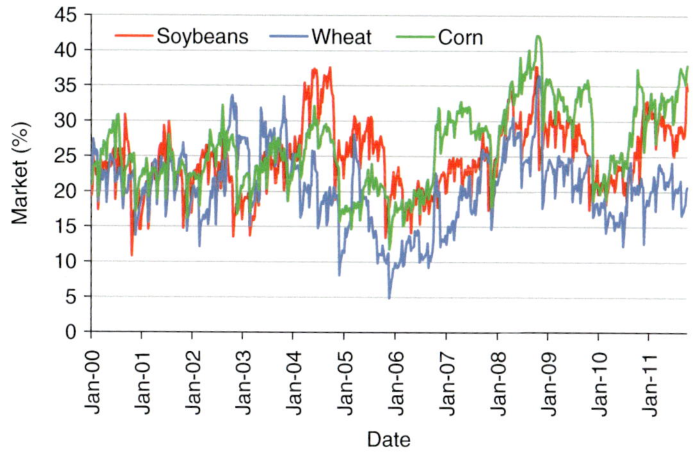

Fig. 6.7. Options percentage of total open interest, grains, 2000–2011.

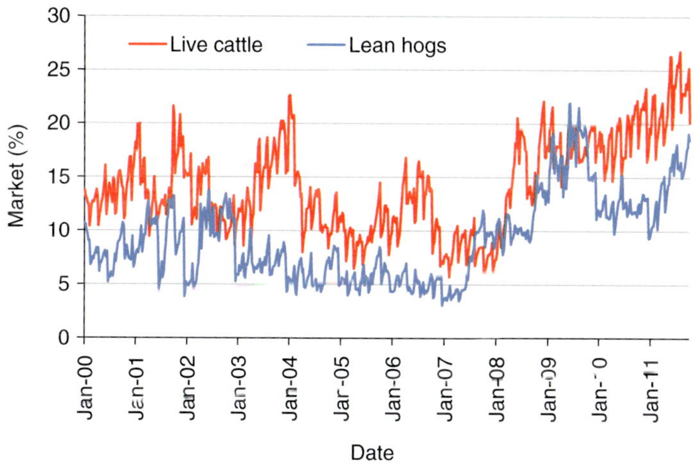

Fig. 6.8. Options percentage of total open interest, livestock, 2000–2011.

Finally, it is important to consider the distribution of open interest across listed calendar months. That is, has the growth in trading been concentrated in the nearby contract months or has there also been an increase in trading activity in deferred contracts? To measure this, the percentage of open interest held in deferred contracts is recorded on the first day of the crop marketing year for the grains and on the first trading day of the calendar year for livestock. In grains, deferred contracts are defined as those that are greater than 1 year from maturity and listed for the subsequent crop marketing year. For livestock, deferred contracts are defined as those that are greater than 6 months from maturity.

The percentage of open interest held in deferred contracts is summarized in the far-right column of Table 6.1. All of the grain contracts showed an increase in the relative amount of open interest held in deferred contracts. For example, in 2000–2003 deferred wheat contracts only contained 2% of the open interest in wheat. By 2009–2011, deferred wheat contracts held 10% of the open interest. Similar increases are shown for corn and soybean futures. The livestock markets show much more muted increases in deferred open interest with lean hogs increasing from 10% to 12% over the sample period and live cattle recording no increase. Collectively, these results suggest that trade – and presumably liquidity – generally increased in deferred futures contracts over this sample period. Improved liquidity in deferred futures contracts could be beneficial to commercial traders desiring to place longer-horizon hedges.

The collective futures and options data clearly suggest that the markets began to grow rapidly around 2004 and continue to show growth in 2011. The rapid growth was in volume and open interest simultaneously, with the ratio of volume-to-open interest remaining relatively constant across the 2000–2011 time period. Likewise, the growth was not isolated to just futures or just options. The share of total open interest held in the options market remained relatively stable for the grain markets. While there was an increase in the role of options in livestock futures markets, the increase was in the latter portion of the sample. The relative amount of open interest in deferred futures contracts increased in the grain markets in 2004–2008 and remained essentially stable in the livestock futures markets. Generally, the futures markets became much larger from 2004 with only modest shifts in these internal measures of market structure.

As noted above, greater trading volume can be associated with greater price volatility (Bessembinder and Seguin, 1993). There is no doubt that uncertainty has increased dramatically in commodity markets over the last decade, and this has been an important contributor to the groundswell in trading volumes. There have also been several historically large structural changes during the same period, and these have also undoubtedly contributed to the increase in market activity. We turn to a discussion of these structural changes.

6.3 Structural Changes

6.3.1 Electronic trading

After floundering with less than 2% of the futures trading volume from 2000 to 2005, electronic trading on commodity markets took hold in 2006 and expanded quickly.[4] In July 2006, electronic trading volume was less than 5% of total monthly volume in soybean futures contracts (Fig. 6.9). Eighteen months later over 80% of the monthly trade had migrated from the trading pit to the electronic platform. In the first 9 months of 2011, only 7% of soybean futures trade was transacted through open outcry in the trading pit. Similar trends are seen for corn and wheat (Fig. 6.10 and Table 6.1).

The livestock futures markets have been a little slower to adopt electronic trading.[5] In the live cattle futures market, trading volume on the

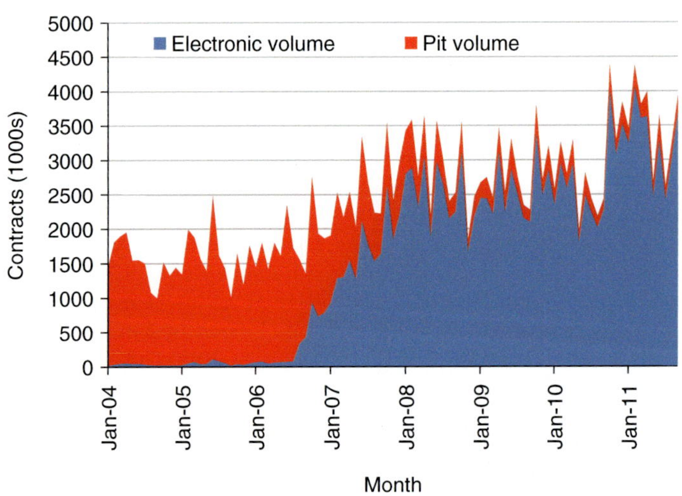

Fig. 6.9. Monthly trading volume, soybean futures, pit and electronic, 2004–2011.

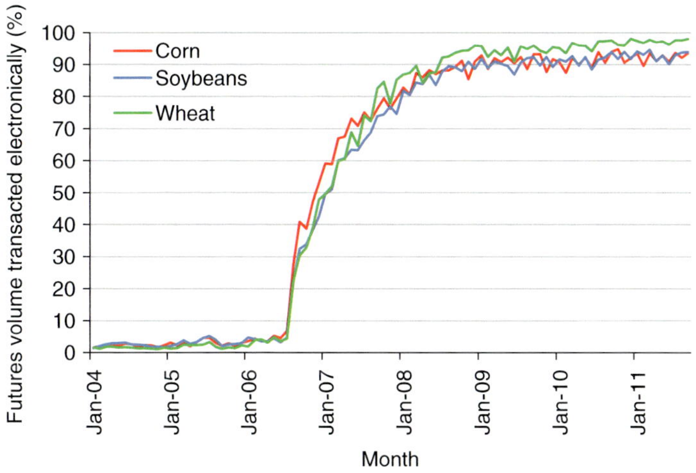

Fig. 6.10. Percentage of futures volume transacted on electronic platform, grains, 2004–2011.

electronic platform was under 5% of the total in April 2007 (Fig. 6.11). The growth was fairly rapid in 2008 and early 2009 as the electronic platform's share of live cattle futures trade rose to over 50%. Growth continues to be steady for the electronic system and the percentage of live cattle futures contracts traded in the pit declined to less than 20% by mid-2011. Lean hog futures show a similar rate of migration to the electronic platform (Fig. 6.12 and Table 6.1).

It is clear that commodity futures markets underwent a fairly dramatic shift in 2006–2008 as a 150-year-old trading mechanism (open outcry pit trading) was largely replaced by an electronic order routing and matching engine. The historic change in how trades are executed certainly could have contributed to the increased trading activity. Moreover, electronic trading may have had a considerable influence on market performance as trading costs likely fell and information transmission improved.

Shah and Brorsen (2011) find that in a side-by-side comparison, electronic order matching at the Kansas City Board of Trade has considerably lower liquidity costs compared to open outcry (floor) trading. Frank and Garcia (2011) find that livestock futures markets also benefited from lower liquidity costs with electronic trading. Studies in financial futures markets suggest that electronic trading can improve efficiency and information transmission (e.g. Ates and Wang, 2005). Moreover, with the lower cost structure inherent in electronic trading, brokerage commissions associated with trading have declined. In the early 2000s, broker commissions from 'discount' brokers were as high as $50 per round turn (buy and sell). Now, fully electronic brokers offer commissions well under $10 per round turn.

Collectively, the move to electronic trading lowered trading costs (liquidity and commission) and, as a result, likely improved information flow. Furthermore, it is possible that overall market efficiency improved, bid-ask spreads narrowed, and the commodity futures markets may have been more fully integrated with other markets, including financial markets.

6.3.2 Market access

Access to futures markets improved dramatically as the trade shifted to an electronic platform. The improved market access stems from two sources. First, the combination of a revolutionary improvement in communication tools (software and hardware) and the rise of electronic trading allowed much easier and direct access to the markets. A potential market participant can open a futures account, deposit and withdraw funds, and trade without ever talking to a broker. Moreover, electronic trading interfaces can be accessed on mobile devices which allow 24-hour access to electronic markets that

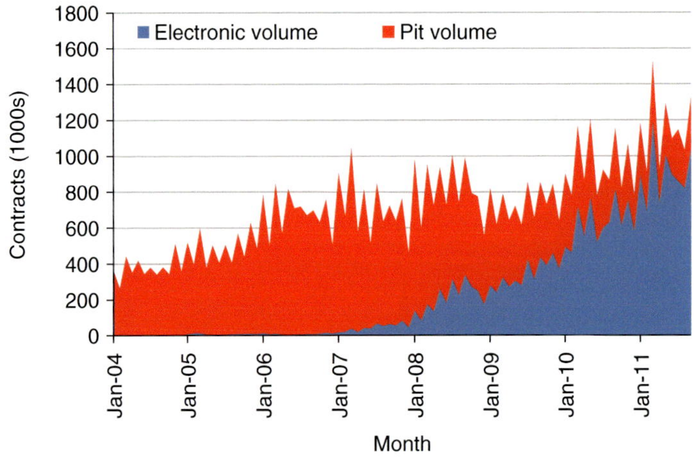

Fig. 6.11. Monthly trading volume, live cattle futures, pit and electronic, 2004–2011.

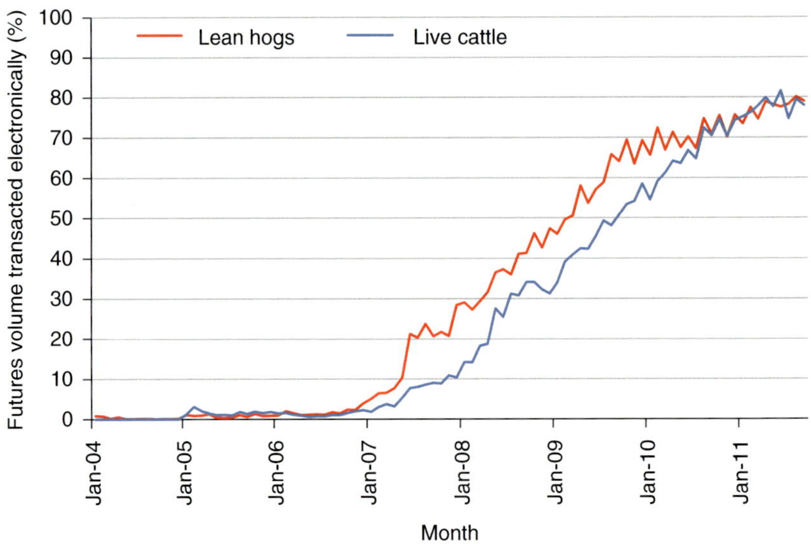

Fig. 6.12. Percentage of futures volume transacted on electronic platform, livestock, 2004–2011.

are open nearly 24 hours a day. New market participants, both domestically and internationally, have visibility to the trading book (bid, offer, and quantity) and receive nearly instant feedback on trade execution. The degree of market transparency and access far exceeds that available under the older broker-based open outcry system. Ease of access at lower cost likely encouraged new participants to enter the market, spawned greater trading by existing participants, and led to potentially more arbitrage-style computer-generated trading.

Second, financial tools were developed that provided easy, but indirect, access to the commodity futures markets. While technological innovation increased direct access to the futures markets, financial innovation provided indirect avenues to participate in commodity futures markets. Specifically, financial instruments were developed by investment banks that gave indirect exposure to a specific commodity futures market or groups of futures markets. Of the exchange-traded products (ETPs), the most common form are exchange-traded funds or ETFs.

ETFs are investment companies that are legally classified as open-end companies. The companies initially sell 'creation units' to a financial institution (e.g. Barclays). The financial institution 'pays' for those creation units with securities that mirror the portfolio that the ETF wishes to target. Then, the financial institution essentially sells shares of the creation units in the secondary market which are the ETF shares available to retail investors. Another common ETP is exchange-traded notes (ETNs). ETNs are similar to ETFs, except the financial institution is selling a debt instrument in the secondary market where the pay-out is indexed to the security bundle being tracked.[6] Regardless of the financial structure, the ETPs share a common goal – provide retail and institutional investors a pay-off tied to the underlying commodity futures market(s).

As an example, the Teucrium Corn Fund (symbol: CORN) is an ETF that tracks corn futures prices. The 'creation units' for this fund are essentially backed by positions in the three front month corn futures contracts. According to the database ETFdb.com, the fund had a notional investment of \$107 million on October 20 2011, which would equate to approximately 3242 corn futures contracts at a price of \$6.60/bushel.[7] A less focused ETF is the Powershares Deutsche Bank Agricultural Fund (symbol: DBA), which holds a portfolio of agricultural commodity futures contracts. On October 20 2011, DBA held \$2.6 billion in assets with an average daily trading volume of 1.5 million shares. DBA allocates 9.4% of the investment funds to corn futures contracts representing approximately 7406 contracts. In total, the ETFdb.com database lists 27 ETFs that are focused on agricultural commodities with a combined notional value of nearly \$3.6 billion (Table 6.2).

Beyond the ETPs that focus on agricultural commodities, there are also broader-based ETPs that mirror popular commodity indices such as the S&P GSCI (see Table 6.3). These ETPs are often heavily weighted toward energy and metals markets; however, their sheer size results in relatively large holdings in agricultural commodities. As an example, the PowerShares Deutsche Bank Commodity Index Tracking Fund (symbol: DBC) has \$5.5 billion under management. So even with just 4.03% allocated to corn futures, the resulting position is equivalent to 7167 contracts.

Importantly, ETPs provide a tool by which retail and institutional investors can essentially trade commodity futures markets as though they are equities. This is sometimes referred to as the 'securitization' or 'equitization' of commodity futures. A retail investor with a standard stock account (e.g. Charles Schwab) can trade corn futures in that account by trading the Teucrium Corn Fund (CORN). Likewise, an institution that is not permitted in their by-laws to hold derivatives positions may be able to gain exposure to agricultural futures by trading in the Powershares Deutsche Bank Agricultural Fund. The development of these exchange-traded instruments has provided market access to a segment of traders that might have otherwise been prohibited or at least reluctant to participate.[8]

It is important to note that not all ETP positions result in long futures positions. Indeed, investors can sell short ETPs in their stock accounts (subject to the standard rules on short selling). Also, a number of 'reverse' or 'inverse' ETPs exist where the creation units are funded by short positions in the tracked market (see Table 6.4). Inverse ETPs are still a small – but growing – part of the ETP universe. As an example, the PowerShares Deutsche Bank Agriculture Short ETN (symbol: ADZ) indexes to a basket of agricultural futures markets such that the share price increases when commodity prices decline. So, an investor can easily get short commodities by simply purchasing one of these 'inverse' ETPs within their traditional stock account. The issuing financial institution will maintain a zero net exposure by taking the same position in the underlying futures markets. It follows that if retail investors are net buyers of a long ETP, the institution is forced to be a net buyer in the underlying futures. Likewise, if retail investors buy an inverse ETP, then the issuing institution must sell the underlying futures. Therefore, the information underlying the buying or selling of commodity ETPs is ultimately transmitted to the underlying commodity futures market.

6.3.3 Passive investment

Several academic studies published over 2004–2007 touted futures market portfolios as viable

Table 6.2. Exchange-traded products (ETPs) tracking agricultural commodities. (Data as of October 19 2011)[a]

Symbol	Name	Assets (1000s $)	Volume (shares)
DBA	PowerShares Deutsche Bank Agriculture Fund	2,560,550	1,729,900
JJG	iPath Exchange Traded Notes Dow Jones – American International Group Grains Total Return Sub-Index ETN Series A	226,785	152,556
JJA	iPath Exchange Traded Notes Dow Jones – American International Group Agriculture Total Return Sub-Index ETN Series A	155,940	48,656
CORN	Teucrium Corn Fund	107,075	130,840
COW	Dow Jones-UBS Livestock Subindex Total Return	96,065	71,827
SGG	iPath Dow Jones-American International Group Sugar Total Return Sub-Index ETN	68,099	49,994
BAL	iPath Dow Jones-American International Group Cotton Total Return Sub-Index ETN	65,262	81,865
RJA	Elements Exchange Traded Notes Rogers International Commodity Index – Agriculture Total Return	49,549	616,836
FUD	UBS E-TRACS Constant Maturity Commodity Index Food Total Return ETN	45,275	11,165
JJS	iPath Dow Jones-American International Group Softs Total Return Sub-Index ETN	36,446	8,534
JO	iPath Dow Jones-American International Group Coffee Total Return Sub-Index ETN	30,456	42,836
GRU	ELEMENTS Exchange Traded Notes Merrill Lynch Commodity Index eXtra Grains Index – Total Return	19,031	72,150
SGAR	iPath Pure Beta Sugar	14,543	5,666
NIB	iPath Dow Jones-American International Group Cocoa Total Return Sub-Index ETN	14,327	16,531
UAG	UBS E-TRACS Constant Maturity Commodity Index Agriculture Total Return ETN	14,250	6,154
AGF	Deutsche Bank Agriculture Long ETN	11,057	7,238
WEET	iPath Pure Beta Grains	7,853	2,648
LSTK	iPath Pure Beta Livestock	7,579	3,407
CAFE	iPath Pure Beta Coffee	7,331	2,781
GRWN	iPath Pure Beta Softs	6,843	2,198
DIRT	iPath Pure Beta Agriculture	6,547	1,863
CTNN	iPath Pure Beta Cotton	6,205	1,645
UBC	UBS E-TRACS Constant Maturity Commodity Index Livestock Total Return ETN	5,159	1,258
CHOC	iPath Pure Beta Cocoa	4,382	2,215
CANE	Teucrium Sugar Fund	2,615	n/a
SOYB	Teucrium Soybean Fund	2,294	n/a
WEAT	Teucrium Wheat Fund	2,282	n/a
	Total	3,573,800	

[a]ETN, exchange-traded note; n/a, not available.

alternative investments. Key among these studies were Gorton and Rouwenhorst (2006) and Erb and Harvey (2006), which claimed 'equity-like' returns to portfolios of commodity futures. These studies also highlighted the diversification benefits relative to traditional asset classes. This academic 'stamp of approval' helped to spur a movement among institutions and pension plans to allocate investment dollars to commodity futures markets. Barclays reports nearly $400 billion invested in commodity-linked investments in early 2011. More dollars are expected

Table 6.3. Exchange-traded products (ETPs) tracking broad commodity indices. (Data as of October 19 2011.)

Symbol	Name	Assets (1000s $)	Volume (shares)
DBC	PowerShares Deutsche Bank Commodity Index Tracking Fund	5,545,984	2,909,512
DJP	iPath Exchange Traded Notes Dow Jones – American International Group Commodity Index Total Return Medium-Term Notes Series A	2,485,630	401,286
GSG	iShares Goldman Sachs Commodity Index Commodity-Indexed Trust	1,334,946	357,573
RJI	Elements Exchange Traded Notes Rogers International Commodity Index – Total Return	649,226	620,054
GCC	GreenHaven Continuous Commodity Index Fund	609,792	208,982
USCI	United States Commodity Index Fund	421,383	65,953
UCI	UBS E-TRACS Constant Maturity Commodity Index Total Return ETN	126,997	36,727
GSP	iPath Standard and Poors' Goldman Sachs Commodity Index Total Return Index ETN	110,174	32,998
GSC	GS Connect Standard and Poors' Goldman Sachs Commodity Index Enhanced Commodity Total Return Strategy Index ETN	70,325	17,097
DJCI	UBS E-TRACS Dow Jones-UBS Commodity Index Total Return ETN	22,653	3,785
BCM	iPath Pure Beta Broad Commodity	10,125	2,237
DPU	Deutsche Bank Commodity Long ETN	6,624	3,123
SBV	iPath Pure Beta Standard and Poors' Goldman Sachs Commodity Index-Weighted	5,075	1,667
GRN	iPath Global Carbon ETN	1,592	591
	Total	11,400,526	

Table 6.4. Inverse or short exchange-traded products (ETPs). (Data as of October 19 2011.)

Symbol	Name	Assets (1000s $)	Volume (shares)
DGZ	Deutsche Bank Gold Short ETN	25,564	370,490
DNO	US Short Oil Fund	16,440	28,183
SZO	PowerShares Deutsche Bank Crude Oil Short ETN	13,485	17,038
BOS	PowerShares Deutsche Bank Base Metals Short ETN	3,655	23,565
PTD	UBS E-TRACS Dow Jones Short Platinum Excess Return ETN	3,428	823
ADZ	Deutsche Bank Agriculture Short ETN	2,275	5,612
DDP	Deutsche Bank Commodity Short ETN	1,325	15,178
	Total	66,172	

to flow into the commodity markets as pension funds increase their allocation to commodity futures markets. For instance, the California State Teachers Retirement System (CALPERS) added nearly $2.5 billion to their commodity allocation in 2010 and other institutions are expected to follow suit (Krishnan and Sheppard, 2010). While these numbers sound large in absolute terms, they represent a fairly small allocation for pension plans. For instance, CALPERS targets a 1% commodity allocation to the S&P GSCI linked investments, with a permissible range of 0.5–3.0%. Pension plans purportedly view these investments as providing an inflation hedge and diversification against their core portfolios of equities and fixed-income investments.

Institutional holdings of commodity-linked investments are most likely to occur through swap agreements with a major financial institution. For instance, CALPERS may simply enter a swap agreement with a major bank where the swap is indexed to the S&P GSCI. However, some institutional money may flow directly into futures positions, ETPs, or traditional open-ended mutual funds. For example, PIMCO's Commodity Real Return Strategy Fund is an open-ended mutual fund that offers 'institutional shares' and requires a \$1.0 million minimum investment. Morningstar reports that this fund has \$22.8 billion under management as of October 20 2011.[9]

In contrast to ETPs, where the funds are often actively traded over relatively short horizons, commodity holdings by institutions and pension funds are generally passive in nature. While there are periodic inflows and outflows for portfolio rebalancing and allocation purposes, the positions generally follow an indexing approach with no active management (in the trading sense). Moreover, passive investments by institutions are long-only. The rather mundane 'buy and hold' strategy pursued by most passive commodity investments is important in understanding potential impacts on the market.

6.4 Trends in Market Composition

With greater market access, passive investments, and new tools to access the futures market, it is not surprising that the mix of market participants has changed as well. The only publicly available data on the changing make-up of market participants is provided by the Commodity Futures Trading Commission's (CFTC) *Commitments of Traders* (COT) report. These reports come in different formats that provide somewhat different views of the trader groups holding reportable positions within commodity futures markets (see Irwin and Sanders, 2012b).

The legacy COT report breaks down open interest into reporting commercial and non-commercial traders as well as non-reporting traders. The imprecise definitions surrounding commercial and non-commercial classifications are well known and make these particular classifications problematic when analyzing changing market participation (Ederington and Lee, 2002; Sanders *et al.*, 2004, 2010). The non-reporting segment – traders with positions less than the predetermined reporting level – is also a mix of trading motives. But it may provide a window on the relative activity of non-professional speculators and small commercial hedgers.

As shown in Figs 6.13 and 6.14, the role of non-reporting traders in the commodity futures markets has been on a consistent downward trend (see also Table 6.5). In the grain futures markets (Fig. 6.13), non-reporting traders comprised as much as 35% of open interest in the soybean, corn, and wheat futures markets in 1995. By 2011, these 'small' traders were only 10% of the participation in these same markets. Livestock futures tell a similar story (Fig. 6.14). Nearly 50%

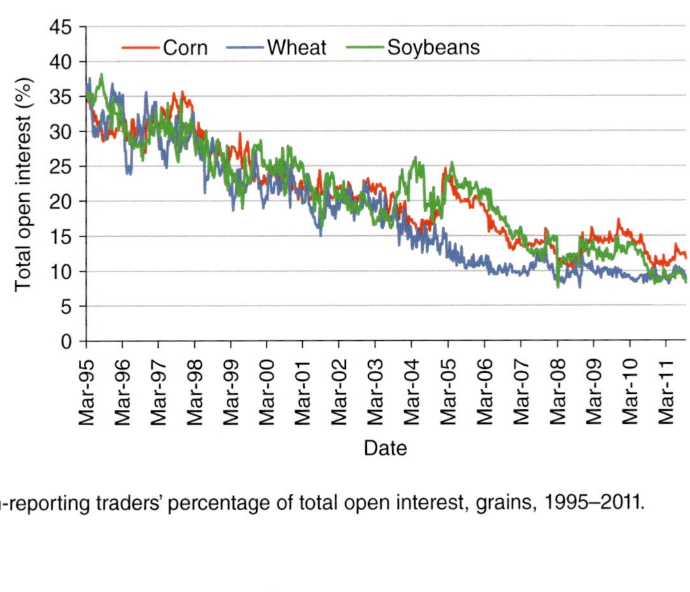

Fig. 6.13. Non-reporting traders' percentage of total open interest, grains, 1995–2011.

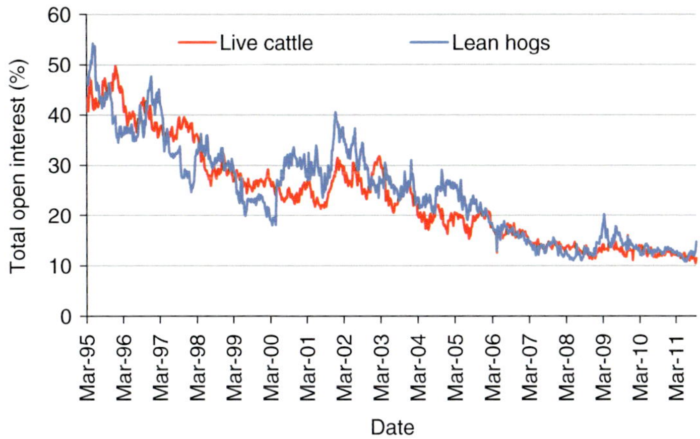

Fig. 6.14. Non-reporting traders' percentage of total open interest, livestock, 1995–2011.

Table 6.5. Percentage of total open interest by trader category, 2004–2011.[a]

Time period	Index traders (%)	Processors and merchants (%)	Swap dealers (%)	Managed money (%)	Other reportables (%)	Non-reporting traders (%)
Panel A: Corn						
Jan. 2004–May 2005	11.1					18.9
June 2006–Dec. 2008	11.6	33.4	12.5	15.3	25.5	13.4
Jan. 2009–Oct. 2011	15.0	29.0	15.1	18.2	24.3	13.4
Panel B: Soybeans						
Jan. 2004–May 2005	9.5					21.8
June 2006–Dec. 2008	13.1	31.6	13.1	18.3	23.0	14.0
Jan. 2009–Oct. 2011	15.3	30.4	14.9	19.3	24.1	11.4
Panel C: Wheat						
Jan. 2004–May 2005	20.5					13.0
June 2006–Dec. 2008	21.9	24.0	21.9	22.3	21.7	10.1
Jan. 2009–Oct. 2011	24.7	22.9	24.3	21.1	22.3	9.5
Panel D: Live cattle						
Jan. 2004–May 2005						19.4
June 2006–Dec. 2008	21.0	26.4	18.8	24.4	16.0	14.5
Jan. 2009–Oct. 2011	17.7	27.3	15.3	25.7	19.0	12.7
Panel E: Lean hogs						
Jan. 2004–May 2005						22.6
June 2006–Dec. 2008	22.0	26.4	19.3	22.1	18.3	14.0
Jan. 2009–Oct. 2011	20.2	25.2	17.1	23.8	20.2	13.7

[a]The time periods are arranged to reflect the starting point for the *Disaggregated Commitments of Traders* (DCOT) reports. The index trader positions are from the *Supplemental Commitments of Traders* (SCOT) reports. The data are not mutually exclusive between the index traders and the other categories.

of the open interest in live cattle and lean hog futures and options markets were held by non-reporting traders in 1995. That percentage has declined markedly and now hovers at just over 10%. In recent years, this trend could be an artifact of static reporting levels in a growing marketplace. But the trend has been in place since the early 1990s – well before the rapid growth documented in Table 6.1. Within the commodity futures markets, this is one of the clearest trends: the relative importance of non-reporting or 'small' traders has diminished markedly over the last decade.

The decline in the relative size of non-reporting traders could suggest that other groups have grown in relative importance. However, the trends among reporting trader groups do not clearly suggest that any single category has grown in importance. The best data to view in this regard are in the CFTC's *Disaggregated Commitments of Traders* (DCOT) report. In this report, the commercial category is further refined into processors/merchants and swap dealers while the non-commercial category is split into money managers and other reportables (not captured in the other groups).

Figures 6.15 and 6.16 show the trends in these groups' percentage of total open interest since June 2006 (when the data became available) for two representative markets. In Fig. 6.15, the market composition for soybeans is shown. While the percentage of reporting open interest does fluctuate somewhat across the groups, there is not a clear trend for a single group either markedly increasing or decreasing as a percentage of open interest. There does appear to be some increase in swap dealer size in soybeans over this interval – but on the flipside, swap dealers' percentage of lean hog open interest declines modestly (Fig. 6.16). Overall, there is not a clear increase or decrease in the relative size of a particular trading group (see Table 6.5). The decline in the relative importance of non-reporting traders is mostly attributed to a fairly steady absolute level of open interest for this group while the overall size of the marketplace increased.

Finally, the CFTC provides a glimpse of the size of index fund positions emanating from passive investments as well as ETPs in their *Supplemental Commitments of Traders* (SCOT). Irwin and Sanders (2012b) have shown that the SCOT report provides reasonable estimates of index positions for the 12 agricultural commodities covered in the report. Public data for all 12 commodities are available beginning in 2006, and data for corn, wheat, and soybeans for the period from 2004–2006 were provided by the CFTC.

Not surprisingly, the participation in these commodity futures markets by CFTC-designated index traders has increased over time. Figure 6.17 shows the percentage of total open interest (long + short positions) held by index traders from 2004 through 2011 for corn, soybean, and wheat futures markets.[10] As documented by Sanders and Irwin (2011a), the initial increase in index trader participation was from 2004 through early 2006. For example, index traders held less than 5% of the open interest in soybeans in early 2004, but that share increased to over 15% by the end of 2005. The corn market shows a similar increase, and both corn and soybean index traders have stabilized at around 15% of the market since hitting that level in 2005. The wheat market displays the same rapid increase in positions through 2004–2005, but the wheat share has stabilized at a higher level of around 25%. Importantly, the participation of index funds has been fairly steady since the relative peaks reached in late 2005.

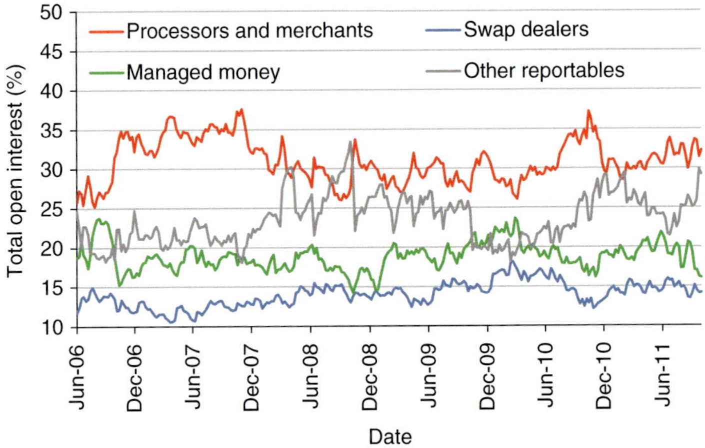

Fig. 6.15. Soybean futures, trader composition, 2006–2011.

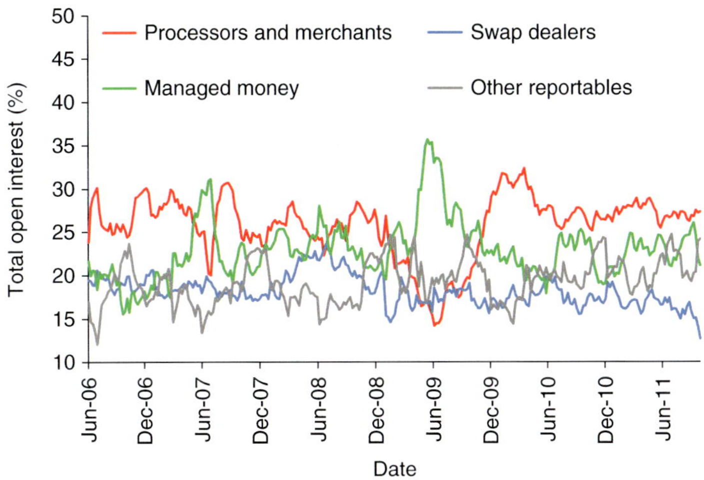

Fig. 6.16. Lean hog futures, trader composition, 2006–2011.

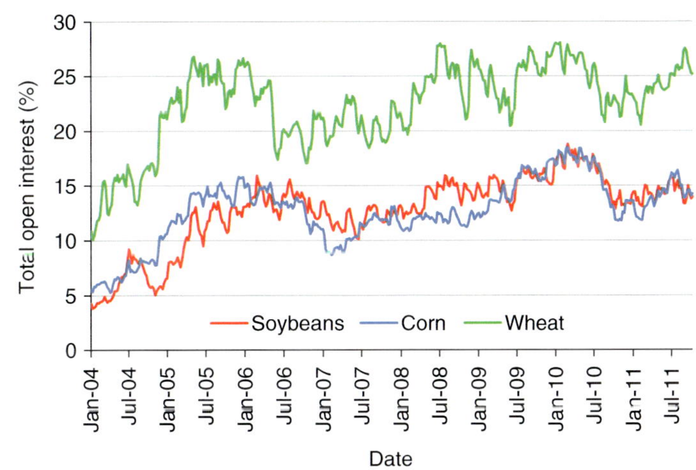

Fig. 6.17. Index traders' percentage of total open interest, grains, 2004–2011.

Unfortunately, the livestock commodity index data are not available prior to 2006, so the initial increase in index positions is not observable. Somewhat surprisingly, the lean hog data show a mild decline in index participation from a high of 25% in 2006 and 2008 to a low of 15% in 2011 (Fig. 6.18). As shown in Table 6.5, index traders' share of live cattle open interest was 21% in 2006–2008 and declined to 17.7% in the 2009–2011 sub-period. It appears that the initial increase in index trader positions in the livestock futures markets was likely at the same time as in grain futures (2004–2005), although the positions may have peaked a bit later. Both relative and absolute levels of long index positions peaked in the livestock futures markets in early 2008 and have since stabilized or even declined.

The view of market participants from the COT database is fairly limited due to the lack of data prior to 2006. However, the following conclusions can be reached based on the data presented. First, the relative importance of small, non-reporting traders has declined uniformly across the grain and livestock markets examined. In the 1990s these traders held as much as 50% of the open interest in some markets and that

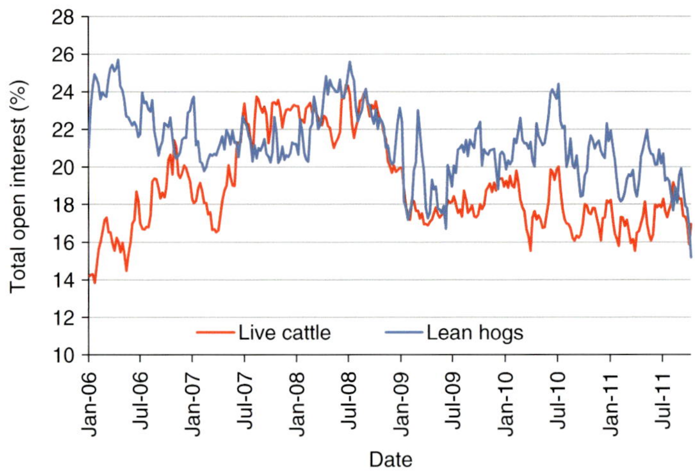

Fig. 6.18. Index traders' percentage of total open interest, livestock, 2006–2011.

share declined to around 10% in recent years. Second, commodity index traders emerged as a major participant in these markets between 2004 and 2006. Index traders comprise approximately 15% of the total open interest in grain and livestock markets, except wheat, where it is closer to 25%. For the 2006–2011 sample, there were no discernible trends in the relative importance of commercials (processors/merchants and swap dealers) or non-commercials (managed money and other reportables) in the markets examined.

It is also important to consider the number of traders and the average size of the reported positions by trader category. These data are compiled for two sub-periods (June 2006–December 2008 and January 2009–October 2011), reflecting the availability of the data and creating consistent samples across markets and measures. The data for long reporting traders are presented in Table 6.6 and short reporting traders are shown in Table 6.7. Importantly, the data are not mutually exclusive. The index trader data are embedded in the other categories because they are taken from different COT reports. Likewise, traders may appear in both Table 6.6 and 6.7 if they have reportable long and short positions.

In Table 6.6, it is clear that the largest positions are generally held by swap dealers. For example, in the corn market from June 2006 to December 2008 there were 19 swap dealers with reportable long positions averaging 19,139 contracts. In that same period, there were 25 index traders with an average long position of 15,998 contracts. While one cannot be certain, it is likely that of the 25 index traders, 19 of them are swap dealers and the other six are dispersed in the managed money and other reportables categories.

Within Table 6.6, a few trends are apparent. First, the number of index traders is fairly consistent at around 25. Second, the number of reportable swap dealers averages fewer than 20 traders. Third, each market increased the number of reportable index traders, swap dealers, and money managers across the sub-periods – which could stem from either more participants or simply from fixed reporting thresholds in a growing marketplace. Fourth, the average position size held by swap dealers declined across the two periods in each market.

Table 6.7 shows the number of traders with reportable short positions and the average size of those short positions across the same sub-periods. For example, from January 2009 to October 2011, there was an average of 315 processors and merchants with reportable short positions holding 2157 contracts. In the same time period, the number of reportable swap dealers was just seven but they had the largest average short position in the corn market at 3900 contracts.

From Tables 6.6 and 6.7 a few numbers are striking. First, the number of short index traders with reportable positions increased by

Table 6.6. Number of long reporting traders and average position size, 2004–2011.[a]

Time period	Index traders	Processors and merchants	Swap dealers	Managed money	Other reportables
	No. of traders (row 1 of each entry) Position size (row 2)				
Panel A: Corn					
June 2006–Dec. 2008	25	271	19	95	115
	15,998	1,423	19,139	2,301	541
Jan. 2009–Oct. 2011	28	225	21	100	116
	15,720	1,309	15,216	2,540	609
Panel B: Soybeans					
June 2006–Dec. 2008	25	107	17	82	74
	6,047	1,010	7,974	1,102	224
Jan. 2009–Oct. 2011	27	98	21	90	82
	6,800	1,171	6,437	1,179	284
Panel C: Wheat					
June 2006–Dec. 2008	25	61	16	59	48
	8,095	839	10,803	1,170	264
Jan. 2009–Oct. 2011	29	64	19	59	64
	7,713	858	9,076	1,156	264
Panel D: Live cattle					
June 2006–Dec. 2008	24	75	15	60	32
	4,825	369	6,438	954	331
Jan. 2009–Oct. 2011	26	80	19	85	39
	4,916	453	5,346	1,090	273
Panel E: Lean hogs					
June 2006–Dec. 2008	22	24	15	42	28
	4,011	783	5,359	710	367
Jan. 2009–Oct. 2011	24	35	17	57	31
	3,595	478	4,005	718	265

[a]The time periods are arranged to reflect the starting point for the DCOT reports. The index trader positions are from the SCOT reports. The data are not mutually exclusive between the index traders and the other categories. The data are also not mutually exclusive between Tables 6.6 and 6.7 as a single trader can have both a reportable long and short position.

at least three (live cattle) to as much as six (soybeans, corn, and wheat) across the sub-periods and the average position size held increased. This is consistent with Table 6.4 and the increase in 'inverse' ETPs. Second, the number of swap dealers also increased in each market, except for lean hogs where there was a notable absence of reporting short positions held by swap dealers. Third, there are fewer reportable short index traders and swap dealers across all markets and time periods than there are with reportable long positions. This is consistent with the notion that these traders are primarily representing long-only index investments in the agricultural markets. Finally, reportable processors and merchants are more prominent on the short side of the market, consistent with classical short hedging by producers.

Clearly, the commodity futures markets have undergone a sea change over the last decade. Trading volume and open interest increased dramatically beginning in 2004 and continues to increase. Undoubtedly, part of the increase in trading activity is linked to the emergence of passive index investments, which also increased markedly from 2004 through 2006. However, trade interest also benefited from a more efficient electronic trading platform that lowered brokerage and liquidity costs at approximately the same time. Electronic trading and subsequent declines in trading costs undoubtedly narrowed arbitrage bands and may have accelerated the use of computer-based or algorithmic trading strategies. Finally, market access was revolutionized by technological innovations (smartphones) and financial innovations (ETPs) that vastly

Table 6.7. Number of short reporting traders and average position size, 2004–2011.[a]

Time period	Index traders	Processors and merchants	Swap dealers	Managed money	Other reportables
	No. of traders (row 1 of each entry) Position size (row 2)				
Panel A: Corn					
June 2006–Dec. 2008	13	339	3	34	109
	1659	2451	4963	1219	385
Jan. 2009–Oct. 2011	18	315	7	41	114
	3262	2157	3900	1465	414
Panel B: Soybeans					
June 2006–Dec. 2008	10	158	4	28	78
	673	1752	1770	542	236
Jan. 2009–Oct. 2011	16	158	6	29	92
	1374	1854	1519	637	226
Panel C: Wheat					
June 2006–Dec. 2008	12	96	6	51	69
	1319	1983	2386	896	290
Jan. 2009–Oct. 2011	18	91	9	67	77
	1776	2009	2094	915	347
Panel D: Live cattle					
June 2006–Dec. 2008	5	136	4	42	27
	431	890	1109	552	330
Jan. 2009–Oct. 2011	8	155	5	38	37
	475	1045	1066	583	258
Panel E: Lean hogs					
June 2006–Dec. 2008	4	44	0	39	31
	363	2137	0	559	219
Jan. 2009–Oct. 2011	9	52	0	33	29
	453	1853	0	513	271

[a]The time periods are arranged to reflect the starting point for the DCOT reports. The index trader positions are from the SCOT reports. The data are not mutually exclusive between the index traders and the other categories. The data are also not mutually exclusive between Tables 6.6 and 6.7 as a single trader can have both a reportable long and short position.

broadened the scope of participation in these markets. It would be amazing if such tectonic shifts did not impact the rational pricing of risk, volatility, liquidity, and storage (spreads). However, it is much less clear if there should or would be an irrational repricing of the underlying commodity.

6.5 Market Impacts

Concerns about irrational pricing impacts have overwhelmingly focused on the positions of long-only passive investors (e.g. USS/PSI, 2009; De Schutter, 2010). Hedge fund manager Michael W. Masters has led the charge that commodity index investment created a massive bubble in commodity futures prices. He has testified numerous times before the US Congress and CFTC with variations of the following argument:

> Institutional Investors, with nearly $30 trillion in assets under management, have decided en masse to embrace commodities futures as an investable asset class. In the last five years, they have poured hundreds of billions of dollars into the commodities futures markets, a large fraction of which has gone into energy futures. While individually these Investors are trying to do the right thing for their portfolios (and stakeholders), they are unaware that collectively they are having a massive impact on the futures markets that makes the Hunt brothers pale in comparison. In the last 4½, years assets allocated to commodity index replication trading strategies have grown from $13 billion in 2003 to $317 billion in July 2008. At the same time,

> the prices for the 25 commodities that make up these indices have risen by an average of over 200%. Today's commodities futures markets are excessively speculative, and the speculative position limits designed to protect the markets have been raised, or in some cases, eliminated. Congress must act to re-establish hard and fast position limits across all markets.
>
> (Masters and White, 2008, p. 1)

In essence, Masters argues that buy-side 'demand' from index funds created a massive bubble in commodity futures prices, with the result that prices, and crude oil prices in particular, far exceeded fundamental values. Irwin and Sanders (2012b) use the term 'Masters Hypothesis' as a short-hand label for this argument. Testing the validity of the Masters Hypothesis is equivalent to testing for the irrational and harmful effects of financialization, so long as one limits the definition of financialization to the rise of index investment in commodity futures markets.

Given the ongoing worldwide debate about the market impact of passive investment, it is not surprising that a burgeoning number of studies have been completed on this topic. Some studies find evidence that commodity index investment increased the level of commodity futures prices.[11] Gilbert (2009) reports evidence of a significant relationship between index fund trading activity and price changes in three commodity futures markets – crude oil, aluminum, and copper. He estimates the maximum impact of index funds in these markets to be a price increase of 15%. In subsequent work, Gilbert (2010) finds evidence of a significant relationship between index fund trading and food price changes. Singleton (2011) estimates a regression model of crude oil futures prices and finds that index investment flows are an important determinant of price changes along with several other conditioning variables. His estimates indicate that a 1 million contract increase in index fund positions in West Texas Intermediate (WTI) crude oil over the previous 13-week period results in a 0.272% increase in nearby crude oil futures prices in the next week.

Alternatively, Brunetti and Buyuksahin (2009) conduct a battery of Granger causality tests and do not find a statistical link between swap dealers positions (a proxy for commodity index fund positions) and subsequent price changes in the crude oil, natural gas, and corn futures markets. Stoll and Whaley (2010) also use a variety of tests, including Granger causality tests, and find no evidence that the position of commodity index traders increased prices in agricultural futures markets. Sanders and Irwin (2010, 2011a, b) report similar results for agricultural and energy futures markets. Buyuksahin and Harris (2011) do not find a statistical link between swap dealers, positions and changes in crude oil futures prices. Irwin and Sanders (2012b) use new data on the positions of index investors in a broad cross-section of commodity futures markets and also fail to find evidence of a link with price movements.

Irwin and Sanders (2011) survey this literature and conclude that the weight of the available empirical evidence tilts decisively against the Masters Hypothesis. They argue that the data and methods used in studies that find evidence of a link between index positions and commodity futures price levels are subject to a number of important criticisms that limit the degree of confidence one can place in their results. In contrast, the results of the studies failing to find a link are robust across combined on- and off-exchange index fund positions, netted or non-netted swap dealer positions, and individual ETF positions, as well as a variety of statistical tests, sample periods, and time horizons. Since the linkage between the level of commodity futures prices and market positions of index funds should be clearly detectable in the data, Irwin and Sanders (2011) argue that no 'smoking gun' has been found with regard to index investment causing a 'massive' bubble.

While most of the attention has been riveted on whether passive index investment caused a massive bubble in commodity futures prices – the Masters Hypothesis – other studies have examined rational impacts of structural changes over the last decade. These studies purport to focus on financialization but generally do not try to disentangle financialization from the other structural changes highlighted earlier in this article.

A first and obvious place that the structural changes may have changed rational pricing in commodity futures markets is risk premiums. The traditional Keynesian risk premium or normal backwardation theory, predicts that hedgers pay speculators to shift the risk that they do not want to bear in the form of downward biased futures prices (i.e. futures prices are systematically below the expected spot price). In this framework, the premium that accrues to the long positions of speculators is a cost associated with the short positions of hedgers. A more flexible version of the theory allows for a time-varying risk premium (e.g. Cootner 1960; Carter *et al.*, 1983; Bjornson and Carter, 1997), where the bias in futures prices changes in relation to the position of hedgers or financial market conditions. There is a long and vigorous debate in the literature about the existence and magnitude of risk premiums, but whatever the level, expanding market participation should decrease premiums, and hence the cost of hedging (Hirshleifer, 1990). And there is indeed some evidence that this happened during the last decade. Hamilton and Wu (2011) present evidence that risk premiums in crude oil futures declined sharply after 2005, coinciding with the increasing participation of passive index investors in the market.

A similar logic can be applied to the volatility of commodity futures prices. Increasing the risk-bearing capacity of the markets may result in reduced price volatility, all else constant. Brunetti and Buyuksahin (2009) find that increasing swap dealer positions are significantly associated with subsequent drops in price volatility for crude oil and natural gas (but not corn) futures markets. Sanders and Irwin (2011b) report a consistent tendency of index trader positions leading reductions in market volatility across a number of position and volatility measures in 12 agricultural and two energy futures markets. The direction of the impact is routinely negative. However, they caution that while index positions lead to lower volatility in a statistical sense, it is possible that trader positions coincide with some other fundamental variable that is actually causing the lower market volatility. Irwin and Sanders (2012b) find mixed evidence that index positions are associated with decreasing volatility.

Another potential avenue of rational impact is market integration. Tang and Xiong (2010, p. 2) argue that commodity markets were not fully integrated with financial markets prior to the development of commodity index investments and, 'The increasing presence of index investors in commodities markets precipitated a fundamental process of financialization amongst the commodities markets, through which commodity prices now become more correlated with the prices of financial assets and each other'. The increased correlation, and hence market integration, implies that commodity futures markets more efficiently reflect shocks to the general economy. Statistical tests confirm that the correlation of commodity futures returns with crude oil returns post-2004 is greater for commodities included in major commodity indices compared to those that are not. Buyuksahin and Robe (2010, 2011) also find increasing correlation between commodity futures and financial returns but attribute the improvement in market integration to hedge funds rather than commodity index funds.

The final avenue of rational impact we will consider here is the market for storage. Classical economic writers such as Cootner (1961) and Weymar (1968) argued that the introduction of futures trading in a commodity market flattens the supply-of-storage curve because the activity of futures speculators increases risk-bearing capacity (similar to the arguments above for risk premiums and volatility). It could therefore be argued that the dramatic increase in futures market participation shifted the supply-of-storage curve to the right so that at any given level of demand for storage, the price of storage (or cost of carry) was lower and inventory higher. Some have argued that the impact actually is better conceptualized as a rightward shift in the demand for storage. Todd Petzel, Chief Investment Officer for Offit Capital Advisors, makes the following interesting argument:

> Seasoned observers of commodity markets know that as non-commercial participants enter a market, the opposite side is usually taken by a short-term liquidity provider, but the ultimate counterparty is likely to be a commercial. In the case of commodity index buyers, evidence suggests that the sellers are not typically other investors or leveraged speculators. Instead, they are owners of the physical commodity who are willing to sell into the futures market and either deliver at expiration or roll their hedge forward if the spread allows them to profit from continued storage. This activity is effectively creating 'synthetic' long positions in the commodity for the index investor, matched against real inventories held by the shorts. We have seen high spot prices along with large inventories and strong positive carry

> relationships as a result of the expanded index activity over the last few years.
>
> (Petzel, 2009, pp. 8–9)

This discussion indicates that one can construct a reasonable argument that financialization and related structural changes increased or decreased the costs of carry in commodity futures markets.[12]

Several studies conduct empirical tests of the impact of passive index investment on the cost of carrying inventories. Spreads between prices for different futures contracts on the same date are examined because theory suggests these spreads provide efficient estimates of the cost of carry (Working, 1948, 1949).[13] Brunetti and Reiffen (2010), Irwin *et al.* (2011), and Garcia *et al.* (2011) conduct various regression tests and do not find a systematic tendency for spreads in corn, soybean, and wheat futures to increase or decrease over time as commodity index positions increase. However, there is some evidence that index trading pushes out spreads during the narrow window when index positions are 'rolled' from one nearby contract to the next. Mou (2010) finds that the rolling of positions by long-only index funds leads to a substantial expansion in spreads in energy and livestock futures markets, but more modest expansion in grain futures markets. Stoll and Whaley (2010) find evidence that spreads increase during roll windows for energy futures but not agricultural futures. Finally, Irwin *et al.* (2011) report that spreads for corn, soybean, and wheat futures increase during roll windows, but the increase is temporary as spreads quickly return to the level prevailing before the roll window.

In sum, the extant literature indicates that the irrational and harmful impacts of financialization and structural change in commodity futures markets over the last decade have been minimal. In particular, there is little evidence that passive index investment caused a massive bubble in commodity futures prices, and therefore the Masters Hypothesis has almost no empirical support. There is intriguing evidence of several other rational and beneficial impacts of the structural changes over the last decade. In particular, the expanding market participation may have decreased risk premiums, and hence the cost of hedging, reduced price volatility, and better integrated commodity markets with financial markets. To date, there is only limited evidence that the changes have permanently impacted the market for storage.

6.6 Summary and Conclusions

Commodity futures and options markets began to grow rapidly around 2004 and continue to show growth in 2011. For example, combined futures and option (delta-adjusted) open interest in CBOT soybeans was relatively stable at an average of 223,000 contracts from 1995 to 2002. By February 2008, open interest had built to a peak of 878,000 contracts and in February 2011 exceeded 1 million contracts. The growth was not isolated to just futures or just options. The share of total open interest held in the options market remained relatively stable for the grain markets. While there was an increase in the role of options in livestock futures markets, the increase was in the latter portion of the sample.

It is well known that greater trading volume can be associated with greater price volatility (e.g. Bessembinder and Seguin, 1993). There is no doubt that uncertainty has increased dramatically in commodity markets over the last decade, and this has been an important contributor to the groundswell in trading volumes. There have also been several historically large structural changes during the same period, and these have also undoubtedly contributed to the increase in market activity.

The first structural change is the fairly dramatic shift in 2006–2008 from a primarily telephone/open outcry trading platform to computer/electronic order matching platform. The historic change in how trades are executed certainly could have driven increased trading activity. Moreover, electronic trading may have had a considerable influence on market performance as trading costs likely fell and information transmission improved.

The second structural change is that access to futures markets improved dramatically as the trade shifted to an electronic platform. A potential market participant can open a futures account, deposit and withdraw funds, and trade without ever talking to a broker. In parallel, new financial tools were developed that provided easy, but indirect, access to the commodity futures markets. While technological innovation increased direct access to the futures markets, financial innovation provided indirect avenues to participate in commodity futures markets. Specifically, financial instruments were developed by investment banks that gave indirect exposure to a specific commodity futures market or groups of futures markets. Of the exchange-traded products (ETPs), the most common form are exchange-traded funds or ETFs.

The third structural change is the entry of new 'financial' participants into the commodity futures arena. Investments that track a commodity index have become an accepted alternative investment for institutions and pension funds. While there are periodic inflows and outflows for portfolio rebalancing and allocation purposes, these new types of positions generally follow an indexing approach with no active management (in the trading sense). Moreover, passive investments by institutions are long-only. Data from the CFTC document the substantial increase in passive index positions in commodity futures markets. For example, index traders held less than 5% of total open interest in soybeans in early 2004, but that share increased to over 15% by the end of 2005. The corn market shows a similar increase, and both corn and soybean index traders have stabilized at around 15% of the market since hitting that level in 2005. The wheat market displays the same rapid increase in positions through 2004–2005, but the wheat share has stabilized at a higher level of around 25%. Importantly, the participation of index funds has been fairly steady since the relative peaks reached in late 2005 and early 2006.

The available literature indicates that the irrational and harmful impacts of these structural changes in commodity futures markets over the last decade have been minimal. In particular, there is little evidence that passive index investment caused a massive bubble in commodity futures prices. There is intriguing evidence of several other rational and beneficial impacts of the structural changes over the last decade. In particular, the expanding market participation may have decreased risk premiums, and hence the cost of hedging, reduced price volatility, and better integrated commodity markets with financial markets.

Acknowledgments

The authors thank Paul Peterson and John Hill of the CME Group and Hongxia Jiao of the University of Illinois for their assistance in collecting the data for this study.

Notes

[1] Original citation: Irwin, S.H. and Sanders, D.R. (2012) Financialization and structural change in commodity futures markets. *Journal of Agricultural and Applied Economics* 44, 371–396. Reprinted by permission of John Wiley and Sons and the Southern Agricultural Economics Association.

[2] This observation only applies to commodity futures markets. There has been a revolution in futures markets due to the rise of financial futures trading (see Lambert, 2010). This article focuses exclusively on commodity futures markets such as corn, soybeans, wheat, cattle, and hogs.

[3] Unless otherwise noted, 'options' refer to options on the corresponding futures contracts. All option open interest is on a delta-adjusted basis and all option volume is in absolute contracts.

[4] In this paper, electronic trading volumes are for futures only. Electronic trading in the options market has been slower to expand due partially to a more complicated order and strategy system. We appreciate the assistance of Paul Peterson and John Hill of the CME Group, Inc. for providing the open outcry and electronic trading volume data and assistance in understanding adoption trends across the grain and livestock markets.

[5] The livestock futures slower move to electronic trading may reflect the more domestic nature of these markets and the more consolidated flow of information among market participants and the cash trade.

[6] For a full discussion of exchange-traded products or ETPs see the SEC website: http://www.sec.gov/answers/etf.htm (accessed January 14 2022).

[7] See http://dbfunds.db.com/Notes/Agriculture/index.aspx (accessed January 14 2022) for full details.

[8] A few cynical market participants have suggested that commodity ETFs are a favorite of traditional stockbrokers because they generate considerable brokerage fees and allow customers access to commodity

futures markets without the customer needing to open a separate brokerage account with a potential brokerage competitor.

[9] See http://etfs.morningstar.com/quote?t=dba (accessed January 14 2022) for full details.

[10] Position data for 2004–2005 were prepared by the CFTC at the request of the US Senate Permanent Subcommittee on Investigations (USS/PSI 2009). We thank the staff of the Subcommittee for allowing us to use these data.

[11] Other studies test for the existence of price bubbles (Einloth, 2009; Phillips and Yu, 2010) or investigate the general impact of speculation in the recent commodity price movements without directly testing for statistical linkages between index fund positions and price movements (Kilian and Murphy, 2010; Juvenal and Petrella, 2011; Lombardi and Robays, 2011). Conclusions are mixed as to whether a price bubble occurred or if speculation was a main driver of prices.

[12] Buyuksahin *et al.* (2008) show that the linkages between nearby and deferred crude oil futures contracts increased from 2001 to 2004. However, it is not clear if this is due to greater liquidity in deferred contracts or changes in the storage market and cost of carry for crude oil.

[13] Garcia *et al.* (2011) show the conditions under which this assumption may be violated.

References

Ates, A. and Wang, G.HK. (2005) Information transmission in electronic versus open-outcry trading systems: an analysis of U.S. equity index futures markets. *Journal of Futures Markets* 25, 679–715.

Bessembinder, H. and Seguin, P.J. (1993) Price volatility, trading volume, and market depth: evidence from futures markets. *Journal of Financial and Quantitative Analysis* 28, 21–39.

Bjornson, B. and Carter, C.A. (1997) New evidence on agricultural commodity return performance under time-varying risk. *American Journal of Agricultural Economics* 79, 918–930.

Brunetti, C. and Buyuksahin, B. (2009) *Is speculation destabilizing?* Working paper, US Commodity Futures Trading Commission. Available at: https://papers.ssrn.com/sol3/papers.cfm?abstract_id=1393524 (accessed January 14 2022).

Brunetti, C. and Reiffen, D. (2010) Commodity index trading and hedging costs. Working paper, Carey Business School, Johns Hopkins University, Baltimore, Maryland. Available at: https://papers.ssrn.com/sol3/papers.cfm?abstract_id=1727723 (accessed January 20 2022).

Buyuksahin, B. and Harris, J.H. (2011) Do speculators drive crude oil futures prices? *Energy Journal* 32, 167–202.

Buyuksahin, B. and Robe, M.A. (2010) Speculators, commodities, and cross-market linkages. Working paper, US Commodity Futures Trading Commission, Washington, DC.

Buyuksahin, B. and Robe, M.A. (2011) Does 'paper oil' matter? Energy markets' financialization and equity-commodity co-movements. Working paper, US Commodity Futures Trading Commission. Available at: https://papers.ssrn.com/sol3/papers.cfm?abstract_id=1855264 (accessed January 14 2022).

Buyuksahin, B., Haigh, M.S., Harris, J.H., Overdahl, J.A. and Robe, M.A. (2008) Fundamentals, trader activity and derivative pricing. European Finance Association, Bergen meetings paper. Available at: https://papers.ssrn.com/sol3/papers.cfm?abstract_id=966692 (accessed November 11 2022).

Carter, C.A., Rausser, G.C. and Schmitz, A. (1983) Efficient asset portfolios and the theory of normal backwardation. *Journal of Political Economy* 91, 319–331.

Cootner, P.H. (1960) Returns to speculators: Telser vs. *Keynes. Journal of Political Economy* 68, 396–404.

Cootner, P.H. (1961) Common elements in futures markets for commodities and bonds. *American Economic Review* 51, 173–183.

De Schutter, O. (2010) Food commodities speculation and food price crises: regulation to reduce the risks of price volatility. Briefing Note 02 by the United Nations Special Rapporteur on the Right to Food. Available at: https://mronline.org/wp-content/uploads/2020/05/20102309_briefing_note_02_en.pdf (accessed January 20 2022).

Domanski, D. and Heath, A. (2007) Financial investors and commodity markets. *Bank for International Settlements Quarterly Review*, March, 53–67.

Ederington, L. and Lee, J.H. (2002) Who trades futures and how: evidence from the heating oil market. *Journal of Business* 75, 353–373.

Einloth, J. (2009) Speculation and recent volatility in the price of oil. FDIC Center for Financial Research Working paper No. 2009-08. Federal Deposit Insurance Corporation. Available at: https://www.fdic.gov/bank/analytical/cfr/2009/wp2009/2009-08.pdf (accessed January 14 2022).

Erb, C.B. and Harvey, C.R. (2006) The strategic and tactical value of commodity futures. *Financial Analysts Journal* 62, 69–97.

Frank, J. and Garcia, P. (2011) Bid-ask spreads, volume, and volatility: evidence from livestock markets. *American Journal of Agricultural Economics* 93, 209–225.

Garcia, P., Irwin, S.H. and Smith, A. (2011) Futures market failure? Working paper, Department of Agricultural and Consumer Economics, University of Illinois at Urbana-Champaign, Champaign, Illinois. Available at: https://papers.ssrn.com/sol3/papers.cfm?abstract_id=1950262&rec=1&srcabs=702281&alg=3&pos=1 (accessed January 20 2022).

Gilbert, C.L. (2009) Speculative influences on commodity futures prices, 2006–2008. Working paper, Department of Economics, University of Trento, Trento, Italy. Available at: https://www.cftc.gov/sites/default/files/idc/groups/public/@swaps/documents/file/plstudy_14_cifrem.pdf (accessed January 14 2022).

Gilbert, C.L. (2010) How to understand high food prices. *Journal of Agricultural Economics* 61, 398–425.

Gorton, G.B. and Rouwenhorst, K.G. (2006) Facts and fantasies about commodity futures. *Financial Analysts Journal* 62, 47–68.

Hamilton, J.D. and Wu, J.C. (2011) Risk premia in crude oil futures prices. Working paper, Department of Economics, University of California, San Diego, California. Available at: https://econweb.ucsd.edu/~-jhamilto/hw4.pdf (accessed January 20 2022).

Hirshleifer, D. (1990) Hedging pressure and futures price movements in a general equilibrium model. *Econometrica* 58, 411–428.

Irwin, S.H. and Sanders, D.R. (2011) Index funds, financialization, and commodity futures markets. *Applied Economic Perspectives and Policy* 33, 1–31.

Irwin, S.H. and Sanders, D.R. (2012a) Financialization and structural change in commodity futures markets. *Journal of Agricultural and Applied Economics* 44, 371–396.

Irwin, S.H. and Sanders, D.R. (2012b) Testing the Masters Hypothesis in commodity futures markets. *Energy Economics* 34, 256–269.

Irwin, S.H., Garcia, P., Good, D.L. and Kunda, E.L. (2011) Spreads and non-convergence in CBOT corn, soybean, and wheat futures: are index funds to blame? *Applied Economic Perspectives and Policy* 33, 116–142.

Juvenal, L. and Petrella, I. (2011) Speculation in the oil market. Working paper 2011-027A. Federal Reserve Bank, St. Louis, Missouri. Available at: https://www.eia.gov/finance/markets/reports_presentations/2012PresentationJuvenalSpeculationInTheOilMarket.pdf (accessed January 20 2022).

Kilian, L. and Murphy, D. (2010) The role of inventories and speculative trading in the global market for crude oil. Working paper, Department of Economics, University of Michigan, May. Available at: https://papers.ssrn.com/sol3/papers.cfm?abstract_id=1578103 (accessed January 14 2022).

Krishnan, B. and Sheppard, D. (2010) Analysis: Commodities supercycle? Pensions loath to commit. Reuters, October 13. Available at: https://www.reuters.com/article/us-commodities-institutions/analysis-commodities-supercycle-pensions-loath-to-commit-idUSTRE69C3CV20101013 (accessed January 14 2022).

Lambert, E. (2010) *The Futures: the Rise of the Speculator and the Origins of the World's Biggest Markets*. Basic Books, Philadelphia, Pennsylvania.

Lombardi, M.J. and Robays, I.N. (2011) Do financial investors destablize the oil price? Working paper series *No. 1346*, European Central Bank. Available at: https://www.ecb.europa.eu/pub/pdf/scpwps/ecbwp1346.pdf (accessed January 20 2022).

Masters, M.W. and White, A.K. (2008) The accidental Hunt brothers: how institutional investors are driving up food and energy prices. Available at: https://www.cftc.gov/sites/default/files/idc/groups/public/@swaps/documents/file/plstudy_31_ahb.pdf (accessed January 14 2022).

Mou, Y. (2010) Limits to arbitrage and commodity index investment: front-running the Goldman Roll. Working paper, Columbia School of Business, New York. Available at: https://papers.ssrn.com/sol3/papers.cfm?abstract_id=1716841 (accessed January 20 2022).

Petzel, T. (2009) Summary comments on the testimony before the CFTC. July 28. Available at: https://www.cftc.gov/sites/default/files/idc/groups/public/@newsroom/documents/file/hearing072809_petzel1.pdf (accessed January 14 2022).

Phillips, P.C.B. and Yu, J. (2010) Dating the timeline of financial bubbles during the subprime crisis. Cowles Foundation Discussion Paper No. 1770, Yale University. Available at: https://cowles.yale.edu/sites/default/files/files/pub/d17/d1770.pdf (accessed January 14 2022).

Sanders, D.R. and Irwin, S.H. (2010) A speculative bubble in commodity futures prices? Cross-sectional evidence. *Agricultural Economics* 41, 25–32.
Sanders, D.R. and Irwin, S.H. (2011a) New evidence on the impact of index funds in U.S. grain futures markets. *Canadian Journal of Agricultural Economics* 59, 519–532.
Sanders, D.R. and Irwin, S.H. (2011b) The impact of index funds in commodity futures markets: a systems approach. *Journal of Alternative Investments* 14, 40–49.
Sanders, D.R., Boris, K. and Manfredo, M. (2004) Hedgers, funds, and small speculators in the energy futures markets: an analysis of the CFTC's Commitments of Traders reports. *Energy Economics* 26, 425–445.
Sanders, D.R., Irwin, S.H. and Merrin, R.P. (2010) The adequacy of speculation in agricultural futures markets: too much of a good thing? *Applied Economic Perspectives and Policy* 32, 77–94.
Shah, S. and Brorsen, W.B. (2011) Electronic vs. open outcry: side-by-side trading of KCBT wheat futures. *Journal of Agricultural and Resource Economics* 36, 48–62.
Singleton, K.J. (2011) Investor flows and the 2008 boom/bust in oil prices. Working paper No. 3116, Graduate School of Business, Stanford University, Stanford, California. Available at: https://www.gsb.stanford.edu/faculty-research/working-papers/investor-flows-2008-boombust-oil-prices (accessed January 14 2022).
Stoll, H.R. and Whaley, R.E. (2010) Commodity index investing and commodity futures prices. *Journal of Applied Finance* 20, 7–46.
Tang, K. and Xiong, W. (2010) Index investing and the financialization of commodities. Working paper, Department of Economics, Princeton University, Princeton, New Jersey. Available at: https://papers.ssrn.com/sol3/papers.cfm?abstract_id=1683135 (accessed January 14 2022).
USS/PSI (United States Senate, Permanent Subcommittee on Investigations) (2009) Excessive speculation in the wheat market. Majority and Minority Staff Report. Available at: https://www.hsgac.senate.gov/imo/media/doc/REPORTExcessiveSpecullationintheWheatMarketwoexhibitschartsJune2409.pdf?attempt=2 (accessed January 14 2022).
Weymar, F.H. (1968) *Dynamics of the World Cocoa Market*. The MIT Press, Cambridge, Massachusetts.
Working, H. (1948) Theory of inverse carrying charge in futures markets. *Journal of Farm Economics* 30, 1–28.
Working, H. (1949) The theory of price of storage. *American Economic Review* 39, 1254–1262.

7

A Reappraisal of Investing in Commodity Futures Markets[1]

New Author Foreword

While working on the OECD report, we were struck by another basic question: Did the hundreds of billions of dollars flowing into commodity index funds really make economic sense in the first place? The answer to that question largely depended on what one believed about risk premiums as a source of returns to long position holders in commodity futures markets. The classical Keynesian form of a risk premium is the payment short hedgers are willing to pay to long speculators in order to offload price risk. We were steeped in classical studies by Telser (1958), Rockwell (1967), Dusak (1973), and Hartzmark (1987) that indicated risk premiums were approximately zero. Consequently, we were skeptical that the boom in commodity index investment rested on a firm economic foundation.

In a famous paper that first appeared in 2004 and was later published in 2006, Gorton and Rouwenhorst challenged the results of the classical studies. They concluded that not only were risk premiums positive in commodity futures markets, but the premiums were large enough to be considered 'equity-like.' This was exactly the kind of justification for commodity funds that impressed institutional money managers. How could a pair of Ivy League economists (Penn and Yale) be wrong? As a result, the paper played an important role in fueling the commodity fund boom of the 2000s.

At about the same time, we kept hearing about something called 'roll yields' and how these were also an important driver of commodity fund returns. Because futures contracts expire on a set schedule, a futures contract cannot be held indefinitely, and long positions have to be 'rolled' out of expiring contracts and into another contract that is further from expiration. It was commonly argued that commodity index investors steadily lose money – or have a negative roll yield – if the futures market is in 'contango' because long ownership is rolled into increasingly higher-priced contracts. Contrarily, it was argued that index positions would gain or have positive roll yield during periods of 'backwardation' by rolling long ownership across futures contracts into lower-priced contracts. While this view was very widespread at the time, we quite frankly thought it was ridiculous and extremely misleading for investors.

With this background, it is not hard to appreciate why we thought it was important to reappraise the case for commodity investments. We collected over five decades of prices for 20 commodity futures markets and found that the monthly return to individual futures markets was statistically indistinguishable from zero, consistent with the classic studies. We found it very puzzling that Gorton and Rouwenhorst reported essentially the same result for individual markets, but a portfolio of the same commodity futures markets had a highly positive return. Huh? It turns out there was some magic with regard to the weighting scheme used by Gorton and Rouwenhorst and it resulted in a positive 'diversification return' even when the individual futures investments had zero returns. Financial economists have a colorful term to describe this mind-bending outcome – 'turning water into wine.'

DOI:10.1079/9781800622104.0007

We then turned our attention to the question of roll yields as a driver of commodity investment returns. It was actually easy to dispense with this nonsensical argument and show that negative or positive roll yields did not represent an actual loss to an investor in their trading account. Otherwise, it would be a blatant violation of market efficiency since the mere switching of contract months would create a virtual money machine. Since money machines do not exist in the real world, roll yields cannot drive actual returns to commodity futures investments.

In some ways, we are proudest of this article since it was published near the peak of the commodity fund boom, and it went against much of the conventional wisdom at the time about such investments. We were definitely swimming against the current. Even more satisfying, our findings have been confirmed multiple times in the last decade. For example, the sample period has been extended to include more recent data and the average returns were essentially unchanged (Main *et al.*, 2018; Irwin *et al.*, 2020). That is, returns to individual futures contracts still appear to be indistinguishable from zero. There is still some mystery surrounding the existence of a diversification return for a portfolio of commodity futures contracts. We remain unconvinced that this is a sound basis for investment. Bessembinder (2018) proved conclusively that roll yield is a myth in commodity futures markets.

The ultimate proof of our results is the poor real-world returns to commodity index funds. A straightforward measure of index investment performance is the return to the GSG (iShares S&P GSCI Commodity-Indexed Trust) exchange-traded fund (ETF), which tracks the Standard and Poor's Goldman Sachs Commodity Index (S&P GSCI). By a wide margin, the S&P GSCI is the most popular index followed by commodity investors. Since the inception of the GSG ETF, the fund's value has declined from $49.25 (on July 21 2006) to $21.72 (on September 6 2022), or a loss of 56% over the last roughly 16 years.[2] Even more remarkable, that time period spans two huge bull markets in commodities, 2007–2008 and the present one. Where's the risk premium?

Despite our enthusiasm for the article, it was disappointing that it never found a wider audience. While it's nice to be first in academic publishing and even nicer to be right, sometimes you bear a cost for saying unpopular things. That is undoubtedly true with respect to the commodity index fund industry. We were seen as friends of the industry for our work showing that index investment did not drive the commodity price spike of 2007–2008. After publishing this article, not so much.

Abstract

Investments into commodity-linked investments have grown considerably over the last 5 years as individuals and institutions have embraced alternative investments. However, unlike investments in equities or real estate, commodity futures markets produce no earnings and, arguably, are not even capital assets. So, the source of returns and the expected returns for commodity futures investments is unclear. This paper examines the history of returns for static long-only futures investments over five decades. The research highlights the following features of commodity futures investments: (i) returns to individual futures markets are zero; (ii) returns to portfolios of futures markets depend critically on the weighting schemes and the embedded trading strategy; and (iii) historical returns are not statistically different from zero and are driven by price episodes such as 1972–1974. Overall, the case for long-only investment in commodities is not as strong as some have argued in recent years.

Key words: futures markets, commodities, investments

JEL categories: D84, G12, G13

7.1 Introduction

As of June 30 2011, the CFTC reported that $181 billion was invested in commodity index funds (CFTC, 2011; see Fig. 7.1).[3] More dollars are expected to flow into the commodity markets over the next 3 years as the California State Teachers Retirement System adds nearly $2.5 billion to their commodity allocation and other institutions continue to allocate portions of their portfolio to commodity investments (Krishnan and Sheppard, 2010). The impact of this investment flow into commodity futures markets has been the subject of intense debate and scrutiny in recent years (e.g. Masters, 2008; USS/PSI, 2009; Pirrong, 2010; Irwin and Sanders, 2011).

Curiously, the source of returns to commodity futures investments does not seem to be as highly disputed as the possible market impact of such investment. Indeed, stalwart companies such as the Vanguard Group present investments in commodity index funds as potentially valuable alternatives (Stockton, 2007) and a

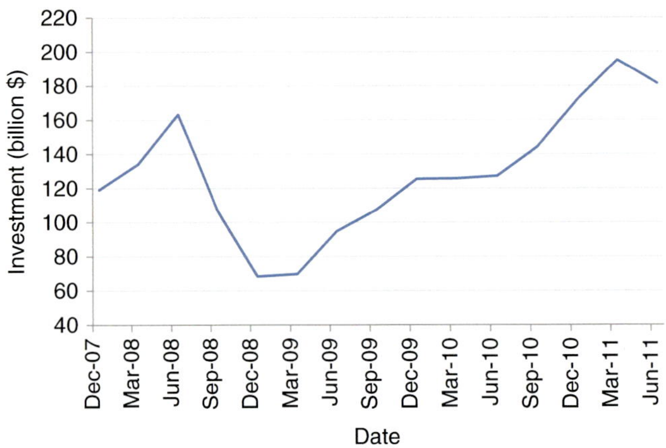

Note: The data represent the notional value of net long positions as reported by the Commodity Futures Trading Commission, *Index Investment Data*.

Fig. 7.1. Commodity index fund investments, December 31 2007–June 30 2011.

report commissioned by the CME Group, Inc. asserts that futures returns stem from 'the existence of these risk premia ... consistent with futures prices' role as biased predictors of expected spot prices' (Abrams *et al.*, 2010, p. 5). The prevalent view within the investment community seems to be that futures positions earn a return for providing risk-transfer services to hedgers.

This view of commodity futures markets – while consistent with the traditional Keynesian risk premium or normal backwardation theory – has been the subject of a heated debate stretching back over 50 years. Some purport to find evidence of substantial risk premiums (e.g. Houthakker, 1957) in the form of downward-biased futures prices – where futures prices are systematically below the expected spot price. This would allow positive returns to accrue to long position holders. Others report evidence of a time-varying risk premium (e.g. Cootner, 1960; Carter *et al.*, 1983; Bjornson and Carter, 1997), where the bias in futures prices changes in relation to the position of hedgers or financial market conditions. However, the bulk of the evidence (e.g. Telser, 1958; Rockwell, 1967; Dusak, 1973; Marcus, 1984; Hartzmark, 1987; Kolb, 1992; Brooks *et al.*, 2011) indicates risk premiums are zero or near-zero. Surveying the literature on the subject, Telser (2000, p. 551) remarked that 'It appears on this evidence that normal backwardation as a theory of futures should be respectfully interred.' Perhaps this conclusion should not come as a major surprise. Black (1976, p. 172) famously noted that 'Commodity contracts, however, are not included in the market portfolio. Commodity contracts are pure bets, in that there is a short position for every long position.' In other words, futures markets are simply side bets on prices, not capital assets.

Despite the historical record, recent academic research has claimed 'equity-like' returns to portfolios of commodity futures while also touting the diversification benefits relative to traditional asset classes (Erb and Harvey, 2006; Gorton and Rouwenhorst, 2006a). Prior to these key studies, some other academics also found evidence of positive returns to long-only futures *portfolios*. In particular, Bodie and Rosansky (1980) found annual excess returns of 9.8% to a collection of 23 markets from 1950 to 1976, Fama and French (1987) reported a geometric annual excess return of 5.4% for 21 markets over 1966–1984, and Greer (2000) documented a 12.2% total annual return from 1970 to 1999 to a futures market portfolio. Gorton and Rouwenhorst (2006a) kick-started the recent commodity investment movement by reporting annualized returns to a portfolio of commodity futures of 10.7%. While a bit more skeptical about the source of returns, Erb and Harvey (2006) also endorsed long-only commodity futures investments. Data compiled by Barclays show that all commodity-linked investments (futures and physicals) grew rapidly during this period as commodity investments were embraced by the financial community (see Fig. 7.2).

In this paper, we re-examine the case for long-only commodity futures investments and find several reasons for caution. Our focus is on the returns generated by static long-only commodity investment, so we do not consider the potential returns to 'tactical' or 'strategic' commodity investments that may actively, but subtlety, manage the individual index components or the switching of contracts (see Seeking Alpha, 2007).[4] Likewise, we do not consider the role of commodity investments within a larger portfolio. It is well known that if long-only index funds have a positive return, then the low correlation with traditional investments would give them a non-zero weight in a mean-variance optimized portfolio (e.g. Elton *et al.*, 1987). In addition, recent research by Daskalaki and Skiadopoulos (2010) shows that the widely touted diversification role of commodities does not necessarily hold up out-of-sample.

Our analysis consists of three parts. First, we investigate the return to individual futures markets going back to the 1960s. Common misconceptions regarding 'contango' and 'roll returns' are dispelled. Second, commodity portfolio returns are analyzed including the 'diversification return.' Third, recent evidence and fund performance is presented.

7.2 Returns to Individual Commodity Futures Markets

A surprising result in recent studies of commodity investment is that the return to individual futures markets is on average zero, yet portfolio returns are positive. Indeed, Erb and Harvey (2006) find that the average return to 12 futures markets from 1982 to 2004 was –1.71% with no individual market producing statistically positive returns. Gorton and Rouwenhorst (2006b) study 36 individual markets and find that 18 had positive returns and 18 had negative returns. None of the individual markets had statistically significant positive returns.

To further investigate this puzzling result, monthly returns were collected for 20 commodity futures markets from 1961 through 2010. Monthly returns were calculated as the percentage change in the nearby futures contract in each month.[5] The average annualized geometric returns by market and decade are shown in Table 7.1 and reflect the returns in excess of the

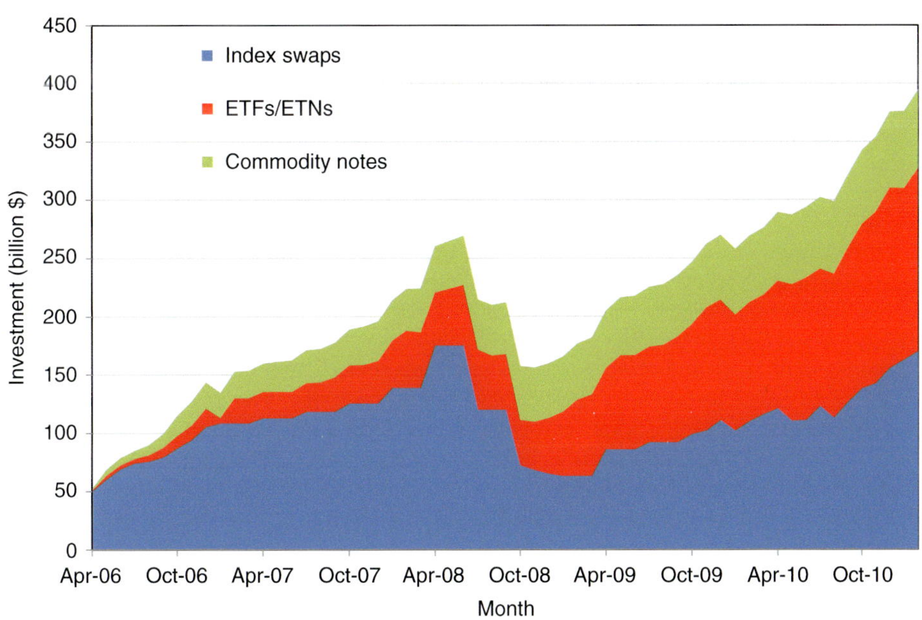

Fig. 7.2. Commodity-linked investments, April 2006–February 2011. (From data compiled by Barclays.)

Table 7.1. Returns (%) to long-only positions in individual US commodity futures markets, 1961–2010.[a]

	Annualized geometric returns (%)							
Market	1960s	1970s	1980s	1990s	2000s	All years	*t*-statistic	*p*-value
Corn	–1.7	2.3	–9.1	–7.9	–6.0	–4.6	–1.40	0.1605
Wheat	–6.0	7.6	–11.2	–6.9	–5.0	–4.5	–1.30	0.1925
Soybeans	4.7	8.4	–10.4	–3.1	13.1	2.2	0.59	0.5550
Soybean meal	15.0	2.4	–7.2	5.4	17.0	6.1	1.41	0.1602
Soybean oil	9.1	23.7	–8.1	–8.7	9.2	4.3	0.99	0.3202
Oats	–4.9	–2.2	–14.3	–13.9	11.3	–5.2	–1.30	0.1925
Rough rice				–11.4	–6.7	–9.1	–1.48	0.1389
Live cattle		4.2	8.3	0.8	0.8	4.2	1.43	0.1543
Lean hogs		15.0	4.8	0.3	–11.3	2.3	0.51	0.6072
Lumber		7.2	–14.1	–3.7	–16.1	–7.7	–1.65	0.0987
Cotton No. 2	–4.6	13.0	6.4	–4.8	–5.0	0.8	0.22	0.8275
Coffee			–0.6	–3.6	–0.8	–1.7	–0.25	0.8024
Sugar No. 11	–7.7	18.8	–24.7	6.4	7.4	–1.1	–0.19	0.8477
Cocoa	–8.6	30.3	–10.4	–15.8	15.3	0.7	0.16	0.8701
Heating oil			6.9	1.8	7.2	5.3	0.91	0.3633
Crude oil				6.4	6.9	6.6	0.91	0.3645
Natural gas				13.0	–34.6	–14.0	–1.29	0.1991
Gold			–11.8	–7.7	14.6	–2.3	–0.76	0.4471
Silver		16.5	–21.5	–4.0	17.4	0.1	0.02	0.9863
Copper	22.5	2.1	4.4	–2.3	18.5	8.6	2.10	0.0364
Average	1.8	10.7	–6.6	–3.0	2.7	–0.5		

[a]Returns represent annualized geometric excess returns to nearby futures contracts and are computed only for complete decades. The *p*-value is for a two-tailed *t*-test that the mean equals zero.

Treasury (T)-bill rate that would be earned on a fully collateralized futures position.[6] The data are only presented for complete decades. For example, a market that started trading in 1978 does not show any data in the 1970s but shows a complete return history for the 1980s. The data are arranged in this manner to keep the markets within each decade consistent and complete.

The average geometric return across all markets and time periods is –0.5% per year. The dispersion of individual market returns is wide. Corn and wheat futures markets had annual average returns of –4.6% and –4.5%, respectively. Copper had the only return (+8.6%) statistically greater than zero at the 5% level (two-tailed *t*-test). Lumber had a marginally statistically significant negative return of –7.7% (*p*-value = 0.0987). Looking at each decade individually, the 1970s was a favorable time period for long commodity futures positions. In that decade, only oat futures had a negative average return. In every other decade, the number of markets with negative returns roughly equals the number with positive returns and the average return is relatively close to zero. Overall, these data are consistent with those presented in the literature and suggest that returns to individual futures markets are essentially zero.

Some researchers (e.g. Till, 2006) explain the cross-market variation in returns with the relative difficulty of storage. Others look at the breakdown of futures returns and pin poor returns for both individual markets and index investments on 'contango' (e.g. Robison *et al.*, 2010). Contango simply refers to the typical situation in commodity futures markets where the term structure of prices is positive – futures contracts further from expiration on a given date have a higher price than those contracts closer to expiration. A contango futures price structure is commonly referred to as a 'normal carry' or 'carrying charge' market.

Because futures contracts expire on a set schedule, one futures contract cannot be held indefinitely. Instead, long-only investors have to 'roll' out of expiring contracts (prior to delivery) and into another contract that is further from expiration. The process of switching or rolling contracts is accomplished with a simultaneous sale of the expiring contract (e.g. March) and purchase of the next contract (e.g. May). There is a misconception among some commodity investors and analysts that the price difference between these two contracts (e.g. March and May) has a direct impact on investment returns. Specifically, it is commonly reported that buy-and-hold index investors will steadily lose money – or have a negative 'roll yield' – if the market is in contango because long ownership is rolled into increasingly higher-priced contracts.

On the flip side, it is purported that long-only positions will gain or have positive 'roll yield' by rolling long ownership across futures contracts that are inverted or in backwardation (i.e. deferred futures contract prices are lower than nearby contract prices). For instance, if a long-only investor owns a March futures contract that can be sold at 505 and replaced with a long May futures contract priced at 500, then it is thought that this generates a positive return because ownership at 505 cents is replaced with ownership at a lower price of 500 cents.

On the surface, this reasoning appears sound and would make sense for a single asset such as common stock on a single company. However, the corn futures market is not a single asset (if it is an asset at all). Instead, it is a series of distinct contracts distinguished by the unique expiration months. March corn futures are not May corn futures. Rolling a long position from March corn to May corn futures is really not any different from rolling a long equity position in the auto industry from General Motors (GM) to Ford. The difference in the prices between the GM stock sold and the Ford stock purchased has no arithmetic bearing on the return to the new long position in Ford. The only thing that determines the realized return is the change in the price of the asset held.

'Contango' *per se* has no direct impact on the realized returns to individual futures contracts. It is simply the structure of the futures market. The term structure of a futures market, whether in contango or backwardation, has no bearing on the cash flows into or out of the futures account associated with a long-only commodity index investment. As Burton and Karsh (2009, p.1) state, 'Negative roll yield does not represent an actual loss of an investor, and in fact, there are no actual losses incurred within the index by rolling from one contract to another.' To be otherwise would be a blatant violation of market efficiency since switching of contract months would create a virtual money machine. More generally, this means that the observed futures return for commodity index investments is not directly affected by the term structure of the underlying commodity futures markets.

It is interesting to note that a number of commodity investments have been introduced that attempt to minimize the impact of contango and roll returns. For example, Goldman Sachs introduced an enhanced GSCI (Goldman Sachs Commodity Index) that 'attempts to boost returns and sidestep the issue of contango' (Seeking Alpha, 2007). However, a careful distinction must be drawn between contango's direct impact on realized futures returns (none) and contango as a market signal. That is, while 'roll' returns are not a direct part of the realized futures return, the market structure could be a market signal. Gorton *et al.* (2007) link high returns with risk premiums that vary inversely with physical inventory levels, which are signaled by futures markets that have an inverted structure or exhibit backwardation. Naturally, this led the authors to introduce a new commodity ETF that exploits this finding by 'concentrating the portfolio in the backwardated portion, or at the least-contango portion of the commodities space' (Crigger *et al.*, 2010).

The efficacy of market term structure (contango or backwardation) as a market signal for higher returns or the existence of risk premiums is not the focus of this paper. Indeed, it is quite possible that an inverted market structure is a proxy for a market situation where long commodity futures positions earn a risk premium. In which case, the 'roll' return is a market signal that suggests a particular strategy. For example, a positive 'roll' return (inverted market structure) could be a *signal* that the realized return in the subsequent period will be positive. This is quite different from the misconception that rolling in an inverted market *creates* a positive realized return.

The balance of the evidence suggests that static long-only positions in individual futures markets have an expected return of zero. The idea that the 'return to roll' is a realized futures return is a fiction. The actual process of rolling a long futures position to a new futures contract does not produce a return, either positive or negative, and if it did this would be equivalent to a futures money machine. So where does the futures return come from?

7.3 Returns to Portfolios of Commodity Futures Markets

The starting point in understanding the source of returns is to dig into the question of how you can get positive returns from assets that have average individual returns of zero. The answer appears to be diversification and the arithmetic properties of geometric returns and the variance reduction inherent in a portfolio. Erb and Harvey (2006) point out the mathematical properties that make diversification the only 'free lunch' in finance (Campbell, 2000). When a portfolio of assets is formed, the portfolio geometric return is equal to the weighted average geometric returns of the components plus the *diversification return*. The diversification return is then the difference between the portfolio's geometric return and the weighted average geometric return of the portfolio's constituents.

An example of the diversification return is shown in Table 7.2. In this example, there are two assets that have geometric or holding period returns of zero. However, the geometric return of the equally weighted portfolio of the two assets is 4%. The 4% is the diversification return – the return to the portfolio is greater than the average geometric return of the two components (0%). Erb and Harvey (2006) refer to this as 'turning water into wine.'

The diversification return is actually an undisputable arithmetic truth stemming from the following two facts: (i) arithmetic averages are always greater than geometric averages and the difference increases with variance; and (ii) the return series for a portfolio of assets will always be less variable than the average variance of the components. So, because the portfolio of assets has a lower variance than the average variance of the components, the geometric mean of the portfolio will be greater than the average geometric mean of the components. Erb and Harvey (2006) show that diversification return (DR) can be expressed as,

$$DR = \frac{1}{2}\left(1 - \frac{1}{k}\right)\bar{\sigma}^2\left(1 - \bar{\rho}\right) \tag{7.1}$$

where k = the number of portfolio components, $\bar{\sigma}^2$ = the average variance of the components, and $\bar{\rho}$ = the average correlation across assets.[7] Hence, the diversification return is an increasing function of asset variability and the number of assets and a decreasing function of the average correlation between the individual asset returns. Commodity futures returns happen to display the characteristics that provide a considerable diversification return. A large number of

Table 7.2. Portfolio diversification return example.

Time period	Price		Return		Equal weighted return (%)
	Asset 1	Asset 2	Asset 1 (%)	Asset 2 (%)	
1	10	10			
2	20	30	100	200	150
3	30	40	50	33	42
4	40	50	33	25	29
5	50	60	25	20	23
6	50	40	0	–33	–17
7	40	10	–20	–75	–48
8	30	20	–25	100	38
9	20	20	–33	0	–17
10	10	10	–50	–50	–50
Arithmetic average			9	24	17
Geometric average			0	0	4
'Diversification return'					4

markets (high 'k') are available that are very volatile (high $\bar{\sigma}^2$) and commodity futures markets tend not to be highly correlated outside of closely linked market segments (low $\bar{\rho}$).[8] For a portfolio of ten markets with average annualized standard deviations of 25% and correlations of 0.20, the diversification return will be 2.25% as calculated using Eqn 7.1.

In order to measure the diversification return found in commodity futures markets, portfolios are formed from 1961 through 2010 using the same data that underlie Table 7.1. When using historical commodity futures returns, there has been some debate as to how to handle the starting date for new contracts as well as contracts that are delisted. For example, Gorton and Rouwenhorst (2006a) use all commodity futures contracts – bringing them into the portfolio when they start trading and removing them if they are delisted – arguing that commodity prices do not have a 'survivorship bias' such as that found in the equity markets. However, Gorton and Rouwenhorst's (2006a) approach leads to the inclusion of markets that may never have achieved sufficient liquidity for active trading and efficient pricing (e.g. butter futures). Perhaps more importantly, allowing contracts to enter as they are listed may bias results if the introduction of contracts to particular sectors is not random. For instance, dairy futures contracts were developed and listed in the mid-1990s as those markets which deregulated and burdensome government-owned inventories were reduced. Conversely, futures markets may be delisted due to very low volume of trade, abnormal pricing performance, or if a structural change in an industry reduces the need for risk-shifting. These factors may not be unrelated to market fundamentals that drive prices. So a more consistent approach to selecting and retaining futures markets used in historic simulations is preferred.

Erb and Harvey (2006) take a more consistent approach by examining a fixed set of 12 markets from 1982 to 2004 that spans the major commodity sectors (energy, metals, grains, livestock, and tropical products). Here we follow a similar approach. Data are collected on ten markets that have continuously traded since 1961 and 20 markets that have traded continuously since 1991. The markets are the same as those listed in Table 7.1, where the ten markets that show complete trade in the 1960s make up the ten-market portfolio and all of the listed markets comprise the 20-market portfolio. Portfolios are constructed using equal weights across markets and rebalanced monthly. The portfolio returns are computed by decade and presented in Table 7.3.

As shown in Table 7.3, the ten-market index had a geometric annual excess return of 3.6% over the 50 years of data. However, over the entire sample, the annual average return is not statistically different from zero at conventional levels

Table 7.3. US portfolio investment returns (%), 1961–2010.[a]

	Annualized geometric returns (%)					
Market	1960s	1970s	1980s	1990s	2000s	All years
US equities (S&P 500)	8.1	8.5	13.8	17.3	1.4	9.7
US T-bills	4.3	6.7	8.4	4.7	2.2	5.2
US T-bonds	2.9	3.5	13.0	8.4	5.5	6.6
Equally weighted index, ten commodities	3.6	15.1	–6.3	–3.6	10.6	3.6
(p-value), two-tailed t-test	(0.2616)	(0.0541)	(0.2235)	(0.2991)	(0.1285)	(0.1520)
'Diversification return'	1.8	4.5	2.1	1.5	3.0	2.8
Equally weighted index, 20 commodities				–0.1	5.9	2.8
(p-value), two-tailed t-test				(0.9695)	(0.2554)	(0.3274)
'Diversification return'				2.9	3.2	3.3

[a]Returns represent excess returns to each portfolio. The ten futures market portfolio includes corn, wheat, soybeans, soybean meal, soybean oil, oats, cotton, sugar, cocoa, and copper. The 20 futures market portfolio includes those in the ten market portfolio plus rough rice, live cattle, lean hogs, lumber, coffee, heating oil, crude oil, natural gas, gold, and silver. The returns to US equities, Treasury (T)-bills, and T-bonds were obtained from the following website: https://www.stern.nyu.edu/~adamodar/pc/datasets/histretSP.xls (accessed February 13 2022). The returns for these markets are total returns and are provided for comparison purposes only.

of significance. By decade, only the 1970s produced returns (15.1%) that were statistically significant at the 10% level (p-value = 0.0541). The 1980s and 1990s produced negative returns. The 20-market portfolio produced a comparable return of 2.8% from 1991 to 2010. For the 20-market portfolio, the returns are not statistically different from zero in aggregate or in any decade.

A closer examination of the pattern of annual returns for the ten-market portfolio is provided in Fig. 7.3, where it is clear that returns in the 1970s were driven by a few select years. Indeed, the three tumultuous years from 1972 to 1974, when the commodity markets underwent dramatic structural shifts, account for 96% of the decade's returns and 68% of all returns during 1961–2010. Over the entire sample, the average annual geometric return is 3.6%. If just 1973 is removed, the return falls to 2.0% and from 1975 forward the annual return is –0.01%. The 1972–1974 episode is clearly an important driver in finding positive portfolio returns. This episode also may be an important driver of the returns reported in the literature (Bodie and Rosansky, 1980; Fama and French, 1987; Greer, 2000). It can be debated if these years should be included in forming expectations for futures years. However, their impact on estimates of historical returns is beyond question.

The 'diversification return' to the commodity portfolios is calculated as the difference between the average geometric return of the component markets and the geometric return of the portfolios. Not surprisingly, the diversification returns behave as suggested by Eqn 7.1. The diversification return for the ten markets averages 2.8% over the 50-year span. The diversification return is higher in volatile decades (1970s and 2000s), and in a given decade it is higher for 20 markets in the portfolio than ten markets. Overall, these results suggest that a portfolio of 20 futures markets will produce a diversification return of around 3% per year. However, as shown in the 1980s and 1990s, a positive diversification return does not assure a positive realized futures return.

Erb and Harvey (2006) argue that the diversification return is closely linked to the portfolio weighting scheme. Simplistically this can be seen in Table 7.2. The two assets have identical initial and terminal values. So, an initially equally weighted portfolio that is not rebalanced or a 'let it run' (no rebalancing) portfolio would have a return of zero even though the equally weighted and rebalanced portfolio has a return

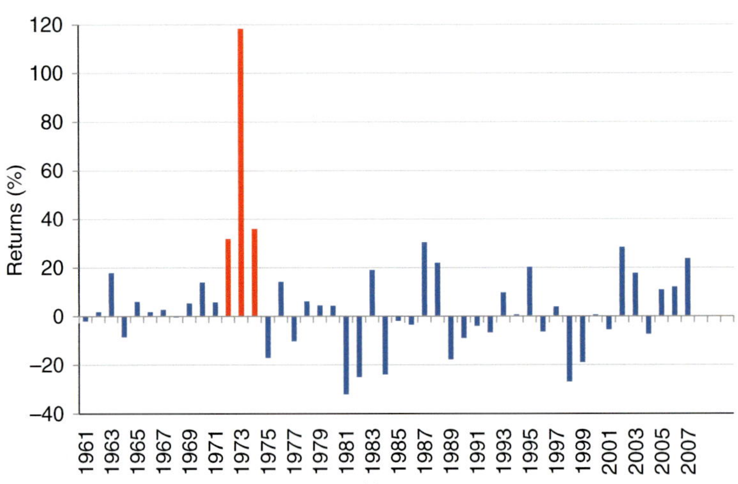

Note: The ten futures market portfolio includes corn, wheat, soybeans, soybean meal, soybean oil, oats, cotton, sugar, cocoa, and copper.

Fig. 7.3. Annual returns to a ten-market equally weighted portfolio of US commodity futures markets, 1961–2010. Returns in the years from 1972 to 1974 account for 68% of all returns during the 1961–2010 period.

of 4%. Erb and Harvey (2006) argue that this loss of diversification return to initially equally weighted portfolios is due to 'covariance drag' or the correlation between an asset's returns and the portfolio weights. Gorton and Rouwenhorst (2006b) dismiss this notion and emphasize that diversification returns are indisputable mathematical properties of the geometric returns of portfolios. Instead, they suggest that an equally weighted and continuously rebalanced portfolio (as opposed to an initially equally weighted and let-it-run portfolio) represents an 'embedded' trading strategy.

Indeed, the portfolio weighting scheme for commodity futures investments is a crucial point. Unlike equities – which have clearly defined market capitalization and a strong economic justification for value-weighted portfolios – futures markets have a zero-market capitalization. Therefore, there is not a clear method to weight a futures market portfolio. Some commercial indices, such as the Standard and Poor's Goldman Sachs Commodity Index (S&P GSCI), attempt to use a value-weighting scheme based on production in the underlying physical markets. However, this approach results in a commodity index that is heavily (> 70%) weighted in the energy markets.

As pointed out by Fung and Hsieh (2002), an equally weighted portfolio is a 'contrarian' strategy because the portfolio is rebalanced by 'selling winners' and 'buying losers.' Conversely, the initially equally weighted and let-it-run portfolios are a momentum-driven allocation strategy where winners naturally increase their weight in the index and losers naturally reduce their portfolio weight. So when evaluating the past performance of a portfolio of commodity futures markets, it is very difficult to disentangle the source of the returns between diversification returns and strategies embedded within a portfolio weighting scheme. The recent performance of some popular commodity index funds further highlights these issues.

7.4 Recent Performance of Commodity Investments

Despite the academic endorsement (e.g. Gorton and Rouwenhorst, 2006a), actual returns to commodity investments have been disappointing. The iShares S&P GSCI Commodity-Indexed Trust is an ETF designed to mimic the performance of the S&P GSCI – one of the most widely followed commodity indices. The ETF was initially offered to the public in July of 2006 at a price near $50 per share. Since then, the share price has generally declined and was $30.19 on September 30 2011 for a total loss of 39.6% over 5 years (Fig. 7.4). The negative

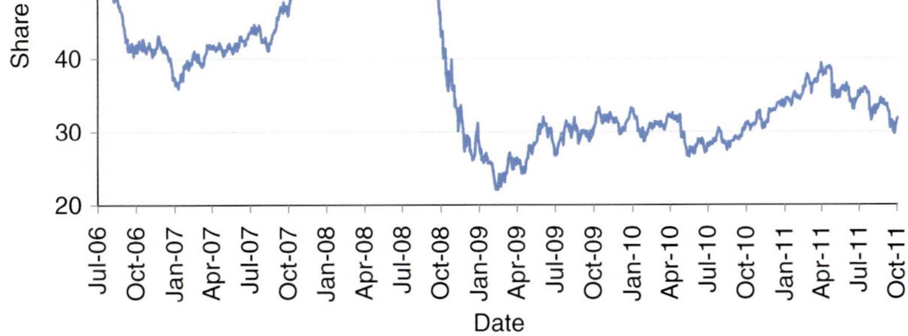

Fig. 7.4. iShares S&P GSCI Commodity-Indexed Trust, 2006–2011. (Source: iShares by BlackRock, no date.)

return occurred over a period of time when there was a general upward trend in overall commodity prices. The disappointing performance of this and similar static long-only commodity funds is partially responsible for the industry's push toward 'enhanced' or 'strategic' funds that attempt to tweak the static funds' portfolio weights or roll-over rules to improve performance.

Recent evidence also shows the differences in performance that can arise from alternative portfolio weighting schemes or embedded trading strategies. Table 7.4 shows the quarterly returns to three long-only ETFs. Consistent with Fig. 7.4, the returns are negative over this particular sample for each fund. However, over the common sample from the second quarter of 2008 through the third quarter of 2011, the GreenHaven Continuous Commodity Index Fund did remarkably better than the funds tracking the S&P GSCI and Dow Jones UBS (DJ-UBS) indices (Murdoch, 2008). The GreenHaven fund tracks the Thompson Reuters Continuous Commodity Index (TR-CCI) – an equally weighted index of 17 commodities – by holding positions in the front 6 months of the futures curve. The fund's relative outperformance over this interval could stem from the holding of contracts other than just the nearby futures (a strategic component of performance) or it could stem from an equal weighting of the 17 markets with daily rebalancing (embedded trading strategy). Regardless, the performance (while still negative) stands in contrast to the funds tracking the S&P GSCI and DJ-UBS, which hold nearby futures only and are weighted more heavily in the energy sector.

The recent performance of funds that track long-only commodity indices is generally negative and certainly lower than the historical returns reported by some researchers (e.g. Gorton and Rouwenhurst, 2006a). Naturally, this can be sample–specific, and the returns may indeed gravitate higher over the longer term. Or, as

Table 7.4. Performance of exchange-traded commodity index funds, 2007–2011. (Source: iShares by BlackRock, no date.)

Quarter ending	GreenHaven CCI	iPath DJ-UBS	iShares S&P GSCI
Mar-07		3.84	3.77
Jun-07		–0.12	2.24
Sep-07		5.95	10.99
Dec-07		4.56	11.83
Apr-08		9.19	9.71
Jul-08	14.04	16.47	28.71
Sep-08	–24.84	–27.93	–27.76
Dec-08	–20.92	–31.72	–46.82
Mar-09	0.14	–6.08	–11.99
Jun-09	4.24	12.38	19.34
Sep-09	4.76	3.82	–1.23
Dec-09	9.80	9.60	7.21
Mar-10	–4.71	–5.68	–2.07
Jun-10	–0.60	–5.57	–10.82
Sep-10	12.96	12.33	7.99
Dec-10	17.01	16.18	13.63
Mar-11	6.92	4.54	9.53
Jun-11	–4.91	–8.02	–8.67
Sep-11	–9.43	–11.62	–11.49
Annualized returns			
January 2007–March 2011		–3.29%	–5.77%
February 2008–March 2011	–1.81%	–10.44%	–16.96%
Expense ratio	0.85%	0.75%	0.75%
Underlying index	TR-CCI	DJ-UBS	S&P GSCI

[a]CCI, Continuous Commodity Index; DJ-UBS, Dow Jones UBS; S&P GSCI, Standard and Poor's Goldman Sachs Commodity Index; TR-CCI, Thompson Reuters Continuous Commodity Index.

found in the presented historical data (Table 7.3 and Fig. 7.3), long-term returns may be driven by periodic upheavals in commodity prices like those seen in 1972–1974 and 2006–2008, which occur very infrequently. In this light, recent negative performance of long-only commodity funds may be viewed as a more representative outcome.

7.5 Summary and Conclusions

Recent academic research claimed 'equity-like' returns to portfolios of commodity futures while also touting the diversification benefits relative to traditional asset classes (Erb and Harvey, 2006; Gorton and Rouwenhorst, 2006a). Not surprisingly, commodity-linked investments have grown considerably over the last 5 years as individuals and institutions have embraced alternative investments and such evidence. However, unlike investments in equities or real estate, commodity futures markets actually produce no earnings and, arguably, are not even capital assets. In this paper, we re-examine the case for long-only commodity futures investments and find several reasons for caution. Our focus is on the returns generated by static long-only commodity indices, and we do not consider the potential returns to 'tactical' or 'strategic' commodity investments that may actively, but subtlety, manage the individual index components or the switching of contracts.

Data on commodity futures returns going back to the 1960s strongly suggest that the returns to individual futures markets are on average zero. Realized futures returns are not directly related to the market term structure (contango or backwardation). The idea that the 'return to roll' is a realized futures return is fiction. That is, the actual process of rolling a long futures position to a new futures contract does not produce a futures return, either positive or negative, and if it did this would be equivalent to a futures money machine. Returns to long-only commodity futures investments therefore must come from the portfolio level.

We find that the return at the portfolio level stems largely from diversification. The 'diversification return' to a portfolio of futures markets is a mathematical fact stemming from the arithmetic of geometric portfolio returns. However, the diversification return alone does not assure positive returns. An equally weighted portfolio of ten commodity futures markets had two decades of negative realized returns (1980s and 1990s) even though the diversification return was reliably positive. The weighting scheme for the portfolio is a critical and confounding factor. There is not a clear, economically justified weighting scheme for commodity futures contracts. All portfolio weighting schemes are an embedded trading strategy. Equally weighted and continuously rebalanced portfolios are contrarian strategies and let-it-run (no rebalancing) portfolios are momentum strategies. It is very difficult to attribute portfolio returns to the various components, which suggests that it may likewise be difficult to form expectations for returns in the future.

These results leave one skeptical about the expected returns of static long-only commodity futures investments. The excess returns from 1961 to 2010 for a ten-market portfolio and the returns from 1991 to 2010 for a 20-market portfolio are near 3% per year but not statistically different from zero. Plus, the returns clearly can be negative for decades (1980s and 1990s) only to be punctuated by positive returns surrounding major commodity upheavals such as during 1972–1974. The typical expense ratio for a long-only index fund is just under 1% per year. If there is a positive expected return to long-only commodity index funds, it is likely to converge to this level over time. Overall, the case for long-only investment in commodities is not as strong as some have argued in recent years.

Acknowledgments

This paper benefited from helpful comments and suggestions by the participants at the 2011 North Central Coordinating Committee (NCCC)-134 Conference on Applied Commodity Price Analysis, Forecasting, and Market Risk Management and seminar participants in the Department of Agricultural and Consumer Economics at the University of Illinois at Urbana-Champaign. We thank Hongxia Jiao for her superb assistance in collecting the return data for this study.

Notes

[1] Original citation: Sanders, D.R. and Irwin, S.H. (2012) A reappraisal of investing in commodity futures markets. *Applied Economic Perspectives and Policy* 34, 515–530. Reprinted by permission of John Wiley and Sons and the Agricultural and Applied Economics Association.
[2] The source for the GSG asset values is: https://www.etf.com/GSG#overview (accessed September 12 2022).
[3] A variety of investment instruments are typically lumped under the heading 'commodity index fund.' Individuals or firms may enter directly into over-the-counter (OTC) contracts with swap dealers to gain the desired exposure to returns from a particular index of commodity prices. Investment funds are also offered whose returns are tied to a commodity index. Exchange-traded funds (ETFs) and structured notes (ETNs) have also been developed to make it even easier to gain commodity exposure. ETFs and ETNs trade on securities exchanges in the same manner as stocks on individual companies. See Engelke and Yuen (2008), Stoll and Whaley (2010), and Irwin and Sanders (2011) for further details.
[4] Erb and Harvey (2006) examine futures markets returns within four potential frameworks: (i) the Capital Asset Pricing Model (CAPM); (ii) a Keynesian risk premium or normal backwardation; (iii) hedging pressure; and (iv) returns associate with 'convenience yield.' Importantly, theories of hedging pressure and those connecting returns to storage signals based on convenience yield do not prescribe static, long-only futures investments. Rather, these theories suggest that risk premiums are time-varying and may accrue to either long or short positions and a dynamic investment strategy is needed to capture them.
[5] Returns are calculated as the percentage change in nearby futures contracts. The nearby contract is defined as the contract closest to delivery month for which the delivery period has not been entered. For most agricultural commodities, the roll-over occurs on the last day of the month proceeding the delivery month (e.g. November 30 for the December contract). For most energy and metals markets, the roll-over occurs near the middle of the month proceeding the delivery month (e.g. November 15 for the December contract).
[6] The data in Table 7.1 and all other tables reflect excess returns. That is, it is implicitly assumed that the futures contract is unlevered, and the notional value of the investment earns the risk-free US Treasury bill rate. The returns are not adjusted for inflation.
[7] Willenbrock (2011) argues that there is no mathematical proof of Eqn 7.1 as presented in Erb and Harvey (2006) and that Gorton and Rouwenhort's (2006a) explanation of the diversification return is faulty. While Willenbrock's (2011) mathematical arguments are convincing, the results are not central to the theme of his paper.
[8] The correlation of commodity futures returns increased substantially after the financial meltdown of September 2008 (e.g. Buyuksahin and Robe, 2011) so the generality of this latter statement is open to debate going forward.

References

Abrams, R., Bhaduri, R. and Flores, E. (2010) Lintner revisited: a quantitative analysis of managed futures in an institutional portfolio. The CME Group, Chicago, Illinois. Available at: http://www.sselectricwv.com/app/webroot/uploads/client_74/files/AlphaMetrix%20-%20CME%20Group%20-%20Lintner%20Revisited.pdf (accessed February 15 2022).

Bessembinder, H. (2018) The 'roll yield' myth. *Financial Analysts Journal* 74, 41–53.

Bjornson, B. and Carter, C.A. (1997) New evidence on agricultural commodity return performance under time-varying risk. *American Journal of Agricultural Economics* 79, 918–930.

Black, F. (1976) The pricing of commodity contracts. *Journal of Financial Economics* 3, 167–179.

Bodie, Z. and Rosansky, V.I. (1980) Risk and return in commodity futures. *Financial Analysts Journal* 36, 27–39.

Brooks, C., Prokopczuk, M. and Wu, Y. (2011) Commodity futures prices: more evidence on forecast power, risk premia and the theory of storage. ICMA Centre Discussion Papers in Finance DP2011-21. International Capital Market Association (ICMA), University of Reading, Reading, UK.

Burton, C. and Karsh, A. (2009) Capitalizing on any curve: clarifying misconceptions about commodity indexing. Credit Suisse Asset Management, LLC. Available at: https://rainmaker-s3-media.s3-ap-southeast-2.amazonaws.com/prod/media/library/CPD/PDFs/Journals/Investment%20Strategy/Volume%204/Number%202/JIS-v4n02_Burton_09.pdf (accessed February 15 2022).

Buyuksahin, B. and Robe, M.A. (2011) Speculators, commodities, and cross-market linkages. Working Paper, US Commodity Futures Trading Commission, Washington, DC. Available at: https://www.cftc.gov/sites/default/files/idc/groups/public/@swaps/documents/file/plstudy_04_iea.pdf (accessed February 15 2022).

Campbell, J.Y. (2000) Diversification: a bigger free lunch. Working Paper, Department of Economics, Harvard University, Cambridge, Massachusetts.

Carter, C.A., Rausser, G.C. and Schmitz, A. (1983) Efficient asset portfolios and the theory of normal backwardation. *Journal of Political Economy* 91, 319–331.

CFTC (Commodity Futures Trading Commission) (2011) Index Investment Data. March 31. Available at: http://www.cftc.gov/MarketReports/IndexInvestmentData/index.htm (accessed February 13 2022).

Cootner, P.H. (1960) Returns to speculators: Telser vs. Keynes. *Journal of Political Economy* 68, 396–404.

Crigger, L., Ludwig, O. and Hougan, M. (2010) US commodity launches contango-killer ETF. *Index Universe.* Available at: http://www.indexuniverse.com/sections/news/7921-us-commodity-funds-launches-contango-killer-etf.html (accessed February 13 2022).

Daskalaki, C. and Skiadopoulos, G. (2010) Should investors include commodities in their portfolios after all? New evidence. Working Paper, Department of Banking and Financial Management, University of Piraeus, Greece. Available at: https://papers.ssrn.com/sol3/papers.cfm?abstract_id=1652699 (accessed February 15 2022).

Dusak, K. (1973) Futures trading and investor returns: an investigation of commodity market risk premiums. *Journal of Political Economy* 81, 1387–1406.

Elton, E.J., Gruber, M.J. and Rentzler, J.C. (1987) Professionally managed, publicly traded commodity funds. *Journal of Business* 60, 175–199.

Engelke, L. and Yuen, J.C. (2008) Types of commodity investments. In: Fabozzi, F.J., Fuss, R. and Kaiser, D. (eds) *The Handbook of Commodity Investing*. Wiley, Hoboken, New Jersey, pp. 549–569.

Erb, C.B. and Harvey, C.R. (2006) The strategic and tactical value of commodity futures. *Financial Analysts Journal* 62, 69–97.

Fama, E.F. and French, K.R. (1987) Commodity futures prices: some evidence on forecast power, premiums, and the theory of storage. *Journal of Business* 60, 55–73.

Fung, W. and Hsieh, D.A. (2002) Asset based style factors for hedge funds. *Financial Analysts Journal* 58, 16–27.

Gorton, G.B. and Rouwenhorst, K.G. (2004) Facts and fantasies about commodity futures. National Bureau of Economic Research (NBER) Working Paper No. 10595. Available at: https://www.nber.org/system/files/working_papers/w10595/w10595.pdf (accessed January 14 2022).

Gorton, G.B. and Rouwenhorst, K.G. (2006a) Facts and fantasies about commodity futures. *Financial Analysts Journal* 62, 47–68.

Gorton, G.B. and Rouwenhorst, K.G. (2006b) A note on Erb and Harvey (2005). Yale ICF Working Paper No. 06-02. Available at: http://depot.som.yale.edu/icf/papers/fileuploads/2595/original/06-02.pdf (accessed February 15 2022).

Gorton, G.B., Hayashi, F. and Rouwenhorst, K.G. (2007) The fundamentals of commodity futures returns. National Bureau of Economic Research (NBER) Working Paper No. 13249. Available at: https://www.nber.org/papers/w13249 (accessed January 14 2022).

Greer, R.J. (2000) The nature of commodity index returns. *Journal of Alternative Investments* 3, 45–52.

Hartzmark, M. (1987) Returns to individual traders of futures: aggregate results. *Journal of Political Economy* 95, 1292–1306.

Houthakker, H.S. (1957) Can speculators forecast prices? *Review of Economics and Statistics* 39, 143–151.

Irwin, S.H. and Sanders, D.R. (2011) Index funds, financialization, and commodity futures markets. *Applied Economic Perspectives and Policy* 33, 1–31.

Irwin, S.H., Sanders, D.R., Smith, A. and Main, S. (2020) Returns to investing in commodity futures: separating the wheat from the chaff. *Applied Economic Perspectives and Policy* 42, 583–610.

iShares by BlackRock (n.d.) iShares S&P GSCI Commodity-Indexed Trust. Available at: https://www.ishares.com/us/products/239757/ishares-sp-gsci-commodityindexed-trust-fund (accessed February 13 2022).

Kolb, R.W. (1992) Is normal backwardation normal? *Journal of Futures Markets* 12, 75–90.

Krishnan, B. and Sheppard, D. (2010) Analysis: commodities supercycle? Pensions loath to commit. Reuters, October 13. Available at: https://www.reuters.com/article/us-commodities-institutions-idUSTRE69C3CV20101013 (accessed February 15 2022).

Main, S., Irwin, S.H., Sanders, D.R. and Smith, A. (2018) Financialization and the returns to commodity investments. *Journal of Commodity Markets* 10, 22–28.

Marcus, A.J. (1984) Efficient asset portfolios and the theory of normal backwardation: a comment. *Journal of Political Economy* 92, 162–164.

Masters, M.W. (2008) Testimony before the Committee on Homeland Security and Government Affairs, US Senate. May 20. Available at: http://hsgac.senate.gov/public/_files/052008Masters.pdf (accessed January 14 2022).

Murdoch, J. (2008) The New GreenHaven Commodity ETF Examined. ETF.com. Available at: https://www.etf.com/sections/features-and-news/652-the-new-greenhaven-commodity-etf-examined (accessed February 13 2022)

Pirrong, C. (2010) No theory? No evidence? No problem! *Regulation* 33, 38–44.

Robison, P., Loder, A. and Bjerga, A. (2010) Amber waves of pain. *Bloomberg Businessweek* July 22. Available at: https://www.bloomberg.com/news/articles/2010-07-22/amber-waves-of-pain (accessed February 15 2022).

Rockwell, C.S. (1967) Normal backwardation, forecasting, and the returns to commodity futures traders. *Food Research Institute Studies* 7, 107–130.

Seeking Alpha (2007) Goldman's new GSCI enhanced commodity ETF: twist on a familiar index. August 8. Available at: http://seekingalpha.com/article/43910-goldman-s-new-gsci-enhanced-commodity-etf-twist-on-a-familiar-index (accessed February 13 2022).

Stockton, K.A. (2007) *Understanding Alternative Investments: the Role of Commodities in a Portfolio*. Vanguard Investment Counseling & Research, London.

Stoll, H.R. and Whaley, R.E. (2010) Commodity index investing and commodity futures prices. *Journal of Applied Finance* 20, 7–46.

Telser, L.G. (1958) Futures trading and the storage of cotton and wheat. *Journal of Political Economy* 66, 233–255.

Telser, L.G. (2000) *Classic Futures: Lessons for the Past from the Electronic Age*. Risk Books, London.

Till, H. (2006) A long-term perspective on commodity futures returns. EDHEC Risk and Asset Management Research Center, EDHEC Business School, Paris. Available at: https://risk.edhec.edu/publications/long-term-perspective-commodity-futures (accessed February 15 2022).

USS/PSI (United States Senate, Permanent Subcommittee on Investigations) (2009) Excessive speculation in the wheat market. Majority and Minority Staff Report. Available at: https://www.hsgac.senate.gov/imo/media/doc/REPORTExcessiveSpecullationintheWheatMarketwoexhibitschartsJune2409.pdf?attempt=2 (accessed January 14 2022).

Willenbrock, S. (2011) Diversification return, portfolio rebalancing, and the commodity return puzzle. *Financial Analysts Journal* 67, 42–49.

8

The 'Necessity' of New Position Limits in Agricultural Futures Markets: the Verdict from Daily Firm-level Position Data[1]

New Author Foreword

A number of regulatory proposals were put forward to 'rein in' commodity speculation after the 2007–2008 price spike. The extension of speculative position limits to all US futures markets for physical commodities became the focal point of the roiling global controversy. Federally mandated limits on speculative positions had been in place for agricultural futures markets for many decades but had not been extended to new commodity futures markets as they were developed, such as crude oil. The 2010 Dodd-Frank Wall Street Reform and Consumer Protection Act (Dodd-Frank) (US Government, 2010) included a provision mandating the US Commodity Futures Trading Commission (CFTC) to develop rules that would extend position limits to futures markets for all physical commodities. This was controversial because the 1936 Commodity Exchange Act (CEA) required the CFTC to demonstrate that position limits are 'necessary' to prevent excessive speculation 'causing sudden or unreasonable fluctuations or unwarranted changes in the price of such commodity' (US House of Representatives, 2019).

The CFTC's first attempt at position limit rules under Dodd-Frank was vacated in 2012 by US District Court Judge Robert Wilkins on grounds that the agency did not establish the 'necessity' of the limits as required by statute. Undeterred, the CFTC proposed new position limit rules in November 2013, with comment periods extended and/or reopened numerous times over the next several years. This second attempt was eventually put on ice in December 2015 after intense opposition from the futures industry. The CFTC ended up 'reproposing' less stringent position limit rules for commodity futures markets in late 2016, but delayed consideration of the rules until the next administration was in place. The last iteration of the position limit rules was finally approved by the CFTC in January 2020. This version stripped out the most objectional features from earlier proposals, but also included some truly tortured logic regarding the necessity issue. In essence, the CFTC said that so long as there was any kind of risk of unwarranted price fluctuations and the contract in question was important enough, then the necessity of position limits could be established.[2] This brings to mind the classic scene from the movie *Dumb and Dumber* (Farrelly, 1994), where Lloyd tells Mary, the love of his life, 'So you're telling me there's a chance?' (fans of the movie will get the reference).

In an interesting twist, we were contacted by a CFTC staffer in the middle of all this and encouraged to submit a comment letter on the proposed rules. Without coming right out and saying it, the staffer clearly wanted to make sure that our research was officially entered into the record so it could be referenced in the final rulemaking. So we dutifully submitted a comment letter in July 2014. The letter certainly did the trick, as the final 2016 rule referenced our research over 90 times.

The preceding discussion indicates that the heart of the political and legal battle over position limits was the question of whether the limits were 'necessary.' Simply put, new regulations on trading should be justified based on empirical evidence, and the more granular the evidence the better. Coincidently, in 2012 we were hired by a large company that developed and offered commodity index products to examine whether this company's positions

DOI:10.1079/9781800622104.0008

had an impact on market prices. Given the political sensitivities surrounding the index speculation controversy, we gave considerable thought to whether we should pursue this consulting project. Since it would give us access to detailed data that had been previously unavailable to academic researchers, we thought it was worth the effort. The company graciously allowed us to use the data for academic research as long as we did not reveal the source of them.

It did not take us long to realize the wisdom of our decision to move forward with the project. The new data could be used to provide a fresh take on the necessity question. In particular, the position data were available on a daily basis and disaggregated down to the level of individual contracts, which contrasted with the public data series on index positions available from the CFTC that were aggregated across contracts at the weekly, monthly, or quarterly horizons. It was possible that index positions measured in this aggregated manner could mask daily price impacts.

We went to work and conducted a battery of statistical tests designed to detect relationships between daily position changes and returns in 12 agricultural futures markets. A unique feature of the position data was that we could precisely track the 'rolling' activity of the private fund, which is the movement of positions from the expiring nearby contract to the first deferred contract. This allowed us to test whether roll trades impacted the calendar spread between commodity futures prices (spread computed as nearby minus deferred contract). One line of argument was that even if index investment did not impact the level of commodity futures prices, it could still cause a permanent decrease in calendar spreads as the impact of rolling cumulated through time. Our statistical tests failed to indicate any evidence linking the fund's trading with market returns. However, there was a small negative relationship between the fund's roll transactions and changes in calendar price spreads, just the opposite of that expected based on price-pressure arguments.

We presented an early version of the article at an invited paper session at the 2013 annual meeting of the Agricultural and Applied Economics Association. The work was well received and we were encouraged to continue developing the article, which was eventually published in 2016. In the end, the new data told the same story – there was little evidence of a systematic relationship between index positions and commodity futures price movements. But this was still an important article because it highlighted that the CFTC flunked the necessity test when it first proposed new speculative position limits.

Abstract

Regulators are proposing new position limits in US commodity futures markets while the actual impact of long-only index funds on futures prices continues to be debated. Researchers have noted the data limitations – frequency and market breadth – associated with using data compiled by the US Commodity Futures Trading Commission (CFTC). This research addresses these shortfalls by using daily position data for a specific long-only index fund. The empirical analysis focuses on the firm-level position data across 13 US agricultural futures markets. The firm-level data are shown to be representative of the overall index fund industry. Empirical tests fail to find any evidence linking the firm's trading with market returns. However, there does appear to be a consistent negative relationship between the firm's roll transactions and changes in calendar price spreads. Notably, the direction of this impact is opposite of price-pressure hypothesis. The results of this study, and others, indicate that a clear verdict can be reached – new limits on speculation in agricultural futures markets are unnecessary.

Key words: agricultural, bubble, commodity, futures market, index funds, Michael Masters, position limits, price, speculation

JEL categories: D84, G12, G13, G14, Q13, Q41

> Excessive speculation ... causing sudden or unreasonable fluctuations or unwarranted changes in the price of such commodity, is an undue and unnecessary burden on interstate commerce in such commodity. For the purpose of diminishing, eliminating, or preventing such burden, the Commission shall ... fix such limits on the amounts of trading ... as the Commission finds are necessary to diminish, eliminate, or prevent such burden.
>
> (Commodity Exchange Act 1936; US House of Representatives, 2019)

8.1 Introduction

The history of US commodity prices includes periods of relatively depressed prices – such as the 1930s and the mid-1980s – and episodes of rapid price increases – such as the early 1970s and the 2007–2008 commodity boom. Periods of historically low and high commodity prices bring about a predictable public outcry against speculators in US futures markets and a corresponding attempt to regulate their activity (Jacks, 2007). Whether commodity prices are viewed as too low by producers or too high by consumers, the blame for undesirable price levels seems to fall conveniently upon the faceless speculator (Petzel, 1981). Catching the ear of politicians, these complaints have led to attempts to regulate speculative trading in US commodity futures markets through restrictive position limits, greater margin requirements, or even outright bans on trading (Working, 1963; Hieronymus, 1977).

Given this backdrop, it should come as no surprise that the historical pattern was repeated when corn, soybean, and wheat futures prices set new nominal price records in 2007–2008. The rapid increase in commodity prices coincided with the emergence of new financial vehicles that provided investors with exposure to indices that track returns in commodity futures markets. These financial investments are packaged in a variety of forms that provide the investor with long-only exposure to an index of commodity prices. Concerns soon emerged among market participants, regulators, and civic organizations that the inflows into new commodity index investments were driving increases in commodity prices. This notion is most commonly associated with hedge fund manager Michael W. Masters and is often referred to as the 'Masters Hypothesis' (Irwin and Sanders, 2012). The Masters Hypothesis argues that unprecedented buying pressure from index investors created massive bubbles in commodity futures prices. In turn, these bubbles were transmitted to spot prices through arbitrage linkages between futures and spot prices. The end result was that commodity prices – and the prices of staple food and energy products – exceeded values warranted by traditional supply-and-demand factors.

Policy makers and other advocates were quick to adopt Masters-like arguments after the 2007–2008 price spikes and pushed for regulations to limit commodity index activity. As a result, the 2010 Dodd-Frank Wall Street Reform and Consumer Protection Act (Dodd-Frank) (US Government, 2010) laid the groundwork for more restrictive speculative limits on commodity futures positions. The Commodity Futures Trading Commission's (CFTC) first attempt at position limits under Dodd-Frank was vacated in 2012 by US District Court Judge Robert Wilkins on grounds that the CFTC in essence did not establish the 'necessity' of the limits as required by the 1936 Commodity Exchange Act (CEA). That is, the CFTC did not show that excessive speculation was causing unwarranted changes in commodity prices (Young *et al.*, 2012). CFTC Commissioner Scott D. O'Malia laid bare the essence of the court's decision: 'the court explicitly stated that the statute unambiguously requires a finding of necessity before establishing position limits' (O'Malia, 2012). As shown in the opening quote to this article, 'necessity' refers to original language in the CEA which grants the CFTC the ability to fix position limits that are a 'necessity' to prevent excessive speculation 'causing sudden or unreasonable fluctuations or unwarranted changes in the price of such commodity' (US House of Representatives, 2019). The CFTC skirted this issue in the proposed rule-making, claiming essentially that Dodd-Frank requires them to implement the new rules irrespective of the 'necessary' conditions in the original CEA. Federal Judge Robert Wilkins clearly disagreed with this omission and indicated that the 'necessity' finding was in fact required (Young *et al.*, 2012).

Undeterred, the CFTC both appealed the court decision and simultaneously formulated new position limit rules in 2013 (Miedema, 2013). While the CFTC ultimately dropped the appeal, the CFTC Commissioners approved new position limit rules in November 2013 (Michaels, 2013). Since then, the CFTC has extended or reopened the comment period numerous times to provide industry participants ample opportunity to weigh in on this controversial rule. The most recent comment period closed on March 28 2015 (CFTC, 2015). How successful the CFTC will ultimately be in establishing the 'necessity' described in the CEA and required by the US District Court remains to be seen. An economist's interpretation of 'excessive speculation' as outlined in the CEA represents a relatively high hurdle. First, the speculation must be 'causing' the price fluctuations. Second, the price changes must be 'sudden' or 'unreasonable' or 'unwarranted.' This definition of excessive speculation seemingly excludes speculation that cannot be shown to *cause* price changes, which implies a possible temporal ordering. Likewise, the CEA description precludes speculation that warrants price changes – that is, informed speculation.

Numerous academic studies in recent years investigate the empirical relationship between commodity index positions and price movements in commodity futures markets. Some find evidence of a positive impact (e.g. Gilbert, 2010; Gilbert and Pfuderer, 2014) but most do not (e.g. Stoll and Whaley, 2010; Hamilton and Wu, 2015).[3] With a few exceptions, prior research relies on data compiled by the CFTC through the *Large Trader Reporting System* (LTRS). These data are made available through two widely used reports, the *Supplemental Commitments of Traders* (SCOT) and the *Disaggregated Commitments of Traders* (DCOT). Prior work that uses these CFTC data suffers from limitations in terms of both the frequency of the data and the availability of data across markets. For example, the SCOT data are relatively accurate measures of commodity index positions (Irwin and Sanders, 2012), but are only available at weekly intervals for 12 agricultural futures markets and exclude important energy and metal futures markets. The DCOT data nets on- and off-exchange index positions, and may therefore substantially underestimate index positions in some markets, especially energy and metals (Irwin and Sanders, 2012). Compiled independently of the LTRS, the CFTC also publishes the *Index Investment Data* (IID) report. The IID are available for all major futures markets and considered the most accurate data available on index positions; but historical data are available only at quarterly and monthly frequencies, which severely limits the number of observations available for statistical tests. Some authors (e.g. Singleton, 2014) have attempted to circumvent these issues by imputing positions for the energy markets from the positions reported for agricultural markets in the SCOT report. Sanders and Irwin (2013) demonstrate how this data-mapping process can lead to unreliable position data and potentially misleading empirical results, which highlights the need for more detailed data.

The objective of this article is to bring new data to bear on the debate over the impact of index funds on food commodity prices. Specifically, daily futures and swaps positions are obtained for a major commodity index fund, which allows for the examination of potential price impacts over daily intervals. The data set spans 22 US futures markets from October 1 2007 through May 30 2012, or a total of 1176 daily observations for each market.[4] This article focuses on the 13 agricultural markets because these markets are directly relevant to the debate about index fund impacts on food commodity prices. This also facilitates comparisons to several previous studies that have examined the 12 agricultural markets included in the CFTC's SCOT report. Sanders and Irwin (2014) use the firm-level data set to examine similar issues in energy futures markets.

The research contributes to the existing literature by bringing a new data set to bear on the empirical debate regarding index funds' impact on prices during the 2007–2008 commodity boom. Importantly, the disaggregated positions provide for a unique examination of trading patterns and potential market impact over daily horizons for both outright and 'roll' transactions, which cannot be analyzed with publicly available data compiled by the CFTC. The data set also includes positions in both futures and swaps markets that are not available in either the SCOT or DCOT reports. Causal linkages between index positions and price changes, if they exist, may be more evident in these data covering both futures and swaps markets. The presented results have important ramifications regarding proposed policy measures intended to limit positions held by index funds in US commodity futures markets.

8.2 Position Data

The position data are collected from a large investment company (the 'Fund') that offers several commodity investment programs. The majority of the Fund's commodity investments are held in a relatively fixed basket of commodity futures to replicate a proprietary index. Detailed data on actual positions held by the Fund in US futures markets are available for 22 US futures markets. The empirical analysis presented here focuses on 13 key agricultural markets: (i) Chicago Board of Trade (CBOT) corn, soybean oil, soybeans, soybean meal, and wheat; (ii) Intercontinental Exchange (ICE) cocoa, cotton, sugar, and coffee; (iii) Chicago Mercantile Exchange (CME) feeder cattle, live cattle, and lean hogs; and (iv) Kansas City Board of Trade (KCBOT) wheat. For each of these 13 markets, complete position data are available for 1176 days from October 1 2007 through May 30 2012.[5]

The position data for the Fund includes futures positions for each market by calendar month contract. In addition to the direct futures positions, 'look alike' swap positions are held in corn, CBOT wheat, soybeans, soybean oil, and cotton. These swaps are constructed to precisely mirror a particular exchange-traded futures contract. The swap positions are smaller than the direct futures positions held in these markets. For example, in corn, the swap position averaged 4613 contracts from February 14 2011 to January 17 2012. Over that same period, the direct futures position was an average of 18,365 contracts. So the swap position represented 20% of the total position. Comparable calculations show that when swap positions are held, the percentage of the total position was 8% for soybean oil, 7% for CBOT wheat, 24% for cotton, and 21% for soybeans. Swap positions were not continuously held in these markets. For instance, on the last day of the data set, May 30 2012, swap positions were only held in three of the five markets. When analyzing the potential impact of positions on market returns, the swap positions are combined with the futures positions to arrive at a total or aggregate position for each market. Notably, this is an improvement over studies that use firm-level daily position data from the CFTC's non-public LTRS (e.g. Buyuksahin, and Harris, 2011; Aulerich *et al.*, 2013), which does not record swaps positions and therefore may not accurately reflect total commodity exposure (Irwin and Sanders, 2012; Sanders and Irwin, 2013; Brunetti and Reiffen, 2014). The data set did not include any instances of a short total position in any market. So the total position in each market is long-only.

This unique data set also provides the ability to distinguish between trading that represents new investment in the Fund and trading that represents roll transactions. Changes in the aggregate long position held by the Fund clearly represent outright buying or selling. However, there are also days with active trading but no change in the overall long position within a market. On those days, the Fund is 'rolling' or transferring long market positions from one calendar maturity month to another. The normal roll transaction is selling nearby contracts and simultaneously buying the next listed contract; thereby, the long position in the nearby contract is transferred to the next active contract.

From the detailed position data, a series is created that represents the number of contracts that are 'rolled' between futures contracts within a market. For example, if the aggregate long position increases by 100 contracts and a total of 100 contracts was traded across calendar months, then there were no roll transactions and the net new investment is represented by the aggregate increase of 100 contracts. If, however, the aggregate long position increases by 100 contracts and 300 contracts trade across the calendar months, then 100 of the contracts traded were used to establish the new position, and 200 total trades (100 sells and 100 buys) represented the rolling or moving of 100 positions across calendar months. The size of 'roll transactions' will be used to analyze the impact on futures spreads. The ability to precisely identify roll transactions for the Fund is a potential improvement over prior research, which has mostly relied on assumed roll 'windows' or aggregate position size as an indicator (Stoll and Whaley, 2010; Aulerich *et al.*, 2013). The data set here provides a detailed and direct measure of rolling activity.

8.3 Position Trends and Characteristics

Figure 8.1 shows the notional value of Fund positions in all 22 US markets that are actively traded. Notional value is simply the net position of the Fund multiplied by the relevant futures contract price. The total position size (futures plus swaps) grows from under \$4.0 billion in 2007 to \$12.0 billion in 2011 and then stabilizes between \$10.0 and \$12.0 billion. As a standard of comparison, the total positions held by the Fund are compared to those reported in the CFTC's IID report. In Fig. 8.2, the total notional value of index positions for US markets reported in the IID are plotted alongside those held by the Fund for each quarter end from December 31 2007 to March 30 2012. Over the sample period, the Fund's total position and that reported in the IID have a positive correlation of 0.86 in levels and 0.97 in differences. The Fund has grown more rapidly than the industry, with the Fund's portion of the industry increasing from 3.0% in late 2007 to a high of 7.6% in 2012.

The Fund's holding on a market-by-market basis are also compared to the 21 markets in the IID that coincide with those traded by the Fund.

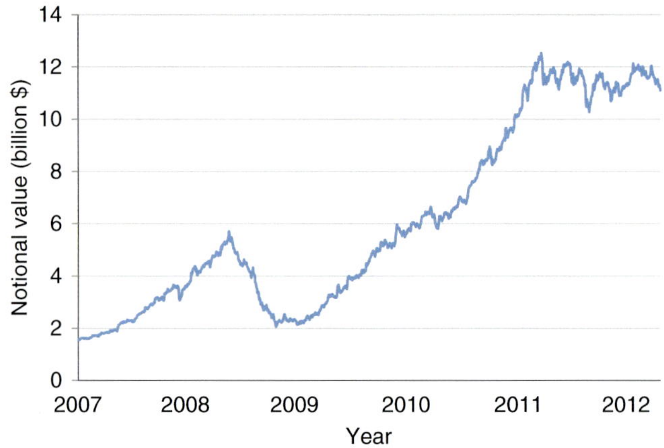

Fig. 8.1. Daily total Fund notional value for 22 US commodity futures markets, October 1 2007–May 30 2012.

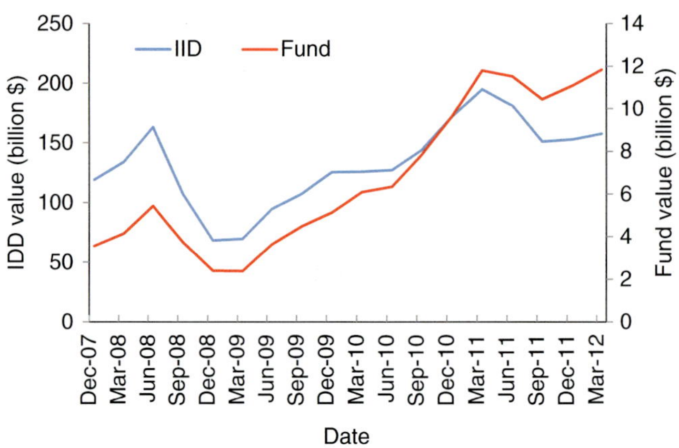

Fig. 8.2. Comparison of quarterly Fund and total *Index Investment Data* (IID) notional value for 21 US commodity futures markets, December 2007–March 2012.

The percentage of index positions held in each market is shown for April 30 2012 in Table 8.1. With regard to allocation across markets, the Fund's holdings are not markedly different from that found in the IID. The top eight holdings for both the Fund and the industry (IID) are the same and account for over 70% of both the Fund and the IID investment allocation. The Fund's agricultural holdings are also compared to those reported in the SCOT report for the nearest date available, May 1 2012 (Table 8.2). On this date, the top five agricultural markets are the same and make up over 70% of the holdings in the 12 SCOT agricultural markets. Notably, across markets in Tables 8.1 and 8.2, the Fund's holdings are fairly consistent at just under 10% of the industry holdings in each market. The exceptions are feeder cattle and soybean meal, which are not included in some of the more popular commodity indices (e.g. S&P GSCI). Overall, the Fund's allocation across markets and aggregate investment flow through time do not differ substantially from that observed for the industry as a whole. In that regard, the Fund's position data should be representative of industry participation and activity in the agricultural futures markets.

The position characteristics for calendar year 2011 are presented in Table 8.3 along with a comparison to descriptive statistics for each futures market. The first column shows the average position size in contracts. The largest number

Table 8.1. Notional values and market allocations of Fund and *Index Investment Data* (IID), April 30 2012.[a]

Market[b]	Fund ($ billions)	Allocation (%)	IID ($ billions)	Allocation (%)	Fund percentage of IID (%)
Crude oil	2.239	19	38.400	25	5.8
Gold	1.508	13	17.400	11	8.7
Soybeans	0.961	8	13.800	9	7.0
Copper	0.823	7	6.300	4	13.1
Natural gas	0.804	7	9.700	6	8.3
Corn	0.764	6	11.900	8	6.4
Heating oil	0.594	5	7.800	5	7.6
RBOB gasoline	0.567	5	9.500	6	6.0
Live cattle	0.544	5	5.600	4	9.7
Sugar	0.497	4	6.500	4	7.6
Silver	0.472	4	5.100	3	9.3
CBOT wheat	0.431	4	7.000	4	6.2
Cotton	0.308	3	3.400	2	9.1
Soybean oil	0.299	3	3.700	2	8.1
Lean hogs	0.278	2	3.100	2	9.0
Coffee	0.266	2	2.900	2	9.2
Soybean meal	0.184	2	0.800	1	23.0
KCBOT wheat	0.097	1	1.300	1	7.4
Feeder cattle	0.091	1	0.600	0	15.2
Platinum	0.076	1	0.600	0	12.6
Cocoa	0.063	1	0.800	1	7.9
Total	11.865	100	156.200	100	7.6

[a]Positions for the industry are based on *Index Investments Data* (IID) reports from the US Commodity Futures Trading Commission (CFTC). Allocations and totals only reflect the US markets displayed in the table.
[b]CBOT, Chicago Board of Trade; KCBOT, Kansas City Board of Trade; RBOB, reformulated blendstock for oxygenate blending.

Table 8.2. Notional values and market allocations of Fund and *Supplemental Commitments of Traders* (SCOT), May 1 2012.[a]

Market	Fund ($ millions)	Allocation (%)	SCOT ($ millions)	Allocation (%)	Fund percentage of SCOT (%)
Soybeans	1,030	22	11,582	20	8.9
Corn	882	19	13,560	23	6.5
Live cattle	535	11	5,344	9	10.0
Sugar	493	10	5,943	10	8.3
CBOT wheat	407	9	6,817	12	6.0
Cotton	306	6	3,255	6	9.4
Soybean oil	293	6	3,245	6	9.0
Lean hogs	280	6	3,126	5	9.0
Coffee	270	6	2,633	5	10.2
KCBT wheat	94	2	1,143	2	8.2
Feeder cattle	89	2	550	1	16.2
Cocoa	67	1	837	1	8.0
Total	4,746	100	58,034	100	8.2

[a]Table 8.2 does not include Fund data for soybean meal because it is not included in the SCOT report.

of contracts was held in the corn futures market at 22,495 contracts, which represents 1.6% of the open interest in that market. The Fund's position averages 2.7% of the open interest across the 14 markets in Table 8.3. The largest relative position is held in Minneapolis Grain Exchange (MGEX) wheat at 5.6% of the open interest (on average) in 2011. The Fund is not an active trader from an outright buying or selling perspective. In 2011, the number of days with a position change in CBOT wheat was 129 out of 252 possible trading days, or 51%. So while trading may occur in bursts, it averages about every other day in CBOT wheat. Position changes are most frequent in corn (161 days) and least common in MGEX wheat (69 days). The relative amount of trading across markets is roughly proportional to the position size in each market which reflects a more frequent need to rebalance larger positions.

The third column in Table 8.3 presents the absolute average daily change in the position for each market in 2011. Since changes in net positions are relatively infrequent, the average is only calculated for days on which there is a change in the position. The change in the aggregate position in each market represents the minimum amount of trading that must have occurred on that day in that market. So if the net position in a market increases from 1000 contracts to 1200 contracts, then a minimum of 200 contracts were bought that day (although not necessarily at the same time). Conversations with Fund management suggest that most trading occurs at the end of the day near the closing price. For 2011, the average change in positions across all markets is 58 contracts. The largest is corn at 244 contracts followed by soybeans (133) and sugar (80). Relative to the average daily volume in each market, the average change in the Fund's position is very small – averaging just 0.1% of daily trading volume across markets. The maximum or largest position change for each market is also shown in Table 8.3 (column 4). Clearly, the Fund does have days with heavy trading. This is especially noteworthy in cotton where the Fund traded 1209 contracts in a single day which represents 5.8% of the average daily trading volume for cotton (20,984). Likewise, in MGEX wheat, the Fund's maximum position change (243) represents 3.5% of average daily volume (6874). Still, even the maximum position changes are generally a small portion of trading volume and average just 1.3% across all markets. It is important to note that the trading does appear to be clustered. The pattern of trading through the month is illustrated in Fig. 8.3, where a majority of the activity occurs at the end of the month when new inflows are most likely to occur.

Further evidence on the characteristics of the Fund's positions is provided in Table 8.4, which shows the Fund's position size along with the average index trader as reported in the SCOT report. The average SCOT index trader's position is calculated as the net long index position in each market divided by the number of reporting long index traders in that market. As a comparison, the average corn position size in 2011 was 22,493 contracts for the Fund, which was larger than that held by the average SCOT index trader (13,484). Indeed, the Fund's average position size is larger than the average index trader in every market except CBOT wheat, where the Fund increases overall wheat exposure by using the CBOT, KCBOT, and MGEX wheat contracts. Interestingly, in only two markets – cotton and sugar – does the Fund's week-to-week position change exceed that of the average SCOT index trader. Nonetheless, the Fund is a relatively large market participant compared to other index traders.

The position data confirm the idea that index traders in general, and the Fund in particular, are not overly active on a daily basis in terms of outright buying and selling. That is, the change in the aggregate position is relatively small while the overall position is relatively large. Not surprisingly then, the Fund must make fairly large, yet somewhat infrequent, transactions to roll or switch long positions from the nearby expiring futures contract to the next.

The frequency and size of the Fund's roll transactions are shown in Table 8.5. On average, the Fund is active in rolling futures positions 70 days/year, or 28% of the trading days. Rolling occurs most frequently in corn (on 96 days) and is least frequent in cocoa (on 37 days). The average roll transaction shifts 5.4% of the position across futures contracts. Given the overall position size that must be rolled, the size of roll transactions is relatively large with the largest relative roll size in cocoa (301 contracts, 11.5% of position), soybean meal (479 contracts, 11.1% of position) and MGEX wheat (319 contracts, 10.5% of position). The maximum roll transactions are indeed quite large with both MGEX wheat and cocoa having maximums that are over 60% of the average position size. Across markets, the average maximum roll is 32.6% of the position size which suggests that nearly a third of the position is sometimes rolled in a single day.

Table 8.3. Fund position levels and characteristics, calendar year 2011.[a]

Market	Fund (no. of contracts)[b]				Futures market (no. of contracts)		Fund's percentage of market		
	Average position size	Days position change (no. of days)	Average position change	Maximum position change	Average open interest	Average daily volume	Position size (%)	Average change (%)	Maximum change (%)
Corn	22,495	161	244	905	1,385,738	313,511	1.6	0.1	0.3
Soybeans	10,851	150	133	625	578,431	179,142	1.9	0.1	0.3
CBOT wheat	5,428	129	36	253	449,685	96,362	1.2	0.0	0.3
KCBOT wheat	4,892	98	22	245	174,531	21,807	2.8	0.1	1.1
MGEX wheat	3,039	69	25	243	54,307	6,874	5.6	0.4	3.5
Soybean meal	6,508	117	40	209	204,162	67,144	3.2	0.1	0.3
Soybean oil	4,302	79	46	590	322,936	95,859	1.3	0.0	0.6
Cotton	4,314	123	68	1,209	161,690	20,984	2.7	0.3	5.8
Live cattle	11,684	154	34	383	337,577	53,701	3.5	0.1	0.7
Feeder cattle	1,441	70	8	62	39,196	6,271	3.7	0.1	1.0
Lean hogs	7,991	153	36	401	240,558	39,563	3.3	0.1	1.0
Coffee	2,844	111	16	120	116,374	20,534	2.4	0.1	0.6
Sugar	15,781	156	80	1,110	581,838	98,033	2.7	0.1	1.1
Cocoa	2,619	73	19	206	165,822	19,635	1.6	0.1	1.0
Average	7,442	117	58	469	343,775	74,244	2.7	0.1	1.3

[a]Minneapolis Grain Exchange (MGEX) wheat is included in the table because complete data were available for 2011. Average position changes and roll size reflect the absolute value of the change to reflect the size (not direction) of the position change. Position changes are only calculated for the days in which there is a non-zero change.
[b]Except for column 'Days position change'.

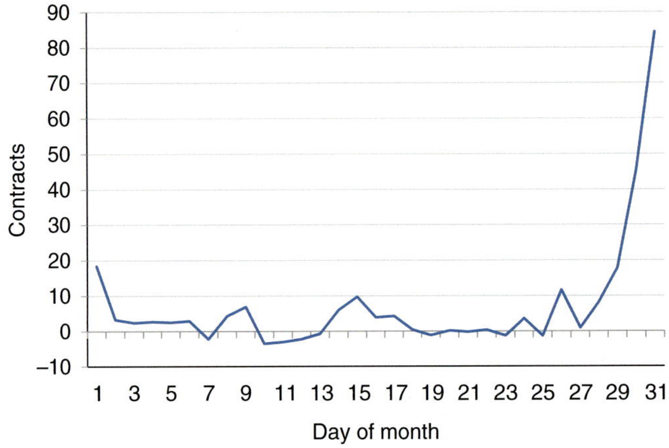

Fig. 8.3. Average Fund net position change by calendar day within the month, 13 US agricultural futures markets, October 1 2007–May 30 2012.

Table 8.4. The Fund's position size (in terms of number of contracts) and position change compared with those of the average index trader in the *Supplemental Commitments of Traders* (SCOT) report, for calendar year 2011.[a]

	Fund (no. of contracts)		Average SCOT trader	
Market	Position size	Position change	Position size	Position change
Corn	22,493	185	13,484	339
Soybeans	10,853	93	6,254	157
CBOT wheat	5,426	67	7,150	187
KCBOT wheat	4,890	40	1842	66
Soybean oil	6,503	56	3,813	122
Cotton	4,348	106	2,003	80
Live cattle	11,685	83	5,517	107
Feeder cattle	1,441	10	481	21
Lean hogs	7,985	78	4,079	93
Coffee	2,846	30	1,582	42
Sugar	15,757	215	7,432	206
Cocoa	2,619	26	1,701	80
Average	8,071	82	4,612	125

[a]The data in Table 8.4 are calculated only on weekly (Tuesday) dates that match up with the release of the SCOT report; therefore, they will differ slightly from those compiled from daily data in Table 8.3. Soybean meal and MGEX wheat are not included in this table because they are not part of the SCOT report. SCOT average position data are calculated as the net long position divided by the number of reporting long index traders.

As shown in Fig. 8.4, the Fund rolls positions primarily between the 8th and 15th day of the calendar month which is consistent with the rest of the industry (Aulerich *et al.*, 2013). Notably, the size of the roll transaction in each market is larger than changes in the outright position, which makes investigating the impact of rolling on market spreads particularly interesting with this data set.

8.4 Empirical Methods and Results

To match up with the Fund's (long) positions, daily log-relative returns, R_t, are calculated using nearby futures contracts adjusted appropriately for contract roll-overs as follows:

$$R_t^1 = \ln\left(\frac{p_t^1}{p_{t-1}^1}\right) * 100 \qquad (8.1)$$

Table 8.5. Fund position levels and roll transaction characteristics (in terms of number of contracts), calendar year 2011.[a]

Market	Futures position	Number of days with roll transaction	Average roll size	Average as a percentage of position (%)	Maximum roll size	Maximum as a percentage of position (%)
Corn	22,495	96	452	2.0	3,324	14.8
Soybeans	10,851	83	352	3.2	2,926	27.0
CBOT wheat	5,428	70	101	1.9	1,050	19.3
KCBOT wheat	4,892	61	330	6.7	1,594	32.6
MGEX wheat	3,039	40	319	10.5	1,875	61.7
Soybean oil	6,508	58	280	4.3	2,552	39.2
Soybean meal	4,302	40	479	11.1	1,756	40.8
Cotton	4,314	59	163	3.8	1,050	24.3
Live cattle	11,684	92	346	3.0	1,160	9.9
Feeder cattle	1,441	95	107	7.4	626	43.4
Lean hogs	7,991	85	185	2.3	1,482	18.5
Coffee	2,844	72	129	4.5	1,089	38.3
Sugar	15,781	96	475	3.0	2,011	12.7
Cocoa	2,619	37	301	11.5	1,919	73.3
Average	7,442	70	287	5.4	1,744	32.6

[a]MGEX wheat is included in the table because complete data were available for 2011. Average position changes and roll size reflect the absolute value of the change to reflect the size (not direction) of the position change. Roll size is only calculated for the days in which there is a non-zero change or roll.

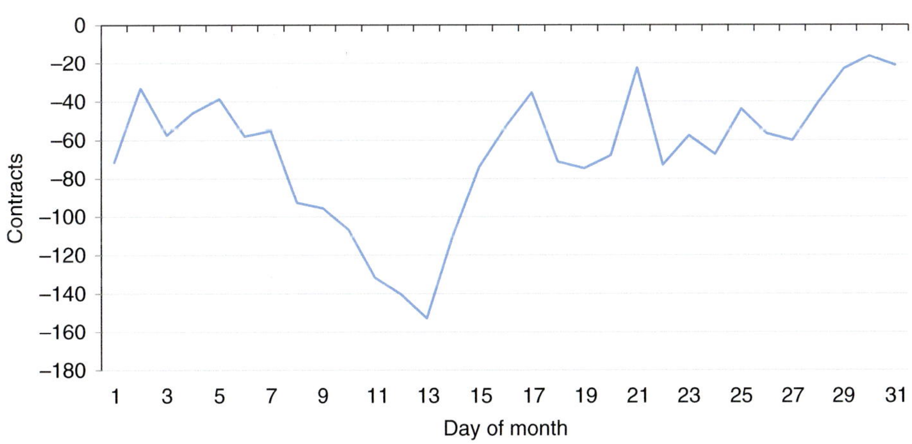

Fig. 8.4. Average Fund roll position change by calendar day within the month, 13 US agricultural futures markets, October 1 2007–May 30 2012.

where p_t^1 is the futures price of the first listed or nearest-to-expiration contract on each trading day. In order to avoid distortions associated with contract roll-overs, p_t^1 in the log-relative price return always reflects the same nearest-to-expiration contract as p_{t-1}^1. Roll-over dates for the 13 markets are set on the 15th of the month prior to the delivery month. The rolling patterns observed in the position data did not appear to be synchronized perfectly across all markets. However, the majority of contract switching generally occurs in the days around the 15th of the month prior to delivery as shown in Fig. 8.4.

Returns for the second or next active futures contract are also calculated as follows:

$$R_t^2 = \ln\left(\frac{p_t^2}{p_{t-1}^2}\right) * 100 \quad (8.2)$$

where $p_t^2 s$ is the settlement price of the second or next actively listed futures contract on each trading day. For example, if the nearby return in corn is calculated using the March futures, then the second listed contract return is calculated using the May contract. The same conventions as described above for switching contracts are used to create a series of daily returns (R_t^2) for the second listed contract for each market.

While some prior researchers have used various absolute measures of the spread between the first and second contract (e.g. differences, price ratios, or percentage of full carry), these measures can be problematic as it is difficult to account for differing storage costs and term structures across markets. Therefore, tests for the impacts of rolling activity focus on a more direct measure of changes in the spread, which is the simple difference in the return between the first and the second listed contracts:

$$\Delta Spread_t = R_t^1 - R_t^2 \quad (8.3)$$

Note that $\Delta Spread_t = R_t^1 - R_t^2 = \ln\left(\frac{p_t^1}{p_{t-1}^1}\right) - \ln\left(\frac{p_t^2}{p_{t-1}^2}\right) = \ln\left(\frac{p_t^1}{p_t^2}\right) - \ln\left(\frac{p_{t-1}^1}{p_{t-1}^2}\right)$ is equivalent to the log-relative change in the price ratio or slope of the futures curve on day t (correctly adjusted for contract switching). As such, it accurately captures the relative movement in the nearby and second-listed futures contracts.[6] The $\Delta Spread$ variable is stationary for all 13 markets. Additionally, the average correlation coefficient across markets for R_t^1, R_t^2 is 0.98; so, using the $\Delta Spread$ variable substantially reduces the variance of the dependent variable in regression models and increases statistical power in time-series tests.

8.4.1 Correlation coefficients

As a first step in testing for possible market impacts, Pearson correlation coefficients are calculated between the change in positions and market returns on the same day (contemporaneous correlation). The lagged correlation is calculated between the change in the net position and the market return the following day. The Pearson correlation coefficients are calculated over 1176 data points in each market. So the correlations have a standard error of $\sqrt{\frac{1}{n-3}}$ or 0.0292 and any correlation that is greater than 0.057 (1.96 × 0.0292) in absolute value is statistically different from zero (5% level, two-tailed t-test).

As shown in Table 8.6, the average contemporaneous correlation across markets is positive.[7] But the relationship is statistically significant in only two of the 13 markets (feeder cattle and lean hogs). So while these two correlations are positive – suggesting that increases in long positions (buying) coincide with upward price movement – they should be interpreted cautiously for a number of reasons. First, the correlations are of a very small magnitude (0.06) and of questionable economic importance. Second, and most important, there are no statistically significant correlations between changes in positions and market returns on the following day. That is, there is no evidence that the buying in these markets precedes a price increase as none of the 1-day lagged correlations are statistically different from zero.

The correlations between roll transactions and spread changes are also shown in Table 8.6. The correlations are calculated in a contemporaneous fashion, as well as with a 1-day lag between the roll position and subsequent spread change. Notably, the average correlation across all markets for both the contemporaneous and lagged correlations is negative. For the contemporaneous correlations, eight correlation coefficients are statistically different from zero at the 5% level and seven of them are negative. Two of the markets – cotton and coffee – continue to show a negative and statistically significant correlation the following day.

The correlation coefficients in Table 8.6 suggest a possible linkage between roll transactions and market spreads. However, the direction of the impact is negative which is the opposite implied by a price pressure effect. Indeed, the negative correlations suggest that when the fund is rolling long positions (selling nearby, buying deferred) the nearby contract's price is actually increasing relative to the deferred contract's price.

8.4.2 Difference-in-means test

Another approach to understanding potential market impacts is to test if returns are different following days where there is active buying (increase in long position) or selling (decrease in

Table 8.6. Correlation coefficients between daily returns and Fund position changes, October 1 2007–May 30 2012.[a]

	Returns		Spreads	
Market	Contemporaneous	1-Day lag	Contemporaneous	1-Day lag
Corn	0.0051	0.0273	−0.1323*	−0.0134
Soybeans	0.0002	0.0124	−0.0475	−0.0314
CBOT wheat	−0.0550	0.0283	−0.0600*	0.0077
KCBOT wheat	0.0484	0.0146	−0.0309	−0.0241
Soybean meal	−0.0074	−0.0317	−0.0166	−0.0090
Soybean oil	0.0273	−0.0069	−0.0133	−0.0133
Cotton	0.0376	0.0454	−0.1512*	−0.0971*
Live cattle	0.0322	0.0451	−0.0507	−0.0562
Feeder cattle	0.0636*	0.0545	0.0759*	0.0328
Lean hogs	0.0667*	−0.0306	−0.0682*	−0.0360
Coffee	−0.0042	0.0440	−0.1040*	−0.0794*
Sugar	−0.0218	0.0385	−0.1934*	0.0011
Cocoa	−0.0046	−0.0223	−0.1146*	−0.0396
Average	0.0145	0.0168	−0.0698	−0.0275

[a]Correlations are computed using all 1176 observations and have a standard error of 0.0292. A star highlights correlations that are statistically different from zero at the 5% level. The 'Returns' column reflects the correlation between changes in the Fund position and daily market returns. The 'Spreads' columns reflect the correlation between futures spreads and the Fund's roll activity.

long position) as compared to days following no activity (no change in the position). The difference-in-mean returns conditioned on market activity can easily be tested within the frame work proposed by Cumby and Modest (1987) because the disaggregated position data allow us to precisely divide the sample into trading and non-trading days for a single large entity. The Cumby-Modest regression is:

$$R_t^1 = \alpha + \beta_1 Buying_{t-1} + \beta_2 Selling_{t-1} + \varepsilon_t \quad (8.4a)$$

where $Buying_{t-1} = 1$ if there is an increase in the long Fund position on day $t - 1$ (0 otherwise) and $Selling_{t-1} = 1$ if there is a decrease in the long Fund position on day $t - 1$ (0 otherwise). In Eqn 8.4a the following day's nearby futures return conditioned on buying ($\alpha + \beta_1$) is statistically different from the unconditional market return (α) if the null hypothesis $\beta_1 = 0$ is rejected using a *t*-test. Likewise, the following day's nearby futures return conditioned on selling ($\alpha + \beta_2$) is statistically different from the unconditional market return (α) if the null hypothesis $\beta_2 = 0$ is rejected. Equation 8.4a is estimated for each market individually using ordinary least squares (OLS) and the Newey-West covariance estimator, which is consistent under general forms of heteroskedastic and serial correlation. It is also estimated across all markets in a pooled estimation using White's estimator to correct for cross market heteroskedasticity.

The behavior of spreads following days with active rolling are investigated in a parallel fashion:

$$\Delta Spread_t = \alpha + \beta_1 Buying_{t-1} + \beta_2 Spelling_{t-1} + \varepsilon_t \quad (8.4b)$$

where $Buying_{t-1} = 1$ if positive roll transactions are transacted (buy nearby/sell deferred) on day *t* (0 otherwise) and $Selling_{t-1} = 1$ if negative roll transactions (sell nearby/buy deferred) are transacted on day *t–1* (0 otherwise). In Eqn 8.4b, the change in the spread (*ΔSpread*) conditioned on buying ($\alpha + \beta_1$) is statistically different from the unconditional change in the spread (α) if the null hypothesis that $\beta_1 = 0$ is rejected using a *t*-test. Likewise, the change in the spread conditioned on selling ($\alpha + \beta_2$) is statistically different from the unconditional market return (α) if the null hypothesis that $\beta_2 = 0$ is rejected.

The estimation results for Eqn 8.4a are presented in Table 8.7 for each market individually as well as a model pooled across all 13 markets. None of the estimated slope coefficients

Table 8.7. Cumby-Modest difference-in-mean return tests for daily Fund positions, October 1 2007–May 30 2012.[a]

	Coefficient estimates						Observations		
Market	No change	*p*-value	Buying	*p*-value	Selling	*p*-value	'No change'	'Buys'	'Sells'
Corn	0.077	0.435	–0.068	0.339	–0.084	0.318	531	321	323
Soybeans	0.081	0.335	–0.019	0.425	–0.006	0.544	570	348	257
CBOT wheat	–0.051	0.601	–0.044	0.969	–0.227	0.339	587	293	295
KCBOT wheat	–0.074	0.351	0.157	0.205	–0.234	0.350	737	212	226
Soybean meal	0.122	0.083	–0.092	0.148	–0.031	0.388	794	217	164
Soybean oil	–0.064	0.369	0.096	0.169	0.092	0.219	742	241	192
Cotton	–0.020	0.831	0.014	0.796	0.028	0.788	619	332	224
Live cattle	–0.041	0.298	0.044	0.150	–0.087	0.488	512	309	354
Feeder cattle	–0.015	0.680	0.034	0.482	–0.048	0.670	783	199	193
Lean hogs	–0.075	0.255	–0.097	0.841	–0.029	0.669	525	349	301
Coffee	0.018	0.812	–0.067	0.568	–0.028	0.725	656	283	236
Sugar	–0.001	0.994	0.049	0.765	0.131	0.492	533	399	243
Cocoa	0.028	0.689	–0.049	0.639	–0.111	0.487	831	193	151
Pooled	0.000	0.997	–0.006	0.928	–0.054	0.433	8,420	3,696	3,159

[a]Buying (selling) is defined as days when there is an increase (decrease) in the long Fund position. The 'No change' column reports the α intercept estimate; the 'Buying' column reports the β_1 slope estimate; and the 'Selling' column reports the β_2 slope estimate. The pooled model is estimated across all markets.

is statistically different from zero at the 5% level. On days following buying and selling, market returns are no different than on days following no change in the position. The result holds true across all individual markets as well as the pooled estimates across markets. The results provide no evidence that market returns are different when conditioned on Fund buying or selling.

Table 8.8 shows the results of estimating Eqn 8.4b when the change in the spread is conditioned on spread buying or selling the previous day. For individual markets, two of the conditional means are statistically different from the unconditional mean at the 5% level (KCBOT wheat and cotton) and another two at the 10% level (live cattle and CBOT wheat). Notably, each of these rejections of the null is associated with a negative impact where positive (negative) roll activity is followed by a negative (positive) change in the calendar spreads. The pooled estimation of Eqn 8.4b shows that on the day after traditional negative roll transactions (sell nearby futures, buy deferred futures), there is a statistically significant and systematic tendency for the nearby contract to gain on the deferred contract by 0.026% (p-value = 0.001). Due to the large number of observations in the pooled model, this provides convincing statistical evidence that futures spreads tend to narrow following the Fund's rolling of long positions. This result differs markedly from the accusation that index funds may cause spreads to widen (nearby futures lose relative to deferred futures). Instead, it suggests the opposite – the market moves towards the Fund's spread trades.

It is also worth noting that the magnitude of spread changes is generally small from a return perspective. Consider the results for CBOT wheat spreads in Table 8.8, where the mean change in the nearby-deferred spread on days with spread selling, or negative roll transactions, was 0.02%. Since the average wheat market price was $6.70/bushel over the sample, two basis points represent less than a quarter of a cent. The impact on a single day – while statistically significant – may be smaller than the bid-ask spread for most markets. An exception is cotton, where the impact of 0.115% on a $0.8688/lb item is $0.001 or $50 per 50,000 lb contract. It is also important to remember that the coefficients in Table 8.8 reflect 1-day impacts. The total economic importance would be greater over a 5-day rolling window as depicted in Fig. 8.4.

Table 8.8. Cumby-Modest difference-in-mean spread tests for daily Fund positions, October 1 2007–May 30 2012.[a]

	Coefficient estimates						Observations		
Market	No roll	p-value	Buying	p-value	Selling	p-value	'No roll'	'Buys'	'Sells'
Corn	–0.013	0.112	0.021	0.532	–0.021	0.765	870	18	287
Soybeans	0.004	0.573	–0.038	0.210	0.002	0.838	889	14	272
CBOT wheat	–0.026	0.007	–0.038	0.753	0.020	0.063	917	18	240
KCBOT wheat	–0.017	0.024	–0.050	0.012	0.000	0.274	969	2	204
Soybean meal	0.012	0.162	0.051	0.572	0.004	0.699	1,054	3	118
Soybean oil	–0.001	0.466	0.017	0.439	–0.001	0.938	999	12	164
Cotton	–0.022	0.233	–0.054	0.721	0.115	0.002	928	13	234
Live cattle	–0.019	0.038	0.056	0.545	0.017	0.077	829	6	340
Feeder cattle	–0.009	0.196	0.199	0.462	–0.012	0.843	884	2	289
Lean hogs	–0.008	0.698	–0.316	0.175	0.033	0.356	834	11	330
Coffee	–0.006	0.005	–0.006	0.975	0.004	0.229	969	10	196
Sugar	–0.021	0.260	0.046	0.618	0.001	0.553	880	11	284
Cocoa	–0.011	0.051	–0.025	0.632	0.058	0.147	1,022	12	141
Pooled	–0.010	0.004	–0.019	0.551	0.026	0.001	12,044	132	3,099

[a]Buying (selling) is defined as days when the Fund is buying (selling) the nearby contract and selling (buying) the deferred contract. The 'No roll' column reports the α intercept estimate; the 'Buying' column reports the β_1 slope estimate; the 'Selling' column reports the β_2 slope estimate. The pooled model is estimated across all markets.

8.4.3 Granger causality tests

We next consider the causal relationship between market returns and the change in Fund positions. Several previous studies use similar procedures (e.g. Stoll and Whaley, 2010). Under the null hypothesis that changes in positions do not Granger cause market returns, the following linear regression is estimated for each market:

$$R_t^1 = \alpha + \sum_{i=1}^{m} \gamma_i R_{t-i}^1 + \sum_{j=1}^{n} \beta_j \Delta Position_{t-j} + \varepsilon_t \quad (8.5a)$$

where return variables are defined as before and $\Delta Position_{t-j}$ is the change in the Fund long position (all contracts) for the market on day $t - j$. The lag structure (m, n) for each market is determined by a search procedure over $m = 30$ and $n = 30$ using OLS and choosing the model that minimizes the Schwartz criterion to avoid over-parameterization. If the OLS residuals demonstrate serial correlation (Breusch-Godfrey Lagrange multiplier test), additional lags of the dependent variable are added until the null of no serial correlation cannot be rejected.

Traditional bivariate causality in a single market, k, is tested under the null hypothesis in Eqn 8.5a that changes in positions cannot be used to predict (do not lead) market returns: $H_0 : \beta_j = 0$ for all j. A rejection of this null hypothesis, using an F-test of the stated restriction provides direct evidence that position changes are indeed useful for forecasting returns in that market. For each market, $\sum_{j=1}^{n} \beta_j$ is calculated as an indicator of the direction of market impact.

Following the lead of Capelle-Blancard and Coulibaly (2011) and Sanders and Irwin (2011), Eqn 8.5a is also pooled and modeled as a system of seemingly unrelated regressions (SUR). Since the error term, ϵ_t, in Eqn 8.5a is correlated across markets, the power of causality tests can be increased by employing a generalized least squares (GLS) estimator within Zellner's seemingly unrelated regression (SUR) framework (see Harvey, 1991, p. 66). Under the SUR approach, GLS parameter estimates are the best linear unbiased coefficient estimates. The efficiency gains over OLS estimates increase with the correlation between the residuals across markets and with the number of equations. To specifically test for a systematic impact across markets, common coefficients are specified for β_j on the lagged position variables across markets.[8]

Using the same specification procedure, an analogous model is estimated and used to test for causality running from the Fund's roll activity to changes in futures market spreads:

$$\Delta Spread_t = \alpha_k + \sum_{i=1}^{m} \gamma_i \Delta Spread_{t-i} + \sum_{j=1}^{n} \beta_j Roll_{t-j} + \varepsilon_t \quad (8.5b)$$

where $Roll_{t-j}$ represents the rolling of positions across calendar months. The standard roll of selling nearby and buying deferred contracts is recorded as a negative quantity (e.g. -500 contracts). The null of no causality is tested again as $H_0 : \beta_j = 0$ for all j.

Table 8.9 shows the test results for the individual markets examining both returns (Eqn 8.5a) and spreads (Eqn 8.5b). Focusing on the estimations for returns (Eqn 8.5a), the (m, n) lag structure that minimized the Schwartz information criterion (SIC) was somewhat trivial with only the soybean meal model containing more than one lag of the position variable. The p-values for the null hypothesis of no causality $H_0 : \beta_j = 0$ for all j in Eqn 8.5a indicate that the null hypothesis is not rejected for any market. The magnitude of the estimated slope coefficients is noticeably small in absolute terms and not statistically different from zero. It is then not surprising that the common coefficients on lagged position changes in the pooled model are not statistically different from zero across this group of markets. Again, there is no evidence of a systematic impact from the Fund's change in position to market returns.

Table 8.9 also shows the results for estimating Eqn 8.5b and testing for causality between the Fund's rolling activity and changes in calendar spreads. There is again some evidence of a causality running from roll transactions to spreads. In particular, the null hypothesis is rejected at the 5% level for two markets (KCBOT wheat and coffee) and at the 10% level for two markets (cotton and live cattle). Importantly, the direction of the impact is negative in these four markets as well as two markets of marginal significance (CBOT wheat and cocoa). Given the number of marginally significant rejections in individual markets and the consistency of the signs, it is not surprising that the pooled model rejects the null of no causality with a p-value of 0.0215. The common coefficient suggests a very

Table 8.9. Granger causality tests that Fund position changes lead market returns, October 1 2007–May 30 2012.

	Returns			Spreads		
Market	*m*, *n*	*p*–value $\beta_j = 0, \forall j$	Estimate $\Sigma \beta_j$	*m*, *n*	*p*–value $\beta_j = 0, \forall j$	Estimate $\Sigma \beta_j$
Corn	1,1	0.4043	0.0425	1,1	0.7969	0.0003
Soybeans	1,1	0.7831	0.0216	2,1	0.7778	–0.0004
CBOT wheat	1,1	0.5883	0.0461	1,1	0.1255	–0.0041
KCBOT wheat	1,1	0.5713	0.0461	12,1	0.0008	–0.0063
Soybean meal	1,2	0.3895	–0.0658	2,1	0.6171	0.0011
Soybean oil	1,1	0.7289	–0.0357	6,1	0.2343	–0.0007
Cotton	1,1	0.1789	0.1000	1,1	0.0686	–0.0390
Live cattle	1,1	0.1591	0.0347	1,1	0.0587	–0.0052
Feeder cattle	1,1	0.2467	0.1877	1,1	0.2305	0.0083
Lean hogs	1,1	0.2337	–0.0494	4,1	0.1757	–0.0126
Coffee	1,1	0.1334	0.2868	1,1	0.0179	–0.0047
Sugar	2,1	0.0980	0.0908	2,1	0.7503	–0.0020
Cocoa	1,1	0.4482	–0.1577	1,2	0.1026	–0.0144
Pooled	2,2	0.6478	0.0054	12,2	0.0215	–0.0019

Notes: The estimated coefficients are scaled by 100. The pooled model is estimated across the 13 markets as an SUR system restricting the β_j, slope parameters to be equal across markets. These restrictions are imposed on the system and the common coefficients are estimated as a single pooled parameter across all 13 markets.

small negative impact, where a 100-contract traditional roll increases the nearby-deferred calendar spread by 0.0019%. While statistically significant, by itself, this would seem to be of doubtful economic importance. Still, the Fund's rolling activity occurs over roughly five days (Fig. 8.4) and the maximum roll within a market is often in excess of 1000 contracts/day. So the cumulative impact may indeed be of economic significance.

Figure 8.5 graphically depicts the average daily roll and average daily change in the futures spread across calendar days for cotton. The negative relationship documented in Tables 8.6, 8.8, and 8.9 for cotton is apparent in the figure. Notably, the direction of this leading relationship is the opposite of what would be found if the Fund's trading were 'pushing around' the spreads. Indeed, the overall spread analysis and results indicate that the Fund is rolling positions when the market gives them the opportunity or is moving 'toward their trade.' The results are consistent with the empirical findings of Aulerich *et al.* (2013) but the reverse of those reported by Brunetti and Reiffen (2014). They are also consistent with a 'sunshine trading' effect (Admati and Pfleiderer, 1991), where large traders essentially pre-announce their intentions and thereby attract potential counterparties, increase liquidity, and lower trading costs (Bessembinder *et al.*, 2012).

8.4.4 Long-horizon tests

The previous three tests are designed to detect the relationship, if any, between daily position changes and returns. Those tests are important because of the uniqueness of this daily data set. However, these tests may have low power to reject the null hypothesis for two reasons. First, the dependent variable in the regressions – the change in commodity futures prices – is well known to be highly volatile. Second, index positions may flow in 'waves' that build slowly, pushing prices higher, and then fade slowly (e.g. Summers, 1986). In this scenario, horizons longer than a day may be necessary to capture the predictive component of index fund positions. Consequently, we implement the long-horizon regression model as described by Valkanov (2003):

$$\sum_{i=0}^{m-1} R^1_{t+i} = \alpha + \beta \sum_{i=0}^{k-1} \Delta Position_{t+i-1} + \varepsilon_{t+1} \qquad (8.6)$$

where all variables are defined as before. In essence, Eqn 8.6 is an OLS regression of a

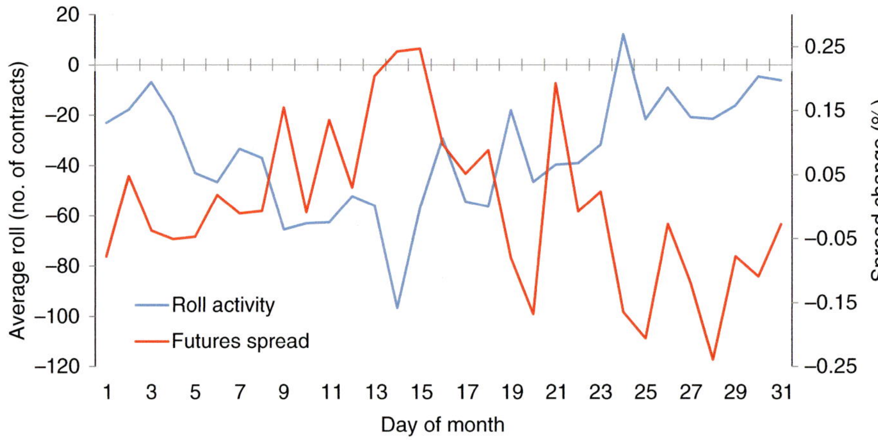

Fig. 8.5. Average Fund roll position by calendar day within the month and the average change in the nearby calendar spread, for cotton, October 1 2007–May 30 2012.

k-period moving sum of the dependent variable at time *t* against an *m*-period moving sum of the independent variable in the previous period, time $t - 1$. If the estimated β is positive (negative), then it indicates a fads-style model where prices tend to increase (decrease) slowly over a relatively long time period after widespread index fund buying (selling). The fads stylization captured in Eqn 8.6 – with a positive β – is consistent with the Masters Hypothesis that position changes can drive bubble-like price behavior in commodity futures prices.

The long-horizon regression (Eqn 8.6) is estimated using the underlying dependent variable of returns and the independent variable of change in positions.[9] Both of these variables are stationary, so the sums are also stationary. Valkanov (2003) demonstrates that the OLS slope estimator in this specification is consistent and converges at a high rate of *T*. The specification in Eqn 8.6 clearly creates an overlapping horizon problem for inference. Valkanov shows that Newey-West *t*-statistics do not converge to well-defined distributions and suggests using the rescaled *t*-statistic, $t/\sqrt{T}$, along with simulated critical values for inference. Valkanov also demonstrates that the rescaled *t*-statistic generally is the most powerful among several alternative long-horizon test statistics.

Recently, Singleton (2014) and Hamilton and Wu (2015) use a variation of this model where $m = 1$ and $k = 13$ weeks. Singleton refers to the 13-week position change as the 'flow' of investment funds and finds considerable predictability between the imputed measure of investment flows and crude oil futures returns. Hamilton and Wu (2015) find that the impact is isolated to crude oil, appears to be sensitive to the lag-length chosen, and does not hold up out-of-sample. As a first step in testing for long-run relationships, we mirror the weekly data frequency used by Hamilton and Wu by setting where $m = 5$ and $k = 65$ days, which essentially equals the 1-week returns and 13-week investment flow identified by Singleton. Additional long-horizon regressions (Eqn 8.6) are estimated over alternative horizons of $m = k = 20, 60, 120,$ and 240 trading days, which approximately correspond to monthly, quarterly, semi-annual, and yearly time horizons. The estimated OLS β coefficients for Eqn 8.6 are shown in Table 8.10 along with the rescaled *t*-statistic. Critical values for the rescaled *t*-statistic (-0.563, 0.595) are taken from Valkanov's (2003) Table 4 for Case 2 and $c = -5.0, \delta = 0.00, T = 750$, and tail values representing the 10% significance level. These represent a conservative case that, if anything, favors a rejection of the null hypothesis that the slope equals zero.

The Singleton case ($m = 5, k = 65$) is shown in the first set of columns. The estimated slope coefficients for this case are noticeably small and the rescaled *t*-statistics do not exceed Valkanov's critical values for any of the markets. Likewise, in all

Table 8.10. Long-horizon regression tests that daily Fund position changes impact returns, October 1 2007–May 30 2012.[a]

	$m = 5, k = 65$		$m = k = 20$		$m = k = 60$		$m = k = 120$		$m = k = 240$	
Market	Slope estimate	Rescaled *t*-statistic	Slope estimate	Rescaled *t*-statistic	Slope estimate	Rescaled *t*-statistic	Slope estimate	Rescaled *t*-statistic	Slope estimate	Rescaled *t*-statistic
Corn	0.0004	0.03	0.0021	0.04	0.0045	0.05	0.0087	0.08	0.0120	0.11
Soybeans	0.0005	0.02	0.0006	0.01	0.0045	0.03	0.0049	0.02	0.0047	0.03
CBOT wheat	0.0001	0.02	0.0002	0.01	0.0013	0.04	0.0006	0.01	–0.0020	–0.03
KCBOT wheat	0.0009	0.05	0.0017	0.03	0.0031	0.04	0.0043	0.04	0.0114	0.07
Soybean meal	–0.0004	–0.01	–0.0047	–0.06	–0.0034	–0.02	–0.0001	0.00	0.0056	0.04
Soybean oil	–0.0002	–0.01	0.0014	0.02	0.0003	0.00	–0.0062	–0.03	–0.0126	–0.09
Cotton	–0.0013	–0.04	0.0013	0.01	–0.0047	–0.02	–0.0072	–0.02	–0.0058	–0.01
Live cattle	0.0001	0.03	0.0011	0.04	0.0015	0.04	0.0018	0.04	0.0018	0.04
Feeder cattle	0.0000	0.00	–0.0001	0.00	0.0023	0.02	0.0042	0.02	0.0061	0.03
Lean hogs	0.0001	0.01	0.0013	0.04	–0.0004	0.00	0.0003	0.00	0.0051	0.04
Coffee	–0.0014	–0.03	–0.0057	–0.05	–0.0112	–0.04	–0.0092	–0.02	–0.0046	–0.01
Sugar	0.0000	0.00	–0.0003	–0.01	0.0029	0.03	0.0057	0.05	0.0059	0.12
Cocoa	0.0001	0.00	–0.0045	–0.02	–0.0072	–0.04	0.0005	0.00	–0.0034	–0.03

[a]This table reports the results of estimating long-horizon regressions between average daily returns and average daily positions held by the Fund. Critical values for the rescaled *t*-statistic (-0.563, 0.595) are taken from Valkanov's (2003) Table 4 for Case 2 and $c = -5.0$, $\delta = 0.00$, $T = 750$, and tail values representing the 10% significance level.

of the other cases ($m = k = 20, 60, 120, 240$), not a single estimated slope coefficient is statistically different from zero. Moreover, among the 65 slope coefficients estimated, 25 (39%) are negative and 40 (61%) are positive, so there is little consistency with regard to the direction of any impact. These results are similar to those reported by Hamilton and Wu (2015) for agricultural markets and provide no evidence that the Fund's market positions impact commodity futures returns over longer horizons. Importantly, the results also indicate that the failure to detect causal linkages between Fund position changes and price changes in earlier tests likely was not due to problems with the statistical power of the tests.

The crux of the argument against index funds – as stated in the Masters Hypothesis – is that a wave of index fund buying from 2007 to 2008 pushed commodity prices well above their true fundamental value. It is notable that long-horizon regressions – which are designed to capture just such a 'fads'-style model – fail to find any such empirical link in the data.

8.5 Summary and Conclusions

After the experience of recent spikes in commodity prices, policy makers are considering additional speculative position limits and other restrictions on futures market participation. Empirical studies examining the linkages between futures market activity and price fluctuations are an important input to the regulatory process (e.g. Stoll and Whaley, 2010; Hamilton and Wu, 2015). This study brings new data to the debate regarding the price impact of long-only index investment in commodity futures markets. Here, high-frequency daily position data for 13 agricultural futures and swaps markets are available from a representative large commodity index fund ('the Fund') from October 1 2007 through May 30 2012. The empirical results provide a fresh perspective on potential market linkages that may not be captured with the more aggregate data sets available from the US CFTC. The Fund's holdings and trading activity suggest that it is representative of long-only index funds in general. Thus, analysis of possible linkages between the Fund's positions and market behavior is likely to capture those of the overall industry.

A battery of statistical tests found no causal relationship between the Fund's outright buying and selling and market returns. Simple correlation tests and Granger causality tests uniformly fail to reject the null hypothesis that changes in positions do not lead market returns in any individual market or across the system of markets. Difference-in-means tests show no statistical difference in market returns on the days after the Fund trades compared to days following no trading. Long-horizon regressions find no evidence that changes in Fund positions exert longer-term pressure on returns in any of the 13 markets. There were no tell-tale signs of any causal linkages between fund position changes and price changes.

Statistically significant findings are documented between the Fund's rolling of long positions across calendar months and changes in futures price spreads. That is, there was consistent evidence of a negative relationship between roll transactions and the change in the nearby-deferred futures spread. In particular, the nearby futures spread narrowed (nearby futures return was greater than the deferred futures return) on days following roll transactions (selling nearby, buying deferred). The result shows up consistently across different statistical tests including Pearson correlation coefficients, difference-in-means tests, and Granger causality tests. Importantly, the directional result is consistently negative across all of the tests. The negative relationship is inconsistent with a price pressure hypothesis but is consistent with a 'sunshine trading' effect where liquidity is actually increased by index fund rolling activity.

In sum, the results of this study add to the growing body of literature showing that buying pressure from index funds was not one of the main drivers of the spikes in food commodity prices in recent years. The results presented here are especially compelling because they are based on daily position data that do not suffer from several of the criticisms that have been leveled against the more commonly used weekly aggregate position data from the CFTC. In particular, the data allow for detailed tests over daily horizons with 13 different agricultural markets and includes both futures and swaps positions.

The empirical evidence presented here and found in prior studies should be relevant inputs into the CFTC's rule-making process.

The 1936 Commodity Exchange Act (CEA) sets what appears to be a high bar for justifying position limits. First, it must be demonstrated under the CEA that position limits are 'necessary' to prevent excessive speculation from 'causing sudden or unreasonable fluctuations or unwarranted changes in the price of such commodity' (US House of Representatives, 2019). Second, position limits must be 'appropriate' in their balance between the prevention of excessive speculation and market manipulation with ensuring sufficient market liquidity and price discovery (Young *et al.*, 2011). The necessary empirical evidence linking 'excessive speculation' to 'unwarranted price changes' is scant. In a comprehensive review, Will *et al.* (2012, p. 18) concluded that 'most empirical studies are unable to confirm that financial speculation has led to an increase in the price levels of agricultural commodities.' From a more legal perspective, Notini (2013, p. 3) argues that:

> The CFTC ignored modern commenter-submitted studies that refute a connection between speculation and price swings. If the CFTC had considered these studies, it might have concluded that the connection between excessive speculation and drastic price movement is an unjustified theory ...
> (Notini, 2013, p. 3)

The research presented here bolsters that conclusion. While no single empirical study is entirely conclusive, the body of empirical evidence is quite convincing. At this point in the policy debate, there is very little evidence that long-only index funds or other speculators are 'causing ... unwarranted fluctuations in price.' Thus, a clear verdict can be reached – new limits on speculation in agricultural futures markets are unnecessary.

Acknowledgments

The authors thank Hongxia Jiao for her assistance in collecting the futures price data. Comments on an earlier version of this article from participants at the invited paper session on Trading Dynamics and Price Behavior in Agricultural Futures Markets at the 2013 annual meeting of the Agricultural and Applied Economics Association are gratefully acknowledged. We are especially indebted to our discussant, Jim Moser, for his many thoughtful comments and suggestions.

Notes

[1] Original citation: Sanders, D.R. and Irwin, S.H. (2016) The 'necessity' of new position limits in agricultural futures markets: the verdict from daily firm-level position data. *Applied Economic Perspectives and Policy* 38, 292–317. Reprinted by permission of John Wiley and Sons and the Agricultural and Applied Economics Association.

[2] See this blog post from Craig Pirrong for an excellent summary of the final position limit rules: https://streetwiseprofessor.com/position-limits-what-a-long-strange-trip-its-been/ (accessed January 23 2022).

[3] Extensive reviews of this rapidly expanding literature are provided by Irwin and Sanders (2011), Will *et al.* (2012), Fattouh *et al.* (2013), Irwin (2013), and Cheng and Xiong (2014).

[4] The proprietary data for this research were provided under the stipulation that it be kept confidential. For simplification, the index fund will simply be referred to as the 'Fund' and detailed position data or statistics that might compromise confidentiality are not presented.

[5] Data are also available for Minneapolis Grain Exchange (MGEX) wheat. However, the series doesn't start until October 30 2009 and is excluded from the time-series models. However, it is included in the tables displaying summary statistics for calendar year 2011.

[6] Please note that we follow market convention and define the underlying calendar spread as the price of the nearby contract minus the price of the deferred contract, or in proportional terms as the nearby contract divided by the price of the deferred contract. This means the order flow pressure of index roll trades should *decrease* the spread. For example, consider a futures curve for a commodity that is in contango, so the deferred contract futures price is higher than the nearby price. The computed spread according to market convention is negative, and roll trades are expected to make the spread become even more negative; and hence, the spread will decrease. If one defines the spread in the opposite manner as the deferred contract futures price minus the nearby contract price, the expected impact of roll trades on the spread are simply reversed.

[7] In Table 8.6 and following tables, the markets are ordered in a fashion that groups like markets (grains, livestock, and softs).
[8] Sanders and Irwin (2011) suggest a more rigorous systems approach to estimating Eqns 8.5a and 8.5b. However, the independent variables only enter the specification at very short lags $(m = 1, n = 1)$ in this case making the systems estimation somewhat trivial.
[9] The long-horizon regressions specified in Eqn 8.6 are not estimated for spreads as most price spreads are bound by storage-related arbitrage conditions. Therefore, it doesn't make much intuitive or economic sense to test for longer-term 'bubbles' in spread relationships.

References

Admati, A.R. and Pfleiderer, P. (1991) Sunshine trading and financial market equilibrium. *Review of Financial Studies* 4, 443–481.

Aulerich, N.M., Irwin, S.H. and Garcia, P. (2013) Bubbles, food prices, and speculation: evidence from the CFTC's daily large trader data files. Working Paper No. 19065, National Bureau of Economics Research, Cambridge, Massachusetts. Available at: https://www.nber.org/system/files/working_papers/w19065/w19065.pdf (accessed Janaury 23 2022).

Bessembinder, H., Carrion, A., Tuttle, L. and Venkataraman, K. (2012) Predatory or sunshine trading? Evidence from crude oil ETF rolls. Working Paper, University of Utah, Salt Lake City, Utah.

Brunetti, C. and Reiffen, D. (2014) Commodity index trading and hedging costs. *Journal of Financial Markets* 21, 153–180.

Buyuksahin, B. and Harris, J.H. (2011) Do speculators drive crude oil futures prices? *Energy Journal* 32, 167–202.

Capelle-Blancard, G. and Coulibaly, D. (2011) Index trading and agricultural commodity prices: a panel Granger causality analysis. *Economie Internationale* 126, 51–72.

Cheng, I.-H. and Xiong, W. (2014) The financialization of commodity markets. *Annual Review of Financial Economics* 6, 419–441.

CFTC (Commodity Futures Trading Commission) (2015) CFTC provides notice of the reopened comment period for its rulemaking proposals on position limits for physical commodity derivatives and aggregation. Press Release PR7126-15. Available at: http://www.cftc.gov/PressRoom/PressReleases/pr7126-15 (accessed January 23 2022).

Cumby, R.E. and Modest, D.M. (1987) Testing for market timing ability: a framework for forecast evaluation. *Journal of Financial Economics* 19, 169–189.

Farrelly, P. (Director) (1994) *Dumb and Dumber*. Film produced by Charles B. Wessler and Brad Krevoy.

Fattouh, B., Lutz, K. and Mahadeva, L. (2013) The role of speculation in oil markets: what have we learned so far? *Energy Journal* 34, 7–33.

Gilbert, C.L. (2010) Speculative influences on commodity futures prices 2006–2008. United Nations Conference on Trade and Development Discussion Paper No.197. Available at: https://unctad.org/system/files/official-document/osgdp20101_en.pdf (accessed January 23 2022).

Gilbert, C.L. and Pfuderer, S. (2014) The role of index trading in price formation in the grains and oilseeds markets. *Journal of Agricultural Economics* 65, 303–322.

Hamilton, J.D. and Wu, J.C. (2015) Effects of index-fund investing on commodity futures prices. *International Economic Review* 56, 187–205.

Harvey, A.C. (1991) *The Econometric Analysis of Time Series*, 2nd edn. MIT Press, Cambridge, Massachusetts.

Hieronymus, T.A. (1977) *Economics of Futures Trading for Commercial and Personal Profit*, 2nd edn. Commodity Research Bureau, New York.

Irwin, S.H. (2013) Commodity index investment and food prices: does the 'Masters Hypothesis' explain recent price spikes? *Agricultural Economics* 44, 29–41.

Irwin, S.H. and Sanders, D.R. (2011) Index funds, financialization, and commodity futures markets. *Applied Economic Perspectives and Policy* 33, 1–31.

Irwin, S.H. and Sanders, D.R. (2012) Testing the Masters Hypothesis in commodity futures markets. *Energy Economics* 34, 256–269.

Jacks, D.S. (2007) Populists versus theorists: futures markets and the volatility of prices. *Explorations in Economic History* 44, 342–362.

Michaels, D. (2013) Traders face curbs on speculation with CFTC vote on new limits. Bloomberg UK. Available at: http://www.bloomberg.com/news/2013-11-05/traders-face-curbs-on-speculation-with-cftc-vote-on-new-limits.html (accessed January 23 2022).

Miedema, D. (2013) Exclusive: U.S. watchdog readies tighter new commodity limits rule. *Chicago Tribune*, October 11. Available at: http://articles.chicagotribune.com/2013-10-11/business/sns-rt-us-commodities-speculation-exclusive-20131011_1_rule-cftc-dodd-frank (accessed 23 January 2022).

Notini, A. (2013) Paper tiger: the validity of CFTC position-limit rulemaking under Dodd-Frank. *Suffolk University Law Review* 46, Rev. 185.

O'Malia, S.D. (2012) Dissenting Statement Regarding the Commission's Appeal of the District Court Decision on the Commission's Position Limit Rule. November 15. Available at: http://www.cftc.gov/PressRoom/SpeechesTestimony/omaliadissentstatement111512 (accessed January 23 2022).

Petzel, T.E. (1981) A new look at some old evidence: the wheat market scandal of 1925. *Food Research Institute Studies* 18, 117–128.

Sanders, D.R. and Irwin, S.H. (2011) The impact of index funds in commodity futures markets: a systems approach. *Journal of Alternative Investments* 14, 40–49.

Sanders, D.R. and Irwin, S.H. (2013) Measuring index investment in commodity futures markets. *Energy Journal* 34, 105–127.

Sanders, D.R. and Irwin, S.H. (2014) Energy futures prices and commodity index investment: new evidence from firm-level position data. *Energy Economics* 46, S57–S68.

Singleton, K.J. (2014) Investor flows and the 2008 boom/bust in oil prices. *Management Science* 60, 300–318.

Stoll, H.R. and Whaley, R.E. (2010) Commodity index investing and commodity futures prices. *Journal of Applied Finance* 20, 7–46.

Summers, L.H. (1986) Does the stock market rationally reflect fundamental values. *Journal of Finance* 41, 591–601.

US Government (2010) H.R. 4173. Dodd-Frank Wall Street Reform and Consumer Protection Act. US Government Printing Office, January 5. Available at: https://www.cftc.gov/sites/default/files/idc/groups/public/@swaps/documents/file/hr4173_enrolledbill.pdf (accessed November 28 2022).

US House of Representatives (2019) 7 U.S.C., United States Code, Title 7 – Agriculture; Chapter 1 – Commodity Exchanges; Section 6a – Excessive speculation. Office of Law Revision Counsel (OLRC), US House of Representatives. Available at: https://www.govinfo.gov/content/pkg/USCODE-2019-title7/html/USCODE-2019-title7-chap1-sec6a.htm (accessed November 28 2022).

Valkanov, R. (2003) Long-horizon regressions: theoretical results and applications. *Journal of Financial Economics* 68, 201–232.

Will, M.G., Prehn, S., Pies, I. and Glauben, T. (2012) Is financial speculation with agricultural commodities harmful or helpful? A literature review of current empirical research. Discussion Paper No. 2012-27. Martin Luther University, Halle, Wittenberg, Germany.

Working, H. (1963) Futures markets under renewed attack. *Food Research Institute Studies* 4, 13–24.

Young, M.D., Gagoomal, P.J. and Kearns, T.S. (2011) CFTC Adopts Final Position Limit Rules for Futures, Options and Swaps on 28 Physical Commodities. Skadden, Arps, Slate, Meagher, & Flom LLP. Available at: https://www.jdsupra.com/legalnews/cftc-adopts-final-position-limit-rules-27819/ (accessed January 23 2022).

Young, M.D., Donley, M.A. and Gagoomal, P.J. (2012) Court Vacates CFTC Position Limit Rules. Skadden, Arps, Slate, Meagher, and Flom LLP. Available at: https://www.jdsupra.com/legalnews/court-vacates-cftc-position-limit-rules-91559/ (accessed January 23 2022).

9

Bubbles, Froth, and Facts: Another Look at the Masters Hypothesis in Commodity Futures Markets[1]

New Author Foreword

The genesis for this article was a conference during the summer of 2014 in Germany. One of us (Scott) had been invited to give a keynote presentation at an international conference on commodity market volatility held at a university in Halle, Germany, which was one of the larger cities in what was once a part of communist East Germany. The other keynote presentation was given by Chris Gilbert, a British economist, who was previously on record arguing that index investment was a significant driver of recent spikes in commodity futures prices. The conference was held in a beautifully refurbished building at the university. It was a little startling to find out that the building served as the Stasi secret police headquarters for the city during the years of communist rule. Apparently, when the building was remodeled, bullet holes were found in the basement walls.

Gilbert's presentation and the questioning during the conference made two things abundantly clear. First, there was still a great deal of confusion about the nature of the controversy surrounding speculation and the commodity price spikes of the previous decade. What exactly was the Masters Hypothesis? Was it about speculators in general, index traders, or both? Was the problem increased price volatility in general or specific bubble episodes? It was obvious that some additional clarity was needed. Second, several new criticisms of our methods had been leveled by Gilbert and his co-authors, and they also needed to be addressed.

We began the article by laying out three basic tenets that were central to the Masters Hypothesis: (i) long-only index investors were directly responsible for driving commodity futures prices higher during price spikes, most notably during 2007–2008; (ii) the deviations from fundamental value during the spikes were economically very large; and (iii) the impact was pervasive across commodity futures markets. While there were numerous aspects of commodity price behavior that were of interest, these tenets were the foundation for the global controversy surrounding index funds.

The remainder of the article addressed the criticisms that had appeared in the literature about the statistical methods we had used in previous studies. Our use of Granger causality tests had come under some valid criticism regarding the potentially low power of such time-series tests in volatile commodity futures markets. Other criticisms revolved around market efficiency issues (i.e. a positive finding would essentially violate the notion of market efficiency as public information on index positions should be impounded instantaneously into futures prices, implying the absence of Granger causality). The other major critique of our bivariate time-series tests was the lack of conditioning variables such as stock market returns or measures of macroeconomic risk.

We addressed each of the major criticisms using updated data and new empirical approaches. As a result, this article contained the most comprehensive set of time-series and cross-sectional tests of any of our published articles. The evidence was once again overwhelming that the Masters Hypothesis came up short in its most basic market predictions.

DOI:10.1079/9781800622104.0009

Ironically, this article was probably the easiest out of all of our published articles to get through the review process. We targeted the *Journal of Agricultural Economics*, which was published by the British Agricultural Economics Society, because this is where several of Chris Gilbert's articles on the subject had been published. We expected that Gilbert would be one of our reviewers, and it would have been surprising if he was not. After a couple of rounds of reviews and revisions, the article published in 2017.

It is an interesting side note that the first half of the title for the article, 'Bubbles, Froth, and Facts,' actually was recycled from one of our earlier papers. We had been forced to delete this part of the title by a journal editor who thought it was too colorful. The reputation of academics as being boring is sometimes well earned. Regardless, we really liked the phrasing, as it rolled off the tongue nicely and sounded a little cheeky. We were delighted when the chance arose to use it again.

Finally, if someone came to us and said they wanted to read only one of our articles, this would be the one we would recommend. The controversy surrounding speculation in commodity futures markets is defined in sharp relief and an exhaustive set of time-series and cross-section statistical tests is performed. In essence, we looked back after almost a decade of research, applied what we had learned, and did our best to detect relationships between index funds and commodity futures price movements. It's almost surprising that after this much effort we uncovered so little evidence of any substantive relationships.

Abstract

The Masters Hypothesis suggests that long-only index funds were the main cause of a massive increase in commodity prices in 2007–2008 and 2011–2012. Central to the Masters Hypothesis are three basic tenets: (i) long-only commodity index funds were directly responsible for driving futures prices higher; (ii) the deviations from fundamental value were economically very large; and (iii) the impact was pervasive across commodity futures markets. There has been a great deal of empirical research on the Masters Hypothesis and commodity market bubbles. However, surprisingly few studies have found evidence that directly supports the main tenets of the Masters Hypothesis. Some have attributed the lack of supporting evidence to the low power of time-series tests, market efficiency issues, and a lack of conditioning variables within models. In this paper, we address each of these issues using updated data and new empirical approaches. Still, price behavior consistent with the Masters Hypothesis is surprisingly difficult to find in the data. This is an important finding given the ongoing policy debate and regulations proposed or being implemented to limit speculative positions in these markets.

Key words: bubble, commodity, futures market, index funds, prices

JEL categories: D84, G12, G13, G14, Q13, Q41

> Institutional Investors are one of, if not the primary, factors affecting commodities prices today ... the current Wheat futures stockpile of Index Speculators is enough to supply every American citizen with all the bread, pasta and baked goods they can eat for the next two years!
>
> (Masters, 2008, Congressional testimony)

9.1 Introduction

This colorful testimony made by Michael Masters (2008) highlights the central claim of the so-called Masters Hypothesis (Irwin and Sanders, 2012): a wave of index fund investment caused irrational and gross mispricing across a wide range of commodities. Masters' claims were quickly adopted by industries – such as the Air Transport Association's anti-speculation campaign dubbed 'Stop Oil Speculation Now' (Blair, 2008) – and it led to other popular movements aimed at curbing speculation (e.g. Better Markets).[2] Some market analysts also supported Masters and claimed that 'excessive speculation' pushed commodity prices during the 2007–2008 period to extremes that were 125% or more over their fundamental value (e.g. Gheit, 2008). More specifically, the Masters Hypothesis and related concerns about speculative impacts rests on the following tenets: (i) long-only commodity index funds were directly responsible for driving commodity futures prices higher; (ii) the deviations from fundamental value were economically very large; and (iii) the impact was pervasive across commodity futures markets. These claims have driven related policy debates about re-regulating speculation in commodity futures markets that are still being actively considered (Moyer, 2016).

Financial theory provides a number of possible mechanisms by which index funds could in fact impact prices. First, the commodity futures markets may not have sufficient liquidity to handle the large position sizes and transactions demanded by index funds. In this case, prices may be temporarily pushed away from fundamental value (see Grossman and Miller, 1988). Second, index funds may create 'noise trader risk' in the marketplace and thus make arbitraging against their positions difficult and price impacts possible (see De Long *et al.*, 1990). Finally, other traders may confuse index fund buying with valuable private information and thus revise their own demands upward, which, in turn, pushes prices higher (Grossman, 1986; Sockin and Xiong, 2015). Given the often-cited reasons for institutional commodity index investments – portfolio diversification, inflation hedges, and capturing long-run risk premiums – none of these theoretical explanations for price impacts seem particularly plausible, especially with regard to generating massive price bubbles. Regardless, the price impact (if any) ultimately becomes an empirical question – one that has seen a surge in academic interest.

The literature on long-only index funds and the possible impact on commodity prices has increased dramatically in recent years. For example, five review articles have appeared recently in the literature (Irwin and Sanders, 2011; Fattouh *et al.*, 2013; Irwin, 2013; Cheng and Xiong, 2014; Will *et al.*, 2016). While some researchers have documented index-related price effects related to market integration (e.g. Tang and Xiong, 2012), changes in risk premiums (e.g. Hamilton and Wu, 2014), and calendar spreads (e.g. Aulerich *et al.*, 2013), the number of studies directly tying commodity index funds to futures price movements is actually surprisingly limited.[3]

Singleton (2014) finds that money flows associated with index investors – along with other macroeconomic variables – help to predict changes in crude oil futures prices. While Singleton's (2014) study is focused only on crude oil over a very specific period, it has been widely cited and serves as an important academic reference in the speculation debate. Mayer (2012) finds a leading causal relationship between index positions and futures prices for soybeans and soybean oil (but not in wheat or corn). Gilbert and Pfuderer (2014) use an instrumental variable approach and find a contemporaneous relationship between index positions and futures prices for soybeans, soybean oil, and Kansas City Board of Trade (KCBT) wheat (but not in corn or Chicago Board of Trade (CBOT) wheat). Tadesse *et al.* (2014) use a seemingly unrelated regression (SUR) estimation approach to show that 'excessive speculation' is a contemporaneous driver of cash price spikes along with fundamental shocks to supply, gross domestic product (GDP), and crude oil. Using a dynamic modeling procedure, Lagi *et al.* (2015) tie the rise of the FAO Food Price Index in 2007–2008 and 2010–2011 to ethanol conversion and investor speculation. In the model, there is no direct measure of speculative positions. Instead, speculative buying is represented by past price changes under the assumption that speculators (index investors) are trend followers.

The authors of some of these papers use their findings to directly call for additional regulatory oversight (e.g. Mayer, 2012; Tadesse *et al.*, 2014; Lagi *et al.*, 2015). Yet they all fall short in some crucial aspect under the basic tenets of the Masters Hypothesis. For example, only Gilbert and Pfuderer (2014) and Mayer (2012) use actual commodity index positions as reported by the US Commodity Futures Trading Commission (CFTC) in their empirical tests. The others use either the broadly defined 'non-commercial' category of futures positions (Tadesse *et al.*, 2014) or attempt to infer index positions in one market, such as crude oil, based on positions held in another market, such as feeder cattle (Singleton, 2014). Prior research has shown that data not specifically classified as commodity index positions or data generated by mapping algorithms can lead to erroneous conclusions (Irwin and Sanders, 2012; Sanders and Irwin, 2013).

None of the above studies provides support across a broad range of commodity futures markets. Gilbert and Pfuderer (2014) as well as Mayer (2012) only find evidence in a few select futures markets, while Tadesse *et al.* (2014) focus on cash markets – not futures. Only Lagi *et al.* (2015) attempt to quantify the economic size of the market impact; but their model focuses on cash prices, and they do not use an actual measure of trader positions. At best, these studies provide circumstantial evidence supporting the Masters Hypothesis. At worst, they are erroneous due to poorly constructed data

(Singleton, 2014) or not using actual position data (Lagi *et al.*, 2015). Either way, none of these studies provides the actual direct evidence needed to link commodity index fund positions to economically large changes in commodity futures prices.

On the other side of the debate, the list of studies that fail to find significant evidence to support the Masters Hypothesis is long (see Fattouh *et al.*, 2013). Many of the studies utilize Granger-style causality tests between weekly futures market returns and commodity index positions reported by the CFTC (e.g. Stoll and Whaley, 2010; Sanders and Irwin, 2011a, b; Rouwenhorst and Tang, 2012; Hamilton and Wu, 2015; Lehecka, 2015). Others test for impacts over long-horizons (Irwin and Sanders, 2012), and some are able to test for impacts using daily CFTC position data (Buyuksahin and Harris, 2011; Aulerich *et al.*, 2013; Brunetti *et al.*, 2016). Still others use detailed position data for a private index fund (Sanders and Irwin, 2014, 2015). Regardless of the horizon, data, or statistical method, these studies uniformly fail to find direct and broad-ranging links between commodity index positions and futures market returns.

Granger-style tests have come under some valid criticism regarding the potentially low power of time-series tests in volatile commodity futures markets. Other criticisms revolve around market efficiency issues (i.e. a positive finding would essentially violate the notion of market efficiency). Specifically, since public information on index positions is impounded instantaneously into futures prices in an efficient market, the absence of Granger causality is merely a consequence of market efficiency (Gilbert and Pfuderer, 2014; Grosche, 2014). The other major critique of bivariate time-series tests is the lack of conditioning variables (Grosche, 2014) such as stock market returns (Singleton, 2014) or measures of macroeconomic risk (Cheng *et al.*, 2015).

In this paper, we address each of the major criticisms of Granger-style tests of the relationship between index fund positions and commodity futures prices using updated data and new empirical approaches. We first examine time-series correlations using weekly index trader positions reported in the CFTC *Supplemental Commitments of Traders* (SCOT) report and nearby futures returns. Special attention is given to the time periods prior to the public release of the SCOT data; thereby, helping to minimize the market efficiency concerns raised by other researchers. Next, cross-sectional correlations are examined using positions from the CFTC *Index Investment Data* (IID) report and nearby futures returns. The cross-sectional correlations between index positions and returns are contemporaneous, which curbs issues associated with market efficiency and eliminates the need for numerous conditioning variables. That is, we assume that the important macroeconomic variables – such as stock market returns, currency movements, and financial risk – impact the cross-section of futures markets in a similar fashion. By focusing on the cross-market correlations over a given period, common control variables are in effect held constant. Collectively, these approaches contribute to the existing literature by addressing many of the concerns expressed with past empirical tests.

9.2 Time-series Analysis

Prior research has used a number of time-series econometric techniques such as Granger causality (e.g. Stoll and Whaley, 2010) and instrumental variables (Gilbert and Pfuderer, 2014) in an attempt to identify linkages between long-only index positions and market behavior. Here we cut to the heart of the Masters Hypothesis – index fund buying causes massive over-valuation in markets. In this interpretation, the Masters Hypothesis leads to three straightforward predictions: (i) large index positions should be associated with high price levels; (ii) large and rapid increases in index positions should be accompanied by increases in prices; and (iii) the prior two findings should be consistent across markets and time. Here we examine the historical correlations that underlie most of the prior empirical studies and then examine some new evidence as it relates to these three predictions.

9.2.1 Time-series data

The CFTC reports the positions held by index traders in 12 agricultural futures markets in the SCOT report. The CFTC's ability to accurately measure index trader positions has some flaws associated with swap activity and internal netting of positions by firms. But the SCOT positions have been shown to be representative of long-only index positions in the agricultural markets where over-the-counter swap activity is minimal (Irwin and Sanders, 2012; Sanders and Irwin, 2013).

The SCOT data are released each Friday in conjunction with the traditional *Commitments of Traders* (COT) report and show the combined futures and options positions as of the previous Tuesday's market close. The weekly data are available for 12 agricultural markets. For eight of the 12 markets (feeder cattle, soybean oil, cocoa, coffee, cotton, sugar, live cattle, and lean hogs) the data are available from 2006 to 2015 for a total of 522 observations. Four of the markets (corn, soybeans, CBOT wheat, and KCBT wheat) have data available from 2004 to 2015 for a total of 626 observations.[4]

Positions are reported as the number of long and short contracts held by index traders as of Tuesday's market settlement. To accurately capture the net buying or selling by index funds, the empirical analysis will focus on net long positions:

$$\begin{aligned} \textit{Net long}_t = \textit{Long open interest}_t \\ - \textit{Short open interest}_t \end{aligned} \quad (9.1)$$

Open interest is the number of open futures contracts or positions held at a point in time. The short open interest held by index investors is generally quite small and net long positions are always positive in the SCOT data set. The change in the net long position is simply calculated as the first difference of net long as defined in Eqn 9.1, and the percentage change in the net long position is calculated as the log-relative change from week $t - 1$ to week t:

$$\textit{Percentage change in net long}_t = \ln\left(\frac{\textit{Net long}_t}{\textit{Net long}_{t-1}}\right) \times 100 \quad (9.2)$$

Finally, the relative size of the index fund position is measured as a percentage of total futures and options open interest:

$$\textit{Percentage of open interest}_t = \left(\frac{\textit{Index trader long open interest}_t}{\textit{Total market open interest}_t}\right) \times 100 \quad (9.3)$$

To correspond with the collection dates for the SCOT data, Tuesday-to-Tuesday log-relative returns are collected for nearby futures contracts. Specifically, R_t^1 is calculated using nearby futures contracts adjusting appropriately for contract roll-overs as follows:

$$R_t^1 = \ln\left(\frac{p_t^1}{p_{t-1}^1}\right) * 100 \quad (9.4)$$

where p_t^1 is the futures price of the first listed or nearest-to-expiration contract on each trading day. In order to avoid distortions associated with contract roll-overs, p_t^1 in the log-relative price return always reflects the same nearest-to-expiration contract as p_{t-1}^1. Roll-over dates for the 12 markets are set on the 15th of the month prior to the delivery month. The futures returns and various measures of index trader positions are used in the following time-series tests.

A potentially important issue when using SCOT time-series data on index trader positions has been raised by Grosche (2014) and Gilbert and Pfuderer (2014). They argue a positive (leading) impact on commodity futures prices would essentially violate the notion of market efficiency. That is, public information on index positions is impounded instantaneously into futures prices, and therefore the absence of Granger causality is simply a consequence of an efficient market. This is a reasonable theoretical argument, but it is applicable in varying degrees across the data.

Here we pay special attention to correlations prior to the public release of the SCOT data, thereby reducing the potential that traders have already reacted to the data and impounded it into current prices. Even then, it is possible that some traders try to anticipate the SCOT index position data (prior to the public release) and trade on that expectation. In that case – depending on how well individual traders can anticipate the actual data – empirical tests may fail to find price impacts even prior to the public release of the SCOT data. While this is possible, empirical tests are still much more likely to uncover price linkages during these and similar periods when the SCOT position data were not available to the public.

9.2.2 Rank order tests

The Masters Hypothesis clearly predicts that large index positions are associated with high prices that have been pushed above fundamental value. Market prices are noisy, and it is notoriously difficult to define fundamental value. Still, as posited by Masters (2008), very high prices over this sample are a direct result of index positions. If so, then there should be some connection between absolute price levels and the position size held by index funds.

To test this assertion, we use a simple Spearman rank-order correlation coefficient. Specifically, the nearby futures prices for all 12 markets for the common sample from 2006 to 2014 are ranked from high to low by week. Long index positions as a percentage of open interest on the corresponding Tuesday are also ranked from high to low. Then, the Spearman rank-order correlation coefficient is calculated between the rankings of (contemporaneous) prices and position size.

The calculated rank correlations are presented in Table 9.1. Ten of the 12 rank-order correlations are statistically different from zero. Of those, six are negative and four are positive. So, over this sample, larger index positions (as a percentage of open interest) show no consistent association with higher-(or lower)-than-average prices.

9.2.3 Time-series correlations

Pearson correlation coefficients are calculated between the change in net long positions and

Table 9.1. Rank correlation test, *Supplemental Commitments of Traders* (SCOT) index position as a percentage of open interest and nearby futures price levels, 2006–2015.[a]

Market	Spearman's rank correlation	*p*-value
Corn	–0.23	0.000
CBOT wheat	–0.06	0.183
KCBT wheat	0.19	0.000
Soybeans	–0.27	0.000
Soybean oil	0.31	0.000
Cotton	–0.51	0.000
Sugar	0.10	0.023
Cocoa	0.35	0.000
Coffee	0.06	0.170
Live cattle	–0.74	0.000
Lean hogs	–0.54	0.000
Feeder cattle	–0.78	0.000

[a]The *p*-value is from a two-tailed *t*-test under the null that the correlation is zero.

market returns for the same week (contemporaneous correlation). The correlations are calculated separately for each year as well as the entire sample to give perspective on the consistency through time. The results are presented in Table 9.2 where the shaded correlations are statistically different from zero at the 5% level using a two-tailed *t*-test.

As shown in the bottom row of Table 9.2, 11 of the 12 contemporaneous correlations calculated for the entire sample are statistically different from zero and positive. Some of the correlations are surprisingly large, such as cocoa and soybeans at 0.25, while most are less than 0.15. Regardless, positive contemporaneous correlations clearly exist between changes in index positions and futures returns. This relationship is undoubtedly the driver behind some of the empirical results that find a contemporaneous market linkage between index funds and market prices (e.g. Gilbert and Pfuderer, 2014; Cheng *et al.*, 2015).

As is well documented in the literature, these linkages essentially disappear when time lags are introduced (e.g. Stoll and Whaley, 2010). Therefore, it is not surprising in Table 9.3 that the correlation between changes in index positions in week t and returns in the following week ($t + 1$) are largely non-existent. Only one of the correlations (corn) is statistically different from zero across the entire sample and it is negative. These correlations underlie the general lack of any causal linkages documented in prior empirical studies (e.g. Stoll and Whaley, 2010) and confirmed by others (Gilbert and Pfuderer, 2014). For the years 2004–2006, when these data were essentially private, there are two statistically significant correlations, but they are negative.

It is worth noting that the contemporaneous correlations in Table 9.2 show very little consistency through time. For instance, soybean oil shows a statistically significant 0.13 correlation across the entire sample; yet the statistically positive correlations are restricted to 2 years, 2007 and 2008. A simple graphical analysis indicates that the years with statistically significant positive correlations do not conform to the notion of an index-induced price bubble. The net long index positions held in soybean oil are plotted in Fig. 9.1 along with the nearby futures prices. The shaded years, 2007–2008, represent those years with statistically significant positive

Table 9.2. Contemporaneous correlations, change in *Supplemental Commitments of Traders* (SCOT) net long index positions and nearby futures returns, 2004–2015.[a]

Year	Corn	CBOT wheat	KCBOT wheat	Soybeans	Soybean oil	Cotton	Sugar	Cocoa	Coffee	Live cattle	Lean hogs	Feeder cattle
2004	0.1556	0.0205	0.1102	0.1626								
2005	0.0220	−0.2345	0.0825	0.1425								
2006	0.1306	0.1936	0.2748	−0.2666	−0.1069	−0.2904	−0.2316	−0.3875	−0.1399	−0.0729	−0.0090	−0.1703
2007	−0.0049	0.0454	0.0324	0.2766	0.4316	0.5620	0.0217	0.0238	0.0264	0.1145	−0.1393	0.0576
2008	0.4672	0.2569	0.1447	0.5620	0.3117	0.2053	0.1608	0.5400	0.4306	0.3412	0.2119	−0.0005
2009	0.2851	0.0194	0.3526	0.3940	−0.0118	0.2400	−0.0569	0.2951	0.2330	0.3258	0.0080	−0.0170
2010	−0.2041	0.1260	0.3635	−0.0211	0.1009	0.0530	−0.1788	0.4204	0.4587	0.3156	−0.0044	0.3268
2011	0.0586	0.3303	0.1430	0.5367	0.2159	−0.0239	0.1581	0.3064	0.2117	0.3111	0.2277	0.4995
2012	−0.0508	−0.2334	0.1758	0.0870	0.0376	0.0673	−0.0522	0.2267	0.2360	0.4220	0.3234	0.2134
2013	0.4342	−0.1053	0.3400	0.0428	−0.1630	0.2332	−0.0815	0.3801	0.2745	0.0473	0.3332	0.0066
2014	0.0297	0.2214	0.0669	0.2286	0.1304	0.1108	−0.5096	0.0448	−0.2261	−0.0152	0.1052	0.0722
2015	0.3116	0.3047	0.4347	0.3776	0.0661	0.2298	0.0127	−0.0562	0.0403	0.3428	0.1666	0.1858
All years	0.1381	0.1050	0.1970	0.2489	0.1335	0.1395	−0.0275	0.2460	0.1448	0.2226	0.1155	0.1264

[a]Shaded correlations are statistically different from zero at the 5% level using a two-tailed *t*-test.

Table 9.3. Lagged correlations, change in *Supplemental Commitments of Traders* (SCOT) net long index positions and nearby futures returns the following week, 2004–2015.[a]

Year	Corn	CBOT wheat	KCBOT wheat	Soybeans	Soybean oil	Cotton	Sugar	Cocoa	Coffee	Live cattle	Lean hogs	Feeder cattle
2004	0.0292	–0.0427	–0.0618	0.1408								
2005	0.1380	–0.1062	0.2322	–0.2135								
2006	–0.1080	0.1161	–0.0039	–0.2577	–0.1311	0.0141	–0.0354	0.2409	–0.1964	–0.2848	–0.3884	–0.2420
2007	–0.3119	–0.0246	0.0671	–0.0258	–0.0349	0.0261	0.0305	0.2694	0.1350	–0.1095	–0.0725	0.0416
2008	0.0792	0.0767	0.0506	0.2372	0.2004	0.3705	0.0139	0.2274	0.1190	0.0447	–0.0588	–0.1957
2009	–0.2447	–0.2554	–0.1656	–0.0133	0.2071	0.1432	–0.1081	–0.1077	–0.1893	0.0920	0.1216	0.2283
2010	–0.1463	–0.2516	0.1377	0.1064	–0.0505	–0.1935	–0.1281	–0.1866	–0.0957	0.0169	–0.1639	–0.0720
2011	–0.2060	–0.1215	0.0802	–0.1060	–0.0936	–0.2982	–0.1227	0.0445	–0.0506	–0.0174	0.1926	–0.0475
2012	–0.2092	0.0055	0.0626	–0.2252	0.0885	–0.0784	0.2988	0.0203	0.1147	–0.1651	–0.1052	0.0024
2013	–0.0159	–0.0537	0.0717	–0.0585	0.1784	0.1019	–0.0969	–0.1126	–0.0901	0.2028	0.5073	0.0602
2014	–0.1005	–0.0539	–0.1515	0.2132	–0.1678	–0.2472	–0.1070	–0.0946	–0.0657	0.0602	0.1956	0.1252
2015	0.0706	0.0384	–0.2306	–0.0108	–0.0149	–0.2094	0.0124	0.1078	0.1414	0.0967	–0.1751	–0.2047
All years	–0.0825	–0.0471	0.0113	0.0129	0.0574	0.0088	–0.0394	0.0533	0.0012	0.0010	–0.0010	–0.0418

[a]Shaded correlations are statistically different from zero at the 5% level using a two-tailed t-test.

contemporaneous correlations (Table 9.2). What is very clear from Fig. 9.1 is how these years in no way resemble an index-driven price bubble. Indeed, from January 2007 through June 2008, the net long position held by index funds increased by a pedestrian 4% (or 2788 contracts) while prices increased by 65%. Then, in the second half of 2008 – in the midst of the financial crises – long-only index positions fell by 27,170 contracts (61%) and prices fell by over 70%. Prices went up and then came down; index positions only came down. It is clear that the positive correlations in 2007–2008 are not consistent with the scenario painted by Masters. Indeed, a visual examination of Fig. 9.1 brings up another striking failure. In the first 3 months of 2010, the net long position in soybean oil increased a remarkable 39,626 contracts (+67%) and yet nearby futures prices actually fell by 4%. This is clearly inconsistent with the Masters Hypothesis.

Overall, the correlation data in Tables 9.2 and 9.3 suggest an almost purely contemporaneous correlation between changes in index positions and market returns. As suggested by Gilbert and Pfuderer (2014) causation may still exist between index positions and returns; however, it may be instantaneous in nature or only detectable over time horizons shorter than a week. Although the studies using shorter horizons – such as daily data – have also failed to find any causal linkages (Buyuksahin and Harris, 2011; Aulerich *et al.*, 2013: Sanders and Irwin, 2014; Sanders and Irwin, 2015; Brunetti *et al.*, 2016), the possibility cannot be ruled out and a further analysis at short horizons is in order. Since the SCOT data are only compiled for the public on a weekly Tuesday schedule, we provide additional insight by examining the correlation between the change in net long positions from Tuesday to Tuesday and the subsequent daily market returns. For example, the change in net long positions from Tuesday to Tuesday may correlate with the market return on the very next day, Wednesday, even though it is not correlated strongly with the return for the entire following week.

The basic data setup is shown in Table 9.4 along with the simple Pearson correlation coefficients. Essentially the weekly change in the index net long position is correlated with each individual day in the subsequent week. The weekly correlations shown in the bottom row of Table 9.3 are disaggregated into their daily components. In Table 9.4, there are just two statistically significant (5% level) positive correlation coefficients in the day following the SCOT compilation with one negative (cotton) and one positive (cocoa). So, not even on the very next day (Wednesday) following the compilation of positions

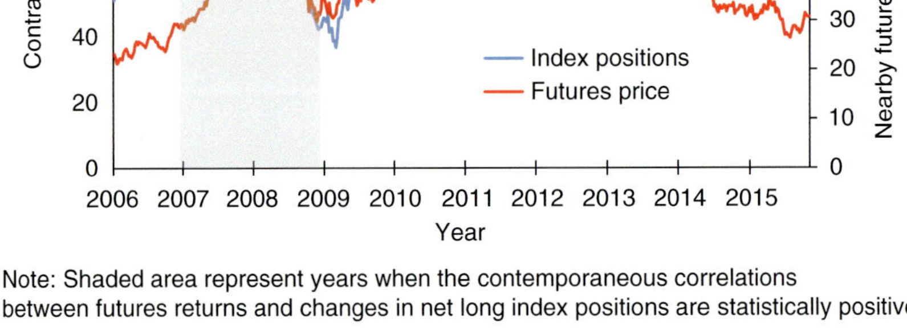

Note: Shaded area represent years when the contemporaneous correlations between futures returns and changes in net long index positions are statistically positive.

Fig. 9.1. Soybean oil, *Supplemental Commitments of Traders* (SCOT) net long index positions and nearby futures prices, 2006–2015.

(Tuesday) is there any evidence of linkages between the change in index positions and prices. On that Wednesday (t +1), the position data are not yet public, so if there were going to be an impact found using these data, it is more likely to be found prior to the public data release (Friday close). The average correlation on that Wednesday is actually negative with nine of the 12 markets having a negative correlation coefficient. There is a total of six statistically significant correlations (four positive, two negative) across the 5 days with no particularly meaningful pattern. The correlations presented in Table 9.4 provide no evidence of price linkages at the daily horizon prior to the public release of the SCOT data.

9.2.4 Extreme moves

Simple weekly correlations may not capture all the key features of the Masters Hypothesis and underlying policy concerns. That is, the real impact of unfettered index buying may be most likely to occur when there are extreme changes in positions. To isolate these impacts, we examine the percentage change in net long positions for each market over rolling 26-week intervals. The half-year period is chosen as it represents a trade-off between often-used horizons (monthly and quarterly) and longer horizon tests (annual). First, the three non-overlapping 26-week windows with the largest increase in fund positions are identified. Then, the cumulative return to nearby futures over that same 26-week window is also calculated. The data are tabulated and presented in Table 9.5.

Consider KCBT wheat in Table 9.5, which has its three largest increases in net long index positions in the 26 weeks ending December 1 2009, February 5 2013, and July 26 2005. The largest increase (December 1 2009) was characterized by a 59.7% increase in net long index positions and a 30.7% cumulative decline in nearby futures prices. The second largest increase (26 weeks ending February 5 2013) had a 53.9% increase in net long positions held by index traders while at the same time futures declined 15.2%. The third largest increase ended July 26 2005 with a 38% increase in index positions and a 3.9% increase in futures. The comparable three largest increases for each market and the corresponding futures returns are shown in the remainder of Table 9.5. Some of the markets – such as cocoa – show relatively large positive futures returns during these episodes. Others – such as CBOT wheat – show negative futures returns during these windows. Most markets show both. Of the 36 windows presented, 15 are positive and 21 are negative.

Table 9.4 Correlations, weekly change in net long *Supplemental Commitments of Traders* (SCOT) index positions and daily futures returns, 2004–2015.[a]

	SCOT private			SCOT public	
Market	Wednesday $t+1$	Thursday $t+2$	Friday $t+3$	Monday $t+4$	Tuesday $t+5$
Corn	−0.0701	−0.0329	−0.0442	−0.1142	−0.0263
CBOT wheat	0.0071	−0.0521	−0.0141	−0.0579	−0.0046
KCBT wheat	−0.0193	−0.0165	0.0495	−0.0149	0.0015
Soybeans	−0.0836	0.0401	0.1171	−0.0760	0.0653
Soybean oil	0.0212	0.0234	0.0126	−0.0714	0.1011
Cotton	−0.0952	−0.0070	0.0931	0.0543	−0.0248
Sugar	−0.0543	0.0398	0.0264	−0.0791	−0.0699
Cocoa	0.0900	0.0033	0.0024	0.0344	−0.0208
Coffee	−0.0794	0.0653	−0.0139	0.0611	−0.0118
Live cattle	−0.0049	0.0180	0.0322	−0.0364	−0.0106
Lean hogs	−0.0525	0.0006	−0.0376	0.0204	0.0407
Feeder cattle	−0.0287	0.0062	−0.0046	0.0389	−0.0768
Average	−0.0308	0.0074	0.0182	−0.0201	−0.0031

[a]Shaded coefficients are statistically different from zero at the 5% level using a two-tailed t-test.

Table 9.5. Largest percentage increases in net long *Supplemental Commitments of Traders* (SCOT) index positions and corresponding futures returns, 2004–2015.[a]

	Largest position increases				Largest position increases		
Market	Ending date	Position change (%)	Futures return (%)	Market	Ending date	Position change (%)	Futures return (%)
Corn	7/6/04	63.1	–4.3	Sugar	2/5/08	59.4	14.9
	5/17/05	61.6	–11.4		7/5/11	45.3	2.8
	5/23/06	58.0	11.6		4/24/12	34.7	–15.4
CBOT wheat	2/15/05	80.1	–12.1	Cocoa	5/12/09	111.5	18.9
	7/6/04	45.6	–20.3		11/17/09	66.6	26.8
	9/6/05	30.3	–16.0		2/26/08	53.9	35.6
KCBOT wheat	12/1/09	59.7	–30.7	Coffee	11/17/09	47.4	–1.2
	2/5/13	53.9	–15.2		12/8/09	45.3	3.2
	7/26/05	38.0	3.9		1/22/13	39.8	–21.9
Soybeans	5/10/05	97.6	20.0	Live cattle	11/17/09	37.9	–3.6
	9/1/09	42.6	30.0		5/16/06	33.4	–13.4
	4/4/06	40.9	–6.9		12/15/09	27.2	–1.7
Soybean oil	6/15/10	67.1	–8.2	Lean hogs	11/9/09	48.3	–14.9
	9/15/09	53.8	7.3		6/3/08	36.5	–9.8
	4/3/12	40.7	11.2		4/4/06	34.0	–27.2
Cotton	4/24/12	48.5	–7.6	Feeder cattle	4/6/10	56.5	16.6
	7/3/06	45.1	–19.4		4/12/11	50.8	17.3
	2/26/08	36.0	21.9		2/3/15	36.2	–5.2
				Average		50.8	–0.7
				No. positive			15
				No. negative			21

[a]Data for 2004 and 2005 are only available for corn, soybeans, CBOT wheat, and KCBOT wheat.

The average 26-week increase in the position size is 50.8% and the average futures return is –0.7%. Just isolating the largest increase for each market, the average 26-week increase in position size is 64.8% and is accompanied by a 1.0% decline in prices.

The parallel data are collected for the three largest decreases in index positions and presented in Table 9.6. With the largest declines in index positions there is somewhat more consistency in terms of the corresponding futures returns. The average 26-week decline in index positions is 49.3% and it is accompanied by a negative 7.9% average futures return. Of the 36 windows, 24 have negative returns and 12 are positive. A closer inspection reveals that ten of the 12 markets have the largest decline in positions during the financial crisis – those windows ending between October 28 2008 (KCBT wheat) and March 17 2009 (soybean oil). The 12 windows corresponding to the financial crisis show an average decline in index positions of 73.7% and a negative futures return of 37.4%. Clearly, during that crisis period there were many other things impacting commodity prices and trader activity. For the 24 non-crisis windows the average index position declined by 37.1% but futures returns were a positive 6.9%. So, ignoring the crisis-related declines, the relationship between positions and returns in Table 9.6 are negative (consistent with Table 9.5). These results simply do not fit with the gross mispricing and overvaluation predicted under the Masters Hypothesis.

9.2.5 Consistency

A Masters Hypothesis proponent could point to the cocoa market as providing some positive proof for the hypothesis. In Table 9.1, cocoa shows a positive rank-order correlation between price levels and the size of the index position. Table 9.2 presents a reasonably consistent and statistically significant positive contemporaneous correlation across time for the cocoa market. The one instance of a single-day leading correlation

Table 9.6. Largest percentage decreases in net long *Supplemental Commitments of Traders* (SCOT) index positions and corresponding futures returns, 2004–2015.[a]

	Largest position decreases				Largest position decreases		
Market	Ending date	Position change (%)	Futures return (%)	Market	Ending date	Position change (%)	Futures return (%)
Corn	11/11/08	–68.0	–54.1	Sugar	2/3/09	–76.6	–23.1
	3/17/15	–30.3	2.2		1/19/10	–34.8	41.7
	2/15/11	–27.5	44.9		1/4/11	–31.8	65.9
CBOT wheat	11/25/08	–41.7	–40.7	Cocoa	11/18/08	–177.6	–25.6
	1/29/13	–28.1	–17.3		12/22/15	–80.0	–2.0
	12/24/13	–24.6	–15.9		4/1/14	–49.4	10.1
KCBOT wheat	10/28/08	–62.9	–47.1	Coffee	3/17/15	–72.3	–33.4
	3/15/11	–44.8	–1.7		1/13/09	–62.8	–27.2
	7/30/13	–36.6	–20.9		1/18/11	–32.5	37.3
Soybeans	12/16/08	–65.5	–61.5	Live cattle	1/27/09	–56.6	–29.2
	9/18/12	–42.0	25.8		1/27/15	–29.9	–6.8
	11/17/15	–26.1	–6.2		1/17/12	–19.0	3.4
Soybean oil	3/17/09	–65.4	–37.6	Lean hogs	2/10/09	–90.8	–40.6
	10/4/11	–29.8	–21.1		1/20/15	–34.7	–35.7
	8/14/12	–23.0	–2.5		3/26/13	–33.7	–2.9
Cotton	2/15/11	–56.2	79.1	Feeder cattle	11/4/08	–62.5	–8.1
	1/13/09	–53.6	–54.3		1/10/12	–38.7	1.6
	2/25/14	–29.4	2.7		6/3/14	–35.6	16.7
				Average		–49.3	–7.9
				No. positive			12
				No. negative			24

[a]Data for 2004 and 2005 are only available for corn, soybeans, CBOT wheat, and KCBOT wheat.

is in the cocoa market (Table 9.4). Likewise, the extreme increases in the net long index positions in the cocoa market are all associated with large increases in prices (Table 9.5).

Perhaps index funds have had a unique impact on the cocoa market for some unknown reason. Or, perhaps, there was something else unique occurring in cocoa such as the attempted market manipulation reported by *The New York Times* (Werdigier and Creswell, 2010). Or maybe it is just a chance finding. Notably, index funds are relatively small players in the cocoa futures market, having the smallest relative position across the 12 markets at just 15% of open interest (compared to 39% in CBOT wheat). Regardless, while it seems most unlikely for there to be an impact in a market such as cocoa – where arbitragers can easily trade the London-based contract against the US-based contract – the data do support that possibility. If presented in isolation, the cocoa market results could be cited as evidence that index funds influence prices. But it should not be viewed in isolation. The Masters Hypothesis and related charges of index-fueled price increases are not presented as a description of the cocoa market or any individual market. Rather, it is presented as an explanation for massive price increases across a broad range of markets. Therefore, evidence to support the underlying notion of an index-driven price bubble must be consistent and pervasive across time and markets. While the results for cocoa are indeed curious and may warrant additional investigation, they by no means provide broad and compelling evidence.

9.3 Cross-sectional Analysis

Time-series tests may have relatively low statistical power with volatile market returns and it is often difficult to specify all relevant conditioning variables across varying time horizons. Precisely for these reasons, cross-sectional methods have become the gold standard for examining returns in stock markets (e.g. Fama and French, 1992). A cross-sectional analysis of commodity futures markets can control for common conditioning variables such as term spreads and other determinants of market returns (see Singleton, 2014). Basically, if there are macroeconomic variables that impact risk premiums or returns, then they should impact each futures market similarly in any given time period. Moreover, if index funds influence market returns, then those markets with the greatest relative increase in index positions over a given period should see the greatest returns. That is, the cross-section of market returns should be positively correlated with the cross-section of index positions. Previous studies that have conducted cross-sectional tests (Sanders and Irwin, 2010; Irwin and Sanders, 2012; Kim, 2015) fail to find evidence that long-only commodity funds impact futures prices.

9.3.1 Cross-sectional data

One of the shortcomings of the SCOT data set is the limited number of markets (12) covered. This limitation is overcome by the IID that is compiled by the CFTC using data from a 'special call' that is independent of the Large Trader Reporting System that underlies the SCOT data (CFTC, 2015). Essentially, the CFTC collects month-end index investments held by swap dealers, institutions, and other index traders. The index investments include swap positions and are non-netted to reflect the total investment in index funds (Sanders and Irwin, 2013). The IID data span 19 major commodity futures markets over the entire sample with three other markets added later. Importantly, IID markets include energy futures markets, which have been at the center of the policy debate about the market impact of index fund investment. The IID are considered the best measure of index positions in commodities by the CFTC, but time-series observations are limited with the data available at only quarterly intervals from December 31 2007 through June 30 2010 and monthly intervals thereafter until the report was terminated following release of the October 30 2015 report. The larger number of markets covered lends the IID well to cross-sectional analysis.

The 19 markets in all IID reports include Kansas City Board of Trade wheat (KC), Chicago Board of Trade wheat (W), corn (C), soybeans (S), soybean oil (BO), feeder cattle (FC), live cattle (LC), lean hogs (LH), cocoa (CC), sugar (SB), coffee (KC), cotton (CT), crude oil (CL), RBOB gasoline (RB), heating oil (HO), natural gas (NG), gold (GC), silver (SI), and copper (HG). Soybean meal (SM) was added in January 2011 and platinum (PL) was added in July 2011. Palladium (PA) data are available from August through December of 2014. Feeder cattle (FC) data were not reported in February or March of 2013. For each cross-section, all of the markets available for that time period are utilized. So the cross-sectional sample size varies from 19 to 22.

One potential concern with cross-sectional analysis is whether there is sufficient variation in the explanatory variable – growth rates in the IID positions across commodities – to estimate market impacts with a reasonable degree of precision. In fact, there is surprising variability across markets with regard to the change in index positions, with increases of 20% and decreases of 20% common within the same quarter. The average range between the largest increase and decrease across quarters is 46% and the average cross-sectional standard deviation is 10.5%. An example can be seen on the horizontal axis of Fig. 9.2 where the percentage change in index positions varies from –29% to +17%. This variability can be exploited to generate relatively precise cross-sectional estimates of index fund impacts in futures markets.

9.3.2 Cross-sectional regressions

Cross-sectional tests of stock returns often introduce firm-specific measures of value such as the price-to-earnings ratio (Fama and French, 1992). Few such measures exist across commodity markets. However, recent research (Gorton *et al.*, 2012; Bhardwaj *et al.*, 2015) indicates that the 'basis' or term structure of the futures markets, an indicator of 'scarcity' or inventory levels across markets, is related to risk premiums. Put simply, futures markets that have more of an inverted market structure – nearby futures higher than deferred futures – will tend to have higher returns. The basic notion is that an inverted market structure is a proxy for tighter inventories and possibly higher risk premiums.

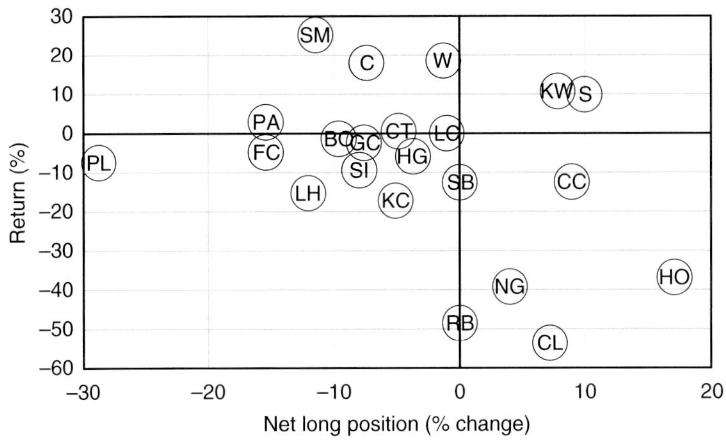

Note: Market abbreviations are as follows (n = 22): Kansas City Board of Trade wheat (KW), Chicago Board of Trade wheat (W), corn (C), soybeans (S), soybean meal (SM), soybean oil (BO), feeder cattle (FC), live cattle (LC), lean hogs (LH), cotton (CT), crude oil (CL), RBOB gasoline (RB), heating oil (HO), natural gas (NG), gold (GC), sugar (SB), coffee (KC), cocoa (CC), silver (SI), copper (HG), platinum (PL) and palladium (PA).

Fig. 9.2. Market returns and percentage change in *Index Investment Data* (IID) index positions, fourth quarter, 2014.

Here, we consider the cross-sectional relationship between market returns and changes in index positions conditioned on the cross-section of basis levels across the markets. The basis variable will essentially control for any time-varying risk premium associated with inventory levels within that market. Following Gorton *et al.* (2012) and Bhardwaj *et al.* (2015), the basis at time t is defined as:

$$Basis_t = \left(\frac{p_t^1 - p_t^2}{p_t^1}\right) \times \left(\frac{T^2 - T^1}{365}\right) \times 100 \qquad (9.5)$$

where p_t^1 is the futures price of the first listed or nearest-to-expiration contract expiring at time T^1 and p_t^2 is the futures price of the second listed contract expiring at T^2, such that Eqn 9.5 represents the basis as a percentage per annum. A positive basis in Eqn 9.5 represents a futures market that is inverted and a negative basis is a normal carry or contango market.

After sorting and ranking commodity futures using basis levels, Bhardwaj *et al.* (2015) show that market returns are greater for those markets with a more positive (inverted) basis structure. Here we directly incorporate this sorting process into the following cross-sectional regression:

$$R_{i,t} = \alpha + \beta_1 \, Position_{i,t} + \beta_2 \, Basis \, rank_{i,t-1} + \varepsilon_t \qquad (9.6)$$

where $R_{i,t}$ is the return in commodity futures market i and period t, $Position_{i,t}$ is the percentage change in the net long index position in market i during period t, and $Basis\ rank_{i,t-1}$ is the basis rank at the start of period t for market i. Note, the basis rank is from 1 to J, where J is the number of markets in the cross-section, and a rank of '1' is the market with the largest basis or most inverted futures market. Also note, the basis rank is calculated on the last day of the prior period following the method of Bhardwaj *et al.* (2015).

Fama and MacBeth (1973) propose a method of estimating Eqn 9.6 that is still widely used in the literature (e.g. Campbell *et al.*, 1997; Peterson, 2009). With the Fama-MacBeth regression procedure, Eqn 9.6 is estimated via ordinary least squares (OLS) regression for each time period $t = 1, 2, 3, \ldots, T$ across the $i = 1, 2, 3, \ldots, N$ markets. The average of the estimated slopes is calculated for the T regressions and the associated standard error is $\sigma_{\bar{\beta}} / T^{1/2}$. The basic estimation strategy is to exploit the information in the cross-section of markets about the relationship between index investment and returns and then treat each cross-section as an independent sample. Ibragimov and Muller (2010) show that the Fama-MacBeth estimator performs well in terms of size and power relative to alternative estimators for the number of cross-section and time-series observations considered here.

The estimations are conducted for annual, quarterly, and monthly horizons. The annual horizons are available from 2008 to 2015, where 2015 is calculated as a partial year ending on the last report as of October 31 2016 (eight cross-sections). The quarterly horizons run from the first quarter of 2008 through the third quarter of 2015 (31 cross-sections). The monthly horizons start in July of 2010 and run through December of 2015 (64 cross-sections). A representative quarterly cross-section is shown in Fig. 9.2 for the fourth quarter of 2014. The returns are plotted along the vertical axis and the percentage change in net long positions is shown on the horizontal axis. The largest increase in index positions was 17% in heating oil (HO), which was accompanied by a 36% decline in prices. The largest decline in index positions was 29% in platinum (PL), which was associated with a negative 7% return. In this particular quarter, there is a –0.34 cross-sectional correlation. That is, on average the markets that had the largest increases in positions had smaller than average returns.

The Fama-MacBeth regression results are shown in Table 9.7. For two horizons there is some statistically significant (5% level) evidence of a negative relationship between changes in index positions and the cross-section of market returns. The estimated coefficient on index positions at both the annual and the quarterly horizons is negative with *t*-statistics of –2.37 (*p*-value = 0.045) and –1.70 (*p*-value = 0.099). The conditioning variable, basis rank, shows a marginally statistically significant coefficient (*p*-value = 0.0835) at the quarterly horizon and with a sign consistent with expectations. Overall, the cross-sectional results show no convincing evidence that relative market returns can be explained by concurrent buying or selling of index funds across markets. If anything, there is some modest evidence of a negative relationship between index buying and the cross-section of market returns. These results are fairly robust

Table 9.7. Fama-MacBeth cross-sectional regressions for index fund impacts in commodity futures markets, 2008–2015.

	Coefficient estimates		
Horizon	Intercept	Index positions	Basis rank
Panel A:			
Annual (2008–2015, n = 8)	–9.8496	–0.1475	0.1042
t-statistic	–1.2800	–2.3700	0.3700
Quarterly (2008–2015, n = 31)	–1.1088	–0.1592	–0.1991
t-statistic	–0.3500	–1.7000	–1.7900
Monthly (2010–2015, n = 64)	–0.3518	0.0447	–0.0189
t-statistic	–0.5000	0.8700	–0.5100
Panel B:			
Annual (2008–2015, n = 8)	–9.0833	–0.0606	
t-statistic	–1.2600	–0.8900	
Quarterly (2008–2015, n = 31)	–2.8913	–0.1865	
t-statistic	–1.1200	–2.0500	
Monthly (2010–2015, n = 64)	–0.5476	0.0366	
t-statistic	–0.9900	0.7500	
Panel C:			
Annual (2008–2015, n = 8)	–8.4765		0.0645
t-statistic	–1.1900		0.2300
Quarterly (2008–2015, n = 31)	–0.2987		–0.1777
t-statistic	–0.1100		–1.5400
Monthly (2010–2015, n = 64)	–0.2508		–0.0139
t-statistic	–0.3600		–0.3800

across models estimated with or without each independent variable (panels B and C, Table 9.7). Fama-MacBeth regressions are a preferred method for examining the cross-section of market returns and they uncover no evidence that index fund positions are associated with relative returns across commodity markets.

9.4 Summary and Conclusions

The Masters Hypothesis – formulated by hedge fund manager Michael W. Masters – suggests that long-only index funds were the main cause of a massive increase in commodity prices that culminated in mid-2008. The crucial policy question related to the Masters Hypothesis is whether or not there were egregious pricing errors in commodity futures markets caused by commodity index funds. Unfortunately, some research has been inadvertently taken as support for an index-fueled price bubble even though it does not establish credible evidence to support the basic tenets of the Masters Hypothesis: (i) long-only commodity index funds were directly responsible for driving futures prices higher; (ii) the deviations from fundamental value were economically very large; and (iii) the impact was pervasive across commodity futures markets. That is, empirical evidence should demonstrate a direct link between long-only commodity index fund positions and commodity futures prices that result in economically large deviations from fundamental value. The few empirical studies that claim to support the Masters Hypothesis tend to fall short on at least two if not all three of these criteria.

On the flipside, numerous empirical studies document a lack of evidence in support of the Masters Hypothesis. Yet these studies can also be rightfully criticized for employing potentially low-power time-series tests, ignoring traders reactions to US CFTC position releases, and not fully accounting for other conditioning variables that may impact market returns. In this paper, we address these issues and present some new empirical evidence. First, we confirm the well-established contemporaneous correlation between changes in the CFTC SCOT index positions and nearby futures returns. However, upon close inspection it is clear that many of the positive correlations stem from years that are far from bubble-like (i.e. both the markets and the SCOT index positions are essentially in a sideways pattern). Second, we show that in periods where there are extremely large changes in the size of SCOT index positions, generally there are not correspondingly large changes in prices. Third, the correlation between index positions and daily market returns is examined. There is no correlation between changes in SCOT index trader positions and daily returns even in the day before the SCOT data are publicly available. Finally, controlling for other variables, cross-sectional analysis of CFTC IID positions fails to find any evidence that the relative returns across markets are related to index buying in those markets. While the analysis may be limited by the available data and perhaps by the straightforward empirical approach, the Masters Hypothesis comes up short on its most basic market predictions.

Market regulation is expensive in terms of both administrative burden and market friction. The Masters Hypothesis has led to a chorus for additional market regulations. The debate over position limits has been long and costly. Indeed, the CFTC is still considering a proposed rule to limit the futures positions held by traders (Moyer, 2016). Recently, the CFTC formed the Energy and Environmental Markets Advisory Committee (EEMAC) to review the empirical evidence and make recommendations regarding positions limits. The EEMAC voted 8-1 to recommend that the CFTC not finalize the proposed position limit rule. Here, we echo Michael Cosgrove – a member of the EEMAC – who, after reviewing the empirical evidence, noted that 'Instead of being obvious, it is undetectable. If we claim that elephants were playing in the backyard then we would expect to see their footprints. The alleged excessive speculation, if it is taking place, is leaving no data footprints' (CFTC EEMAC, 2016, p. 7).

Acknowledgments

A version of this paper was presented at the 2015 NCCC-134 Conference on Applied Commodity Price Analysis, Forecasting, and Market Risk Management. The authors acknowledge the helpful comments made by the conference participants and also anonymous referees.

Notes

[1] Original citation: Sanders, D.R. and Irwin, S.H. (2017) Bubbles, froth and facts: another look at the Masters Hypothesis in commodity futures markets. *Journal of Agricultural Economics* 68, 345–365. Reprinted by permission of the John Wiley and Sons and the Agricultural Economics Society.
[2] Better Markets (available at: https://bettermarkets.com/ (accessed February 22 2022)), 1825 K Street NW, Suite 1080, Washington, DC.
[3] An alternative approach is to directly test for bubbles in commodity futures prices. There is a burgeoning literature applying recently developed bubble tests to commodity futures prices (e.g. Gilbert, 2010; Gutierrez, 2013; Etienne *et al.*, 2014, 2015a, b; Brooks *et al.*, 2015; Figuerola-Ferretti *et al.*, 2015). While the results of these studies are mixed, few have found evidence of large and long-lasting bubbles consistent with the basic tenets of the Masters Hypothesis.
[4] The CFTC collected these additional data for selected grain futures markets over 2004–2005 at the request of the US Senate Permanent Subcommittee on Investigations and these data are used in the present analysis.

References

Aulerich, N.M., Irwin, S.H. and Garcia, P. (2013) Bubbles, food prices, and speculation: evidence from the CFTC's daily large trader data files. Working Paper No. 19065. National Bureau of Economics Research. Available at: https://www.nber.org/system/files/working_papers/w19065/w19065.pdf (accessed Janaury 23 2022).

Bhardwaj, G., Gorton, G. and Rouwenhorst, G. (2015) Facts and fantasies about commodity futures ten years later. ICF Working Paper No. 15-18, Yale International Center for Finance, New Haven, Connecticut. Available at: https://papers.ssrn.com/sol3/papers.cfm?abstract_id=2610772 (accessed February 2 2022).

Blair, P. (2008) Stop oil speculation now: the airlines disgraceful political agenda. *Capitalism Magazine.* Available at: http://capitalismmagazine.com/2008/07/stop-oil-speculation-now-the-airlines-disgraceful-political-agenda/ (accessed February 2 2022).

Brooks, C., Prokopczuk, M. and Wu, Y. (2015) Booms and busts in commodity markets: bubbles or fundamentals? *Journal of Futures Markets* 35, 916–938.

Brunetti, C., Büyükşahin, B. and Harris, J.H. (2016) Speculators, prices and market volatility. *Journal of Financial and Quantitative Analysis* 51(5), 1545–1574.

Buyuksahin, B. and Harris, J.H. (2011) Do speculators drive crude oil futures prices? *Energy Journal* 32, 167–202.

Campbell, J.Y., Lo, A.W. and MacKinlay, A.C. (1997) *The Econometrics of Financial Markets*. Princeton University Press, Princeton, New Jersey.

Cheng, I.-H. and Xiong, W. (2014) The financialization of commodity markets. *Annual Review of Financial Economics* 6, 419–441.

Cheng, I.-H., Kirilenko, A. and Xiong, W. (2015) Convective risk flows in commodity futures markets. *Review of Finance* 19, 1733–1781.

CFTC (Commodity Futures Trading Commission) (2015) *Index Investment Data*. Available at: http://www.cftc.gov/MarketReports/IndexInvestmentData/index.htm (accessed February 22 2022).

CFTC EEMAC (CFTC Energy and Environmental Markets Advisory Committee) (2016) Report on EEMAC's 2015 review and consideration of the CFTC's proposed rule on position limits. Commodity Futures Trading Commission (CFTC), February 25. Available at: https://www.findknowdo.com/sites/default/files/2016/02/25/00/Final%20Approved%20EEMAC%20Report%202-25-2016%20EMBARGOED.pdf (accessed October 6 2022.)

De Long, J.B., Shleifer, A., Summers, L.H. and Waldman, R.J. (1990) Noise trader risk in financial markets. *Journal of Political Economy* 98, 703–738.

Etienne, X.L., Irwin, S.H. and Garcia, P. (2014) Bubbles in food commodity markets: four decades of evidence. *Journal of International Money and Finance* 42, 129–155.

Etienne, X.L., Irwin, S.H. and Garcia, P. (2015a) Price explosiveness, speculation, and grain futures prices. *American Journal of Agricultural Economics* 97, 65–87.

Etienne, X.L., Irwin, S.H. and Garcia, P. (2015b) $25 spring wheat was a bubble, right? *Agricultural Finance Review* 75, 114–132.

Fama, E.F. and French, K.R. (1992) The cross-section of expected stock returns. *Journal of Finance* 47, 427–465.

Fama, E.F. and MacBeth, J.D. (1973) Risk, return, and equilibrium: empirical tests. *Journal of Political Economy* 81, 607–636.

Fattouh, B., Kilian, L. and Mahadeva, L. (2013) Financial speculation in the oil markets and the determinants of the price of oil. *Energy Journal* 34, 7–33.

Figuerola-Ferretti, I., Gilbert, C.L. and McCrorie, J.R. (2015) Testing for mild explosivity and bubbles in LME non-ferrous metals prices. *Journal of Time Series Analysis* 36, 763–782.

Gheit, F. (2008) Testimony before the Subcommittee on Oversight and Investigations of the Committee on Energy and Commerce, US House of Representatives. June 23. Available at: https://www.google.com/books/edition/Speculative_Investment_in_Energy_Markets/mCo6_xKYnQMC?hl=en&gbpv=1&dq=-fadel+gheit+testimony&pg=PA50&printsec=frontcover (accessed January 14 2022).

Gilbert, C.L. (2010) Speculative influences on commodity futures prices 2006–2008. United Nations Conference on Trade and Development, Discussion Paper No.197. Available at: https://unctad.org/system/files/official-document/osgdp20101_en.pdf (accessed January 23 2022).

Gilbert, C.L. and Pfuderer, S. (2014) The role of index trading in price formation in the grains and oilseeds markets. *Journal of Agricultural Economics* 65, 303–322.

Gorton, G., Hayashi, F. and Rouwenhorst, K.G. (2012) The fundamentals of commodity futures returns. *Review of Finance* 17, 35–105.

Grosche, S.-C. (2014) What does Granger causality prove? A critical examination of the interpretation of Granger causality results on price effects of index trading in agricultural markets. *Journal of Agricultural Economics* 65, 279–302.

Grossman, S.J. (1986) An analysis of 'insider trading' on futures markets. *Journal of Business* 59, S129–S146.

Grossman, S.J. and Miller, M.H. (1988) Liquidity and market structure. *Journal of Finance* 43, 617–633.

Gutierrez, L. (2013) Speculative bubbles in agricultural commodity markets. *European Review of Agricultural Economics* 40, 217–238.

Hamilton, J.D. and Wu, J.C. (2014) Risk premia in crude oil futures prices. *Journal of International Money and Finance* 42, 9–37.

Hamilton, J.D. and Wu, J.C. (2015) Effects of index-fund investing on commodity futures prices. *International Economic Review* 56, 187–205.

Ibragimov, R. and Muller, U.K. (2010) *t*-statistic based correlation and heterogeneity robust inference. *Journal of Business and Economics Statistics* 28, 453–468.

Irwin, S.H. (2013) Commodity index investment and food prices: does the 'Masters Hypothesis' explain recent price spikes? *Agricultural Economics* 44, 29–41.

Irwin, S.H. and Sanders, D.R. (2011) Index funds, financialization, and commodity futures markets. *Applied Economic Perspectives and Policy* 33, 1–31.

Irwin, S.H. and Sanders, D.R. (2012) Testing the Masters Hypothesis in commodity futures markets. *Energy Economics* 34, 256–269.

Kim, A. (2015) Does futures speculation destabilize commodity markets? *Journal of Futures Markets* 35, 696–714.

Lagi, M., Bar-Yam, Y., Bertrand, K.Z. and Bar-Yam, Y. (2015) Accurate market price formation model with both supply-demand and trend-following for global food prices providing policy recommendations. *Proceedings of the National Academy of Sciences of the United States of America* 112, E6119–E6128.

Lehecka, G.V. (2015) Do hedging and speculative pressures drive commodity prices, or the other way round? *Empirical Economics* 49, 575–603.

Masters, M.W. (2008) Testimony before the Committee on Homeland Security and Governmental Affairs, US Senate. May 20. Available at: http://hsgac.senate.gov/public/_files/052008Masters.pdf (accessed January 14 2022).

Mayer, J. (2012) The growing financialisation of commodity markets: divergences between index investors and money managers. *Journal of Development Studies* 48, 751–767.

Moyer, L. (2016) Commodity agency advised to drop plan to limit futures contracts. *The New York Times*, February 24. Available at: http://www.nytimes.com/2016/02/25/business/dealbook/commodities-agency-advised-to-drop-plan-to-limit-futures-contracts.html?_r=0 (accessed February 2 2022).

Peterson, M.A. (2009) Estimating standard errors in finance panel data sets: comparing approaches. *Review of Financial Studies* 22, 435–480.

Rouwenhorst, K.G. and Tang, K. (2012) Commodity investing. Yale ICF Working Paper No. 06-12. Yale International Center for Finance (ICF), New Haven, Connecticut. Available at: https://papers.ssrn.com/sol3/papers.cfm?abstract_id=2079664 (accessed February 2 2022).

Sanders, D.R. and Irwin, S.H. (2010) A speculative bubble in commodity futures prices? Cross-sectional evidence. *Agricultural Economics* 41, 25–32.

Sanders, D.R. and Irwin, S.H. (2011a) New evidence on the impact of index funds in U.S. grain futures markets. *Canadian Journal of Agricultural Economics* 59, 519–532.

Sanders, D.R. and Irwin, S.H. (2011b) The impact of index funds in commodity futures markets: a systems approach. *Journal of Alternative Investments* 14, 40–49.

Sanders, D.R. and Irwin, S.H. (2013) Measuring index investment in commodity futures markets. *Energy Journal* 34, 105–127.

Sanders, D.R. and Irwin, S.H. (2014) Energy futures prices and commodity index investment: new evidence from firm-level position data. *Energy Economics* 46, S57–S68.

Sanders, D.R. and Irwin, S.H. (2015) The 'necessity' of new position limits in agricultural futures markets: the verdict from daily firm level position data. *Applied Economic Perspectives and Policy* 38, 292–317.

Singleton, K.J. (2014) Investor flows and the 2008 boom/bust in oil prices. *Management Science* 60, 300–318.

Sockin, M. and Xiong, W. (2015) Informational frictions and commodity markets. *Journal of Finance* 70, 2063–2098.

Stoll, H.R. and Whaley, R.E. (2010) Commodity index investing and commodity futures prices. *Journal of Applied Finance* 20, 7–46.

Tadesse, G., Algieri, B., Kalkuhl, M. and von Braun, J. (2014) Drivers and triggers of international food price spikes and volatility. *Food Policy* 47, 117–128.

Tang, K. and Xiong, W. (2012) Index investment and financialization of commodities. *Financial Analysts Journal* 68, 54–74.

Werdigier, J. and Creswell, J. (2010) Trader's cocoa binge wraps up chocolate market. *The New York Times*, July 24. Available at: https://www.nytimes.com/2010/07/25/business/global/25chocolate.html (accessed February 2 2022).

Will, M.G., Prehn, S., Pies, I. and Glauben, T. (2016) Is financial speculation with agricultural commodities harmful or helpful? A literature review of empirical research. *Journal of Alternative Investments* 18, 84–102.

10

Mapping Algorithms, Agricultural Futures, and the Relationship between Commodity Investment Flows and Crude Oil Futures Prices[1]

New Author Foreword

It is hard to overstate the influence of Singleton's (2014) article on the debate about index funds and commodity futures prices. The first working paper version of the article appeared in May 2010, and since then the article has been cited over 650 times according to Google Scholar. Even more important, the work provided instant credibility to the argument that index speculation was a major culprit in the commodity price spike of 2007–2008 and it has been used over and over by those pushing for more restrictive limits on speculation in commodity futures markets. It certainly was a coup of sorts for the anti-speculation crowd to have someone of Singleton's stature supporting their position. He was a distinguished professor of finance at Stanford and editor of the *Journal of Finance*, the leading finance journal in the world. That is the pedigree of a true intellectual heavyweight.

Despite the pedigree, we were skeptical from the outset about the main finding in Singleton's article that commodity index investment had a large and statistically significant impact on West Texas Intermediate (WTI) crude oil futures prices. It was inconsistent with nearly all of our research, and we were beginning to accumulate a fairly impressive set of results. One of us had to be very wrong. As we discussed in Chapter 5 (this volume), Singleton used a simple algorithm to map known index positions in agricultural futures to unknown index positions in crude oil futures. We labeled this method the 'Masters algorithm,' once again, after Michael W. Masters who pioneered its use. The algorithm seemed sensible enough at first glance. If index investment was concentrated in one of the major commercial indexes, such as the Standard and Poor's Goldman Sachs Commodity Index (S&P GSCI), then the size of index positions across markets should be in fixed proportions through time. As long as the proportions were known, one could infer positions in one market from those in another. At least that was the theory.

We used the *Index Investment Data* (IID) report from the US Commodity Futures Trading Commission (CFTC) in a 2012 article (see Chapter 5, this volume) to show that the Masters algorithm produced inaccurate estimates of positions in crude oil. Specifically, the Masters algorithm indicated that index positions in crude oil futures spiked in 2008 almost exactly in parallel with crude oil prices, while the IID positions showed that this never happened, with crude oil index positions actually declining during the price spike. This explained why Singleton's finding of an economically large impact of index positions on crude oil futures prices was faulty. However, this was not the entire story. To convince the skeptics (if that were possible), we knew we needed to explain precisely why the Masters algorithm produced estimates of index positions in crude oil futures that were so wrong.

We were extremely fortunate that a brilliant PhD student in agricultural economics at Illinois, Lei Yan, became interested in working on this problem. After writing a term paper on the topic for a graduate seminar in futures market research, we encouraged him to pursue the work further. He decided this was a good topic for one of his dissertation essays and we were in business. With Lei leading the way, we dug in and went to work figuring

DOI:10.1079/9781800622104.0010

out what went haywire with the Masters algorithm estimates. We first replicated Singleton's regression model and discovered two important findings: (i) the relationship between estimates of index positions and crude oil futures prices was limited to just the second half of 2008; and (ii) the relationship reversed from positive to negative after the end of Singleton's original sample period. This alone did not inspire confidence in the original results.

Next, we decomposed index position estimates based on the Masters algorithm and found the startling result that the estimates were driven by feeder cattle – one of the unique agricultural futures markets underlying the mapping algorithm. Using the exact same regression framework as Singleton, we showed that changes in (actual) index positions for feeder cattle had a significant impact on crude oil futures prices, obviously a spurious result. In the end, a surge in agricultural index investment during 2007–2008, most likely due to the rise of specialized agricultural index funds, was not matched by a surge in crude oil index investment. Hence, an idiosyncratic spike in agricultural index positions during 2007–2008, together with a coincidental spike in crude oil prices, caused the spurious relationship between estimated index positions and crude oil futures prices in Singleton's study. Mystery solved.

We may have solved the mystery of why Singleton's results were wrong, but that was just the beginning of a nearly 3-year journey trying to get the paper published. We tried a top-level finance journal and were once again rejected. So we decided to change strategy and submitted to the top agricultural economics journal, the *American Journal of Agricultural Economics*. The reviews came back quite positive, but in a bizarre twist, the editor in charge of the submission said he wanted a completely different paper. That was extremely disappointing. Rather than destroy the paper, we sent it off to what was fast becoming one of our go-to journals – *Energy Economics*. The first-round reviews were again quite positive, and we resubmitted the paper in short order, expecting that acceptance would not be too far off. Boy, were we in for a surprise! The second-round reviews came back and indicated that the editor had pulled in a completely new reviewer who was not included in the first round. This was more than a little unusual. Our best guess was that the editor was quite nervous about publishing a paper that eviscerated the findings of a top financial scholar and he wanted to have another senior person in the field pass judgment before accepting the article. We managed to satisfy the additional reviewer after considerable work and the article was finally published in 2018.

It will come as no surprise that we think this is an important piece of research in the literature on commodity speculation. Singleton was not the only one that used mapping algorithms to estimate index positions in energy futures markets, and the results in all of these studies are suspect if not downright faulty as a consequence. In fact, if one excludes from the literature studies that use mapping algorithms, only a handful of studies are left that find any evidence of a significant impact of index funds on the level of commodity futures prices.

As a final note, it is fascinating how much an entire literature can turn on what seems to be a reasonable assumption. In this case, the assumption was that commodity index investment followed fixed proportions through time. In the words often attributed to Sherlock Holmes, 'Elementary, my dear Watson.'

Abstract
Several studies employ mapping algorithms to infer index positions in WTI crude oil futures from positions in agricultural futures and report an economically large and statistically significant impact of index positions on crude oil futures prices. In this article, we provide direct evidence that the apparent impact of index investment based on mapping algorithms is spurious. Specifically, an idiosyncratic spike in agricultural index positions during 2007–2008, coupled with the spike in oil prices, causes the spurious impact of index investment on crude oil futures prices found in these earlier studies.

Key words: commodity futures, crude oil, index investment, mapping algorithm

JEL categories: D84, G13, Q13, Q41

10.1 Introduction

Commodity futures prices increased substantially over 2003–2008, with the crude oil futures price hitting a record high of $147/barrel in mid-2008. As prices soared, concerns emerged that the record price rise was driven by the increasing participation of financial investors.[2] Hedge fund manager Michael W. Masters is a leading proponent of the view that commodity index investment was the main driver of the spike in commodity futures prices. In a series of testimonies and reports, Masters argues that index inflows from institutional investors imposed strong buying pressure and created a massive bubble in commodity futures prices, most notably in the crude oil market (e.g. Masters and White, 2008). This argument has become widely known as the 'Masters Hypothesis' (Irwin and Sanders, 2012). Masters-like arguments were quickly adopted by some policy makers and other advocates to push for regulations to limit commodity index activity. As called for in the 2010 Dodd-Frank Wall Street Reform and Consumer Protection Act, the US Commodity Futures Trading Commission (CFTC) proposed regulations implementing limits on speculative futures and swaps positions in December 2013 and proposed a revised set of rules in December 2016. The European Securities and Markets Authority (ESMA) also published new regulatory rules on commodity derivatives with a focus on ancillary activity and position limits, which took effect in January 2018.[3]

Empirical research has examined the impact of index investment on crude oil prices based on different index position measures. Some researchers use positions from the CFTC *Disaggregated Commitments of Traders* (DCOT) and *Index Investment Data* (IID) reports or private index funds and find no significant impact on futures prices for WTI crude oil (e.g. Buyuksahin and Harris, 2011; Sanders and Irwin, 2011, 2014; Irwin and Sanders, 2012; Brunetti *et al.*, 2016). While informative, these index position measures are subject to a netting problem or limitations on data frequency, sample length, and representativeness. Other researchers infer index positions in WTI crude oil using mapping algorithms and report an economically large and statistically significant impact on oil prices (e.g. Mayer, 2012; Singleton, 2014; Cheng *et al.*, 2015). Figure 10.1 demonstrates the large correlation between index positions inferred from a mapping algorithm and WTI crude oil futures prices during 2007–2008. A mapping algorithm is a method of estimating index positions for individual futures markets such as WTI crude oil based on index positions in agricultural futures markets that are known from the weekly CFTC *Supplemental Commitments of Traders* (SCOT) report. The best-known mapping algorithm – the Masters algorithm – infers index positions in WTI crude oil from a few agricultural commodities (Kansas wheat, feeder cattle, and soybean oil). The algorithm implicitly assumes a constant relationship in index positions between WTI crude oil and the agricultural commodities.

Fig. 10.1. Nearby prices of WTI crude oil (left scale) and index positions from the Masters algorithm (right scale), weekly, June 13 2006–December 29 2015.

In this article, we provide direct evidence that the relationship between index positions based on mapping algorithms and futures prices for WTI crude oil is spurious. We re-estimate Singleton's (2014) (henceforth 'SNG') model and introduce a dummy variable for 2008 and find that the forecasting power of index positions based on the Masters algorithm is limited to 2008, especially the second half of 2008. To analyze the sensitivity of SNG's results, we extend the analysis to a post-sample period from January 19 2010 through December 29 2015 and find that index positions based on the Masters algorithm become negatively significant and all the conditional variables lose significance, contradicting the alleged impact of index investment. This suggests that the relationship identified in the earlier period is unstable and does not persist out-of-sample. We then consider two alternative measures of index positions from the iShares S&P GSCI Commodity-Indexed Trust and a large private index fund. These results show no significant impact of index positions on futures returns for WTI crude oil. Altogether, the evidence suggests that the seemingly large impact of index positions based on mapping algorithms is spurious.

To discover why a spurious relationship may arise, we provide an anatomy of mapping algorithms and explore the inaccuracy of index position measures based on mapping algorithms. We show theoretically that in order to replicate a commodity index, the positions of any two commodities should maintain annually fixed ratios. The Masters algorithm implicitly assumes fixed-ratio relations between WTI crude oil and agricultural commodity futures. Using index positions from the SCOT and IID reports, we develop a formal test and empirically reject the underlying fixed-ratio relations. Furthermore, compared with IID positions – the most accurate available – the Masters algorithm provides poor estimates of index positions for WTI crude oil in both direction and magnitude. Index positions from the Masters algorithm show a clear spike during 2007–2008 while the IID measure does not. Decomposition shows that the spike in index positions based on the Masters algorithm is largely driven by positions in feeder cattle – one of the unique agricultural markets underlying this mapping algorithm. Within the same regression framework, we show that 13-week changes in index positions of feeder cattle have a significant impact on crude oil prices, which is obviously spurious. In sum, an idiosyncratic spike of agricultural index positions during 2007–2008, together with the coincidental spike in crude oil prices, causes the spurious relationship between index investment and crude oil prices found in previous studies that employ mapping algorithms.

10.2 Literature Review

A large number of academic studies have examined the impact of financial index investment on commodity prices.[4] Theoretical models suggest several pathways for financial index investment ('financialization') to impact commodity futures prices. First, the flow demand of index investment may be larger than available liquidity due to the large position sizes of index investors, and this flow may temporarily push prices away from fundamental value (e.g. Grossman and Miller, 1988). Second, competition from index investment may reduce risk premiums earned by long speculators in commodity futures markets (e.g. Acharya *et al.*, 2013; Hamilton and Wu, 2014, 2015). Third, increased integration of commodity and financial markets brought about by index investment may result in increased exposure of commodity futures prices to financial shocks that increase prices (e.g. Etula, 2013; Basak and Pavlova, 2016). Fourth, other traders may confuse index buying with valuable private information and thus revise their own demands upward which, in turn, pushes commodity futures prices higher (Sockin and Xiong, 2015).

Depending on the way that index positions are measured, empirical research on the impact of index investment on crude oil prices falls into three groups.[5] The first set of studies uses long positions of swap dealers from the CFTC DCOT report as a measure of index positions and find no evidence of significant impacts of swap dealer positions on crude oil futures prices (e.g. Buyuksahin and Harris, 2011; Sanders and Irwin, 2011; Brunetti *et al.*, 2016). While a large fraction of index investment is placed through swap contracts, net swap dealer positions in energy futures markets may be a poor approximation of total index positions because of the large offsetting non-index swap business conducted in these markets.[6] The second group of studies uses index positions from the CFTC IID report or private index funds, also finding no significant impact of index positions on crude oil prices (e.g. Irwin and Sanders, 2012; Sanders and Irwin, 2014). These measures are direct and, in the case of the IID, generally accurate, but subject to limitations on frequency, sample length, and potentially representativeness of private index fund positions. The third group of studies relies on mapping algorithms to estimate crude oil index positions from agricultural index positions, which are available from the weekly CFTC SCOT report. There are two different but related mapping algorithms – the Masters algorithm and the weighted-average algorithm. The Masters algorithm infers crude oil index positions from a few agricultural commodities that are unique to a particular index (Masters, 2008). Using this algorithm, Singleton (2014) finds an economically large and statistically significant influence of index positions on crude oil futures prices. The weighted-average algorithm derives index positions in crude oil from the aggregate index positions of all 12 SCOT agricultural commodities, with initial period prices as the weight. Using the weighted-average algorithm or a close variant, Mayer (2012) and Cheng *et al.* (2015) find significant impacts of index investment flows in several commodity markets including crude oil.[7,8]

A few studies have questioned the accuracy of index positions generated from mapping algorithms. Irwin and Sanders (2012) and Sanders and Irwin (2013) argue that index positions based on the Masters algorithm are overestimated and directionally wrong in 2008 for WTI crude oil. Hamilton and Wu (2015) extend SNG's analysis to agricultural markets and find no evidence that index flows could help predict returns on agricultural commodity futures, and the significant impact identified in WTI crude oil largely disappears beyond SNG's sample period. Likewise, Sanders and Irwin (2014) find that crude oil position flows for a large private index fund do not predict crude oil futures returns, but agricultural position flows do. These studies, however, are only suggestive because they fail to explain why the estimate of index positions from mapping algorithms is inaccurate and how exactly this may cause a spurious relationship between index investment and crude oil futures prices.

10.3 Impact of Index Investment on Crude Oil Prices

In this section, we investigate the impact of index positions from different sources on crude oil futures prices. First, we replicate SNG's results using the Masters algorithm to estimate index positions and examine the sensitivity of the results to prediction horizons, market events, and change of sample period. Second, we consider two alternative measures of actual WTI crude oil index positions and compare the results.

10.3.1 Impact of index positions from the Masters algorithm

We follow SNG's framework and consider the following forecasting regression model,[9]

$$ERm_{t+1}^{oil} = \alpha + \beta X_t + \gamma IIP13_t + \varepsilon_t \qquad (10.1)$$

where ERm_{t+1}^{oil} is the 1-week realized excess return of WTI crude oil on futures contracts that expire in m months from t to $t+1$, X_t is the set of control variables that are known up to t, and $IIP13_t$, defined as $IIP_t - IIP_{t-13}$, is the 13-week change in commodity index positions (million contracts) based on the Masters algorithm.[10,11] The 13-week (or 3-month) horizon is initially selected to be consistent with SNG, who argues that the impact of index investment flows may take place over longer periods than one week. More specifically, if the estimated slope coefficient, γ, is positive (negative) it indicates a process where crude oil expected returns tend to increase (decrease) slowly over a relatively long time period after widespread index fund buying. The fads- or bubble-like process captured in Eqn 10.1 is consistent with the notion that index investment flows in waves that build slowly, pushing expected returns higher and then slowly fading (e.g. Summers, 1986).

To facilitate comparison, we follow SNG's variable definitions as closely as possible. For the 1-week realized excess return, ERm_{t+1}^{oil}, we consider contracts that have maturities from 1 month to 24 months. This allows us to check whether index positions have similar effects on prices of nearby and deferred contracts. Excess returns are based on rolling futures positions on the tenth business day of the month preceding the delivery month. Whenever rolling occurs, returns are adjusted to exclude price jumps across contracts (see SNG's Appendix). The sample spans from June 13 2006 through December 29 2015. We partition the sample into two parts: (i) June 13 2006–January 12 2010, the same as in SNG's original study; and (ii) January 19 2010–December 29 2015 as a post-sample period. Futures prices are provided by the Commodity Research Bureau.

Control variables in X_t include: (i) *RSP* and *REM*, the 1-week returns (in percent) on the S&P 500 Index and the Morgan Stanley Capital International (MSCI) Emerging Markets Asia Index collected from the Bloomberg database and assumed to reflect alternative investing opportunities faced by index investors; (ii) *REPO*, the 1-week change in overnight repo positions on Treasury bonds held by primary dealers in trillion dollars, which measures the flexibility of balance sheets of large financial institutions;[12] (iii) *MMS*13 and *OI*13, the 13-week changes in managed-money spread positions and total open interest (both in million contracts) from the CFTC DCOT report; (iv) *AVB*, the 1-week change in average basis, where the basis of a contract with maturity T_i at time t is defined as $basis_t^{T_i} = \left(F_t^{T_i} / S_t\right)^{1/i} - 1$, and the average basis is calculated for maturities $i \in \{1, 3, 6, 9, 12, 15, 18, 21, 24\}$; and (v) ERm_{lag}^{oil}, the lagged dependent variable introduced to control potential autocorrelation in excess returns.

Table 10.1 reports the sample means and standard deviations of predictor variables and excess returns for the two sample periods. Not surprisingly, the summary statistics for the June 13 2006–January 12 2010 sample are very similar or even identical to those reported in SNG's Table 2. The average 1-week excess return on the S&P 500 Index is –0.083% in SNG's sample period because of the market meltdown in 2008, while the MSCI Emerging Markets Asia Index provides a positive average return (0.153%) due to its rapid recovery from the crisis. In the post-sample period, the US stock market performs better with an average 1-week excess return (0.194%) versus the Asian stock markets (–0.008%). The average 1-week changes of repo positions on Treasury bonds held by primary dealers are –0.003 and zero million dollars for the two periods, suggesting tight constraints faced by large financial institutions. The average 13-week changes in index positions are 0.041 and –0.018 million contracts for the two sample periods, indicating that index positions were growing before 2010 but declining afterwards (see Fig. 10.1). The average 13-week changes in managed-money spread positions and total open interest are both positive, reflecting the growth of spread trading and expansion of the WTI crude oil futures market. The average 1-week excess returns for WTI crude oil are positive across maturities in SNG's sample period and become negative in the post-sample period because of the large drop in crude oil prices since the summer of 2014.

Table 10.2 shows the correlations between 1-week excess returns and the contemporaneous and lagged predictor variables in both sample

Table 10.1. Sample means, standard deviations of 1-week excess returns on futures positions of WTI crude oil and predictor variables.[a]

	SNG's sample period: June 13 2006–January 12 2010		Post-sample period: January 19 2010–December 29 2015	
	Mean	SD	Mean	SD
Predictor variables				
RSP	–0.083	2.997	0.194	2.110
REM	0.153	4.994	–0.008	2.666
REPO	–0.003	0.078	0.000	0.056
*IIP*13	0.041	0.087	–0.018	0.057
*MMS*13	0.001	0.044	0.006	0.046
*OI*13	0.010	0.100	0.019	0.129
AVB	0.000	0.777	0.002	0.219
Excess returns				
*ER*1	0.001	5.816	–0.270	4.324
*ER*3	0.119	5.444	–0.215	4.101
*ER*6	0.146	5.106	–0.171	3.848
*ER*12	0.163	4.743	–0.151	3.401
*ER*24	0.186	4.315	–0.146	2.807

[a] *RSP* and *REM* are the 1-week returns (in percent) on the S&P 500 Index and the MSCI Emerging Asia Index. *REPO* is the 1-week change in overnight repo positions on Treasury bonds held by primary dealers in trillion dollars. *IIP*13 is the 13-week change in index positions (million contracts) from the Masters algorithm. *MMS*13 and *OI*13 are the 13-week changes in managed-money spread positions and total open interest (million contracts). *AVB* is the 1-week change in average basis. *ERm* is the 1-week excess return (in percent) on WTI crude oil futures contract that expires in *m* months.

Table 10.2. Correlations between 1-week excess returns on futures positions of WTI crude oil and contemporaneous and lagged predictor variables.[a]

	RSP	*REM*	*REPO*	*IIP*13	*MMS*13	*OI*13	*AVB*
Panel A: SNG's sample period: June 13 2006–January 12 2010							
				Contemporaneous predictors			
*ER*1	0.39	0.37	0.11	0.26	0.20	0.16	−0.24
*ER*3	0.43	0.43	0.08	0.26	0.19	0.16	−0.22
*ER*6	0.44	0.45	0.05	0.26	0.17	0.16	−0.19
*ER*12	0.44	0.46	0.03	0.26	0.15	0.14	−0.16
*ER*24	0.41	0.45	0.03	0.25	0.12	0.13	−0.13
				Lagged predictors			
*ER*1	0.13	−0.10	−0.20	0.26	0.20	0.12	−0.35
*ER*3	0.13	−0.09	−0.20	0.27	0.18	0.13	−0.34
*ER*6	0.15	−0.10	−0.19	0.27	0.17	0.13	−0.32
*ER*12	0.17	−0.11	−0.19	0.26	0.15	0.12	−0.26
*ER*24	0.16	−0.14	−0.18	0.25	0.13	0.11	−0.19
Panel B: Post-sample period: January 19 2010–December 29 2015							
				Contemporaneous predictors			
*ER*1	0.53	0.37	−0.05	−0.07	0.02	0.06	−0.41
*ER*3	0.56	0.39	−0.05	−0.07	0.02	0.06	−0.34
*ER*6	0.58	0.41	−0.05	−0.07	0.02	0.05	−0.29
*ER*12	0.60	0.44	−0.05	−0.06	0.02	0.05	−0.24
*ER*24	0.62	0.47	−0.05	−0.06	0.02	0.04	−0.16
				Lagged predictors			
*ER*1	−0.14	0.00	−0.03	−0.13	0.06	0.05	−0.02
*ER*3	−0.14	0.01	−0.02	−0.13	0.06	0.05	−0.03
*ER*6	−0.14	0.01	−0.02	−0.13	0.05	0.05	−0.04
*ER*12	−0.13	0.01	−0.02	−0.13	0.05	0.04	−0.05
*ER*24	−0.11	0.01	−0.02	−0.12	0.06	0.04	−0.06

[a]*RSP* and *REM* are the 1-week returns (in percent) on the S&P 500 Index and the MSCI Emerging Asia Index. *REPO* is the 1-week change in overnight repo positions on Treasury bonds held by primary dealers in trillion dollars. *IIP*13 is the 13-week change in index positions (million contracts) from the Masters algorithm. *MMS*13 and *OI*13 are the 13-week changes in managed-money spread positions and total open interest (million contracts). *AVB* is the 1-week change in average basis. *ERm* is the 1-week excess return (in percent) on WTI crude oil futures contract that expires in *m* months.

periods. Correlations in SNG's sample period (panel A) are quite similar to those shown in SNG's Table 1. The 13-week changes in index positions consistently have a positive correlation with contemporaneous and future excess returns. However, these correlations become negative in the post-sample period (panel B). Similarly, the excess returns have very different correlations with other predictor variables in SNG's original sample versus the post-sample period, especially with lagged predictors in which many correlation coefficients are close to zero.

Table 10.3 presents coefficient estimates and *t* values (in parentheses) of the model in Eqn 10.1 for SNG's sample period. Standard errors and *t*-statistics are computed based on Newey and West's (1994) procedure to allow for serial correlation and conditional heteroskedasticity. We report results for selected contracts with maturity equal to 1, 3, 6, 12, and 24 months and compress intercepts to save space. Predictors are standardized by dividing by sample standard deviations. By doing so, coefficient estimates can be interpreted as the impact on excess returns of a one standard deviation change in the predictor and their magnitudes reflect the relative importance of predictors in explaining 1-week ahead excess returns for WTI crude oil.

The most notable finding by SNG, and confirmed in Table 10.3, is that the 13-week changes in index positions from the Masters algorithm have an 'economically large and statistically significant effect' (Singleton, 2014, p. 313) on 1-week excess returns. This result is not surprising given the relationship illustrated in Fig. 10.1 between the price of nearby WTI contracts and index positions based on the Masters algorithm. In particular, the figure shows a strong co-movement between futures prices and index positions during 2007–2008, so that increases in index positions over the preceding 3 months predict higher crude oil prices. For example, a one standard deviation increase in *IIP*13 raises the 1-week excess return on the 1-month contract by 1.792% holding all other variables constant.[13]

Following SNG, the estimated impact can be compared to the standard deviation of *ER*1 (5.816%, Table 10.1) in order to assess the absolute size of the effect. Since the index impact is slightly more than 30% of the weekly standard deviation of returns, it is reasonable to conclude that the estimated impact is large in an absolute sense. Relative to other predictor variables, *IIP*13 also has the largest positive impact. This is true for excess returns on contracts across all maturities. The rest of the predictor variables also show statistically significant

Table 10.3. Coefficient estimates for 1-week excess return forecasting for WTI crude oil with index positions from the Masters algorithm, June 13 2006–January 12 2010.[a]

	RSP	*REM*	*REPO*	*IIP*13	*MMS*13	*OI*13	*AVB*	R_{lag}	R^2_{adj}
*ER*1	0.804	−0.935	−1.217	1.792	1.385	−0.716	−1.948	−0.706	0.30
	(3.26)	(−2.16)	(−4.02)	(4.16)	(6.28)	(−3.61)	(−17.20)	(−2.39)	
*ER*3	0.816	−0.788	−1.152	1.647	1.172	−0.524	−1.764	−0.781	0.29
	(2.71)	(−1.63)	(−3.18)	(2.91)	(5.18)	(−1.85)	(−13.82)	(−2.60)	
*ER*6	0.874	−0.883	−1.050	1.532	1.019	−0.482	−1.477	−0.581	0.27
	(2.54)	(−1.64)	(−2.43)	(2.75)	(4.57)	(−1.25)	(−9.71)	(−1.92)	
*ER*12	0.980	−1.017	−0.959	1.414	0.841	−0.429	−1.030	−0.432	0.25
	(2.50)	(−1.62)	(−1.94)	(2.87)	(3.92)	(−1.02)	(−6.20)	(−1.34)	
*ER*24	0.911	−1.121	−0.798	1.254	0.669	−0.365	−0.608	−0.250	0.21
	(2.48)	(−1.87)	(−1.79)	(3.06)	(3.43)	(−1.07)	(−3.23)	(−0.75)	

[a]Independent variables include: *RSP* and *REM*, the 1-week returns (in percent) on the S&P 500 Index and the MSCI Emerging Asia Index; *REPO*, the 1-week change in overnight repo positions on Treasury bonds held by primary dealers in trillion dollars; *IIP*13, the 13-week change in index positions (million contracts) from the Masters algorithm; *MMS*13 and *OI*13, the 13-week changes in managed-money spread positions and total open interest (million contracts); *AVB*, the 1-week change in average basis; R_{lag}, lagged excess return; R^2_{adj}, the adjusted R-squared measure of goodness of fit. Dependent variable *ERm* is the 1-week realized excess return (in percent) on WTI crude oil futures contract that expires in *m* months. Independent variables are standardized by dividing by standard deviations. Standard errors are adjusted by Newey-West's (1994) approach and *t*-statistics are shown in parentheses.

forecasting power for excess returns on WTI crude oil at least in this particular period. The absolute values of coefficient estimates decrease in maturity, suggesting that it is more difficult to predict excess returns on contracts that expire at more distant horizons. The same explanation applies to the decreasing pattern of adjusted R^2 across maturities. In sum, using the same data and sample period, we obtain virtually the same finding as SNG (i.e. index positions seem to have an economically large and statistically significant impact on futures prices for WTI crude oil).

We now consider sensitivity of the estimation results for SNG's original sample period to the selection of a 13-week change in index positions based on the Masters algorithm. Hamilton and Wu (2015) use a simplified version of SNG's model and find that the predictability of crude oil returns is maximized for a 12-week change in notional index positions (dollars). Here, we estimate model Eqn 10.1 for index position changes over the previous 1–52 weeks ($IIPm$, $m = 1, \cdots, 52$). Figure 10.2 reports the adjusted R^2 and coefficient estimate of $IIPm$ with a 95% confidence interval in predicting 1-week excess returns on the 1-month contract. Maximal predictability in terms of adjusted R^2 is achieved when $m = 19$, which is different from Hamilton and Wu (2015) because we include conditioning variables and express index positions in contracts instead of dollars. The coefficient estimate on $IIPm$ is significantly positive for 10–30-week changes in index positions. While there is evidence of a significant impact of index positions for a fairly wide range of lags, there is also considerable sensitivity in the degree of predictability, particularly for lags less than 10 weeks compared to 10–30-week lags. SNG's finding that index positions have a significant impact on futures prices for WTI crude oil is somewhat sensitive to the length of interval in which index position changes are calculated.

We next consider how much of the significant predictability of crude oil returns in SNG's sample is influenced by events during 2008. Hamilton and Wu (2015) report that the predictability of crude oil returns using index positions from the Masters algorithm can be traced to the first phase of the 2008 Great Recession. To further investigate this possibility, we introduce a dummy variable to model Eqn 10.1,

$$ER m_{t+1}^{oil} = \alpha + \beta X_t + \gamma IIP13_t + \theta IIP13_t * D08_t + \varepsilon_t \quad (10.2)$$

where $D08_t$ is equal to 1 for 2008 and 0 otherwise. The interaction term allows us to isolate the differential impact of $IIP13$ on excess returns in 2008 relative to the rest of the sample period. Table 10.4 shows estimation results for model Eqn 10.2. All predictor variables have quite similar estimates and levels of significance as in Table 10.3. The main difference is that the 13-week change in index positions, $IIP13$, becomes less significant or insignificant, especially for distant contracts. In contrast, the interaction term is highly significant for excess returns on all contracts, suggesting that the predictive power of $IIP13$ is mainly associated with 2008. In terms of magnitude, the impact of index positions is much larger in 2008 than other periods. For example, a one standard deviation increase in $IIP13$, on average, raises 1-week excess returns on the nearby contracts by 2.74% in 2008, which is about 4.5 times as large as the impact over the rest of the period (0.605%).

To further refine the analysis, we specify the dummy variable for the first and second half of 2008 and present estimation results in Table 10.5. Note in panel A the coefficient estimate on $IIP13$ is larger and more significant than the interaction term $IIP13 * D08$. Referring back to Fig. 10.1, this means that index positions had limited forecasting power for crude oil returns during the most dramatic upward phase of the crude oil price spike. We find just the opposite result when the dummy is specified for the second half of 2008 (panel B). None of the coefficients is significant for $IIP13$ outside of the second half of 2008, but the interaction term is highly significant for excess returns on all contracts. The coefficient estimate on $IIP13 * D08$ is also the largest of any estimates for SNG's sample period. For example, a one standard deviation increase in $IIP13$, on average, raises 1-week excess returns on nearby contracts by 3.948% in the second half of 2008. Our results confirm Hamilton and Wu's (2015) finding that the predictability of crude oil returns using index positions from the Masters algorithm is actually limited to the onset of the Great Recession and the rapid decline in crude oil prices in the second half of 2008.

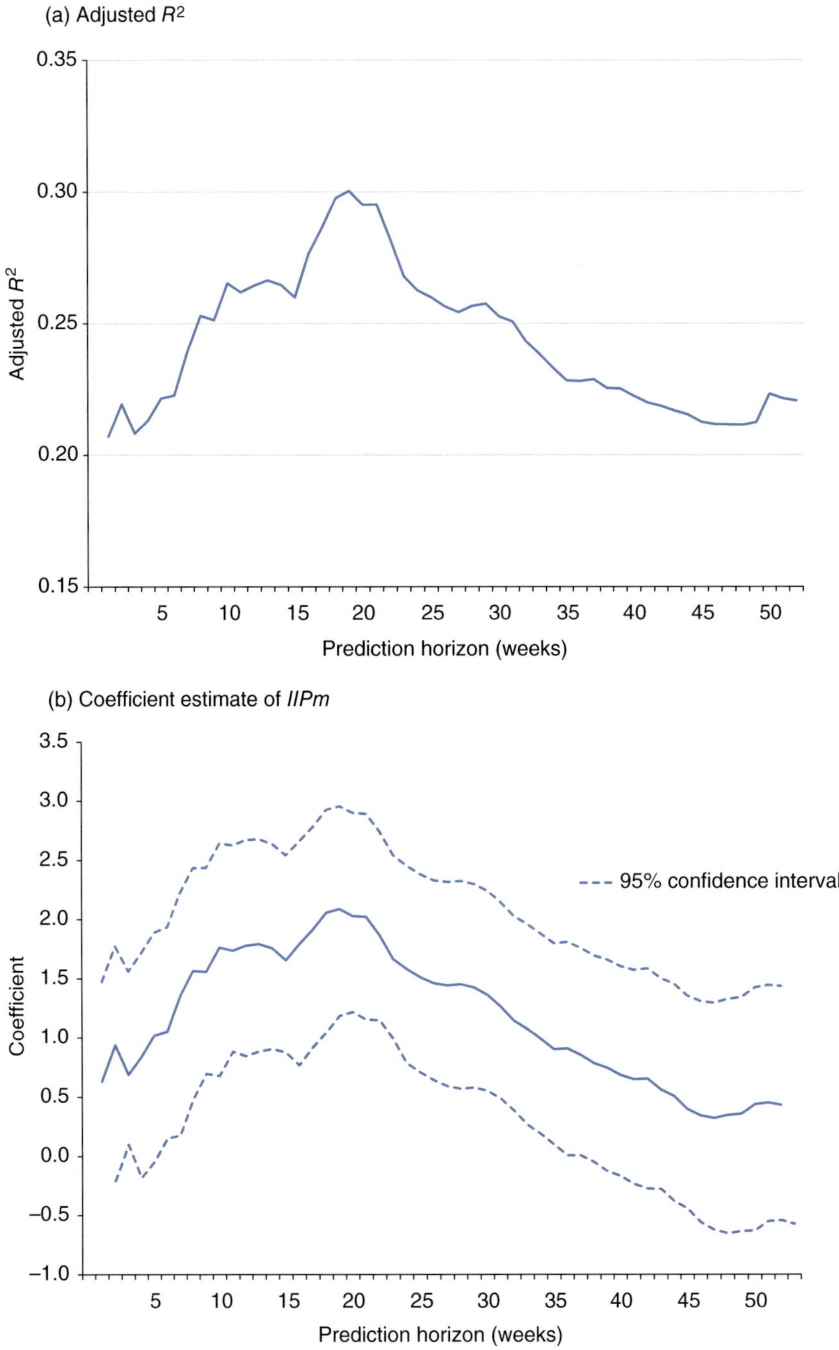

Fig. 10.2. Adjusted R^2 and coefficient estimates in predicting crude oil returns on the 1-month contract as a function of m-week changes in index positions from the Masters algorithm ($m = 1, 2, \ldots, 52$).

Table 10.4. Coefficient estimates for 1-week excess return forecasting for WTI crude oil with index positions from the Masters algorithm and the dummy for 2008, June 13 2006–January 12 2010.[a]

	RSP	*REM*	*REPO*	*IIP*13	*IIP*13×*D*08	*MMS*13	*OI*13	*AVB*	R_{lag}	R^2_{adj}
*ER*1	0.868	−0.879	−1.283	0.605	2.135	1.187	−0.688	−1.989	−0.900	0.33
	(−2.73)	(−1.64)	(−3.47)	(−1.81)	(−4.58)	(−5.65)	(−3.22)	(−13.05)	(−2.76)	
*ER*3	0.883	−0.722	−1.217	0.536	1.997	0.984	−0.495	−1.800	−0.978	0.32
	(−3.60)	(−1.96)	(−4.14)	(−1.93)	(−5.72)	(−7.57)	(−3.43)	(−16.36)	(−4.57)	
*ER*6	0.940	−0.815	−1.116	0.481	1.892	0.839	−0.454	−1.509	−0.775	0.30
	(−2.86)	(−1.60)	(−2.91)	(−1.57)	(−5.67)	(−4.39)	(−2.17)	(−9.61)	(−2.95)	
*ER*12	1.043	−0.947	−1.024	0.418	1.795	0.665	−0.404	−1.057	−0.620	0.27
	(−2.77)	(−1.54)	(−2.07)	(−1.27)	(−3.65)	(−3.51)	(−1.25)	(−6.00)	(−2.15)	
*ER*24	0.961	−1.061	−0.859	0.355	1.619	0.507	−0.342	−0.625	−0.413	0.24
	(−2.69)	(−1.73)	(−1.85)	(−1.16)	(−3.59)	(−2.85)	(−1.21)	(−3.45)	(−1.37)	

[a] Independent variables include: *RSP* and *REM*, the 1-week returns (in percent) on the S&P 500 Index and the MSCI Emerging Asia Index; *REPO*, the 1-week change in overnight repo positions on Treasury bonds held by primary dealers in trillion dollars; *IIP*13, the 13-week change in index positions (million contracts) from the Masters algorithm; *MMS*13 and *OI*13, the 13-week changes in managed-money spread positions and total open interest (million contracts); *AVE*, the 1-week change in average basis; R_{lag}, lagged excess return; R^2_{adj}, the adjusted R-squared measure of goodness of fit. Dummy variable *D*08 is equal to 1 for 2008 and 0 otherwise. Dependent variable *ERm* is the 1-week realized excess return (in percent) on WTI crude oil futures contract that expires in *m* months. Independent variables are standardized by dividing by standard deviations. Standard errors are adjusted by Newey-West's (1994) approach and *t*-statistics are shown in parentheses.

Table 10.5. Coefficient estimates for 1-week excess return forecasting for WTI crude oil with index positions from the Masters algorithm and the dummy for the first and second half of 2008, June 13 2006–January 12 2010.[a]

	RSP	*REM*	*REPO*	*IIP*13	*IIP*13×*D*08	*MMS*13	*OI*13	*AVB*	R_{lag}	R^2_{adj}
Panel A: *D*08 = 1 for the first half of 2008										
*ER*1	0.820	−0.893	−1.220	1.637	0.973	1.383	−0.623	−1.964	−0.749	0.30
	(4.10)	(−2.55)	(−4.98)	(5.03)	(3.15)	(7.95)	(−3.70)	(−21.48)	(−3.30)	
*ER*3	0.836	−0.737	−1.155	1.478	1.061	1.169	−0.422	−1.782	−0.835	0.30
	(2.35)	(−1.38)	(−2.99)	(2.67)	(2.23)	(4.83)	(−1.28)	(−12.69)	(−2.83)	
*ER*6	0.894	−0.831	−1.054	1.365	1.050	1.016	−0.381	−1.495	−0.638	0.28
	(2.50)	(−1.43)	(−2.27)	(2.68)	(1.92)	(4.55)	(−0.84)	(−9.15)	(−2.18)	
*ER*12	1.002	−0.961	−0.964	1.246	1.068	0.836	−0.327	−1.047	−0.493	0.25
	(2.52)	(−1.49)	(−2.03)	(2.67)	(1.79)	(3.81)	(−0.73)	(−5.90)	(−1.69)	
*ER*24	0.932	−1.065	−0.804	1.086	1.070	0.663	−0.263	−0.623	−0.314	0.22
	(2.52)	(−1.76)	(−1.84)	(2.75)	(1.76)	(3.32)	(−0.73)	(−3.21)	(−1.05)	
Panel B: *D*08 = 1 for the second half of 2008										
*ER*1	0.853	−0.997	−1.319	0.374	3.574	1.062	−1.012	−1.956	−0.875	0.33
	(2.16)	(−1.60)	(−3.34)	(0.57)	(3.32)	(3.99)	(−5.04)	(−12.04)	(−2.60)	
*ER*3	0.861	−0.833	−1.243	0.433	3.056	0.893	−0.774	−1.767	−0.928	0.31
	(2.67)	(−1.73)	(−3.30)	(0.69)	(3.45)	(5.17)	(−5.03)	(−12.90)	(−3.24)	
*ER*6	0.916	−0.921	−1.136	0.418	2.805	0.761	−0.711	−1.478	−0.718	0.30
	(3.73)	(−2.30)	(−3.29)	(0.78)	(3.79)	(6.39)	(−4.51)	(−11.81)	(−3.15)	
*ER*12	1.017	−1.050	−1.038	0.411	2.528	0.605	−0.636	−1.027	−0.552	0.27
	(3.61)	(−2.28)	(−2.54)	(0.77)	(3.58)	(4.46)	(−3.90)	(−6.40)	(−2.68)	
*ER*24	0.935	−1.153	−0.865	0.423	2.088	0.472	−0.536	−0.600	−0.336	0.23
	(2.81)	(−2.22)	(−1.77)	(0.84)	(2.63)	(2.40)	(−2.34)	(−3.50)	(−1.26)	

[a]Independent variables include: *RSP* and *REM*, the 1-week returns (in percent) on the S&P 500 Index and the MSCI Emerging Asia Index. *REPO*, the 1-week change in overnight repo positions on Treasury bonds held by primary dealers in trillion dollars. *IIP*13, the 13-week change in index positions (million contracts) from the Masters algorithm. *MMS*13 and *OI*13, the 13-week changes in managed-money spread positions and total open interest (million contracts). *AVB*, the 1-week change in average basis. R_{lag}, lagged excess return. Dummy variable *D*08 is equal to 1 for the first and second half of 2008, respectively. Dependent variable *ERm* is the 1-week realized excess return (in percent) on WTI crude oil futures contract that expires in *m* months. Independent variables are standardized by dividing by standard deviations. Standard errors are adjusted by Newey-West's (1994) approach and *t*-statistics are shown in parentheses.

Finally, we estimate model Eqn 10.1 using the post-sample period of January 19 2010 through December 29 2015. The estimation results are presented in Table 10.6 and provide a strikingly different picture of predictive relationships. The coefficient estimate on *IIP*13 is negative and statistically significant at all return horizons. This implies increasing commodity index positions predict decreasing crude oil returns, just the opposite of the alleged impact of index investment and the estimated impact in SNG's original sample. This is actually not surprising based on the post-2010 trends found in Fig. 10.1. After 2010, index positions generated from the Masters algorithm trend downward sharply while crude oil prices rise until mid-2014. Perhaps even more surprising, all the other predictor variables are insignificant in the post-sample period and lose their ability to predict excess returns for WTI crude oil. The disappearance of forecasting power for all conditioning variables suggests that the significant effects identified in SNG's period are sample-specific and do not persist for an extended period.

10.3.2 Impact of index positions from alternative sources

Given previous criticisms about the accuracy of the Masters algorithm (e.g. Irwin and Sanders 2012; Sanders and Irwin 2013) and the sensitivity of the results documented above, it is natural to ask whether other measures of index positions in WTI crude oil futures yield similar results. We consider two alternative measures of index positions, both based on actual trading records of index funds in WTI crude oil futures. The first is from the iShares S&P GSCI Commodity-Indexed Trust (GSG), which is designed to track the performance of the S&P GSCI Total Return Index, by all accounts the most widely tracked price index for commodity index investment. SNG examines quarterly GSG WTI crude oil positions and reports that the broad trends in crude oil positions from GSG and the Masters algorithm are similar. The inception date for GSG is July 10 2006, less than a month after the start date of the sample for this study. Assets under management of the GSG rose quickly after inception and reached a peak of $2.1 billion in April 2011. The GSG index positions for WTI crude oil are calculated as the product of total net assets in dollars and dollar weights of WTI crude oil in the index divided by futures prices.[14] The second measure of index positions comes from a large private commodity index fund ('the Fund') that replicates a proprietary commodity index. Sanders and Irwin (2014) show that the Fund's allocations across markets and investment flows through time do not differ substantially from that observed as a whole for the commodity

Table 10.6. Coefficient estimates for 1-week excess return forecasting for WTI crude oil with index positions from the Masters algorithm, January 19 2010–December 29 2015.

	RSP	*REM*	*REPO*	*IIP*13	*MMS*13	*OI*13	*AVB*	R_{lag}	R^2_{adj}
*ER*1	−0.764	0.464	−0.209	−0.674	0.426	−0.051	−0.138	−0.294	0.060
	(−1.48)	(1.03)	(−0.66)	(−2.72)	(1.26)	(−0.16)	(−0.43)	(−0.77)	
*ER*3	−0.739	0.483	−0.177	−0.620	0.356	−0.023	−0.159	−0.287	0.060
	(−1.43)	(1.08)	(−0.59)	(−2.68)	(1.11)	(−0.08)	(−0.58)	(−0.82)	
*ER*6	−0.687	0.446	−0.167	−0.571	0.319	−0.036	−0.164	−0.244	0.060
	(−1.39)	(1.02)	(−0.59)	(−2.70)	(1.06)	(−0.13)	(−0.65)	(−0.79)	
*ER*12	−0.577	0.400	−0.138	−0.498	0.290	−0.053	−0.159	−0.203	0.050
	(−1.32)	(0.99)	(−0.54)	(−2.82)	(1.12)	(−0.20)	(−0.72)	(−0.82)	
*ER*24	−0.362	0.301	−0.100	−0.384	0.279	−0.091	−0.164	−0.217	0.050
	(−1.01)	(0.90)	(−0.46)	(−2.76)	(1.31)	(−0.42)	(−0.91)	(−1.07)	

Notes: Independent variables include: *RSP* and *REM*, the 1-week returns (in percent) on the S&P 500 Index and the MSCI Emerging Asia Index; *REPO*, the 1-week change in overnight repo positions on Treasury bonds held by primary dealers in trillion dollars; *IIP*13, the 13-week change in index positions (million contracts) from the Masters algorithm; *MMS*13 and *OI*13, the 13-week changes in managed-money spread positions and total open interest (million contracts); *AVB*, the 1-week change in average basis; R_{lag}, lagged excess return; R^2_{adj}, the adjusted R-squared measure of goodness of fit. Dependent variable *ERm* is the 1-week realized excess return (in percent) on WTI crude oil futures contract that expires in *m* months. Independent variables are standardized by dividing by standard deviations. Standard errors are adjusted by Newey-West's (1994) approach and *t*-statistics are shown in parentheses.

index investment industry, and as a result, argue that the Fund's position data are representative of index participation and activity in commodity futures markets. These data are available beginning on February 13 2007.

Figure 10.3 shows the three standardized measures of index positions for WTI crude oil in SNG's sample period. The variables are standardized by subtracting the mean and then dividing by the sample deviation for comparison. The trends are broadly consistent for index positions from the Masters algorithm, the GSG, and the Fund, which is confirmed by high pairwise correlations above 0.8. There is a spike during 2007–2008 in index positions for the Masters algorithm, which is less pronounced or not present in the GSG or Fund measures. Figure 10.4 compares the two alternative measures of WTI index positions on a quarterly basis to IID positions, the most accurate available (Irwin and Sanders, 2012). The two alternative measures are consistent with the IID measure during SNG's sample period, providing support for the use of the variation of index positions from the GSG or the Fund as representative of the variation in total index positions in WTI crude oil.[15]

Using the GSG and Fund crude oil index positions, we re-estimate model Eqn 10.1 for SNG's sample period and present the results in Table 10.7. The contrast with the estimation results based on index positions from the Masters algorithm (Table 10.3) is striking. The size of coefficient estimates on *IIP*13 is reduced about two-thirds with the GSG and Fund position measures and none of the estimates across horizons are statistically significant. Coefficient estimates for the conditioning variables are somewhat larger and maintain similar significance levels as before. The fit of the regression is uniformly smaller due to the insignificance of index position changes. Clearly, there is no significant evidence that index position changes help predict excess returns for WTI crude oil using the GSG and Fund data.[16]

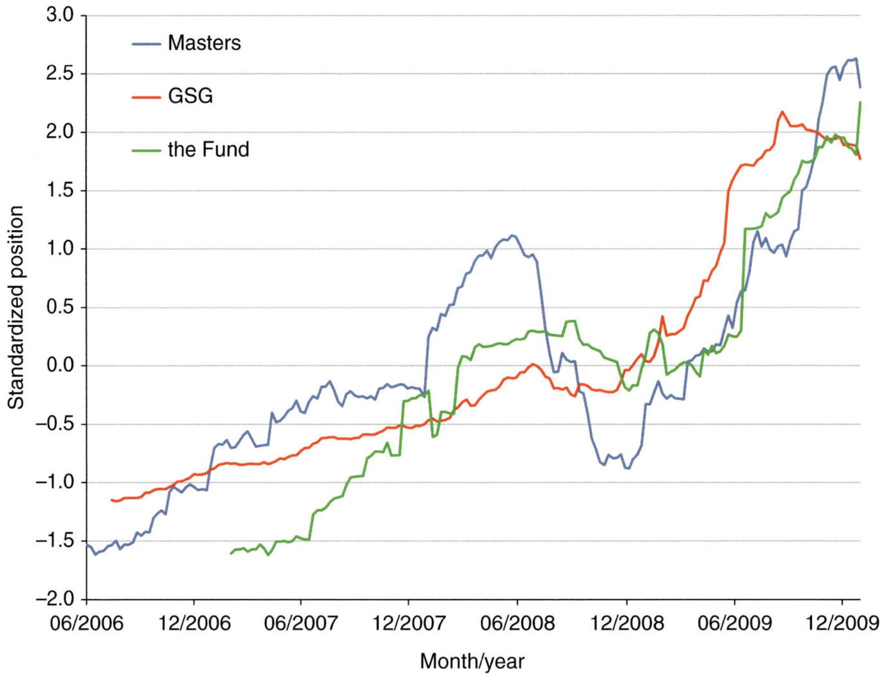

Note: The variables are standardized by subtracting the mean and then dividing by the sample deviation for comparison.

Fig. 10.3. Standardized index position measures from the Masters algorithm, the iShares S&P GSCI Commodity-Indexed Trust (GSG), and the Fund, weekly, June 13 2006–January 12 2010.

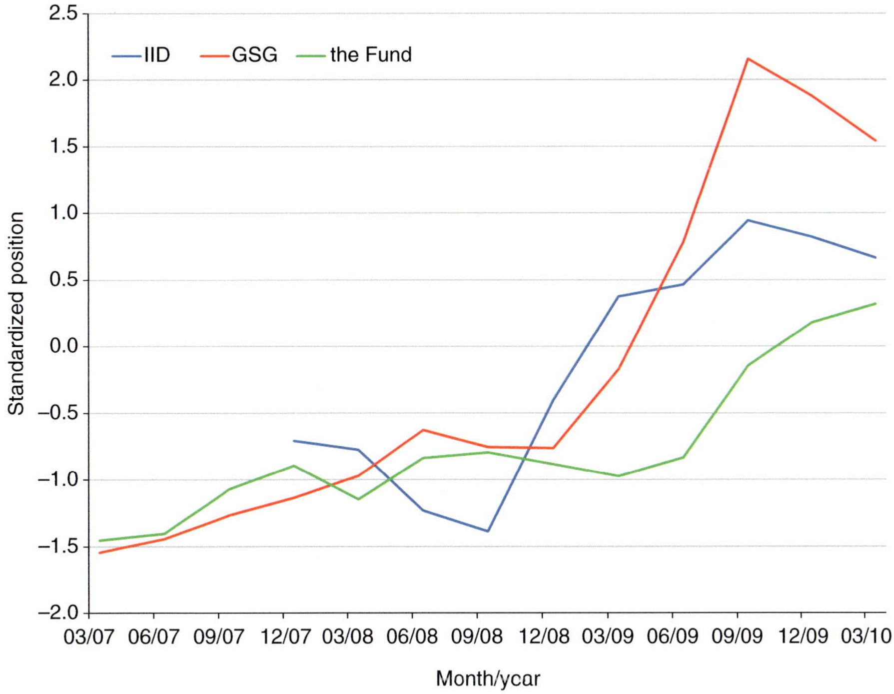

Note: The variables are standardized by subtracting the mean and then dividing by the sample deviation for comparison.

Fig. 10.4. Standardized index position measures from the *Index Investment Data* (IID) report, the GSG, and the Fund, quarterly, 2007Q1–2010Q1.

The estimation results presented in this section lead us to concur with Hamilton and Wu (2015) that 'the correlation identified by Singleton thus has no success at describing data since his paper was written and indeed seems not to have captured a stable predictive relation even within the sample that he analyzed' (p. 203). In view of these sample sensitivities and the fact that the relationship with crude oil futures disappears with the two alternative measures of index positions, it is reasonable to conclude that mapping crude oil positions from agricultural futures is unreliable due to substantial measurement errors. This raises the obvious question of the source of inaccuracy in mapping algorithms, and by implication, the spurious estimation of the relationship between index investment flows and crude oil futures prices found in several previous studies (e.g. Mayer, 2012; Singleton, 2014; Cheng *et al.*, 2015).

10.4 Mapping Algorithms

This section investigates the accuracy of mapping algorithms and the source of the spurious relationship between index investment flows and crude oil futures prices. First, we show theoretically that investing in a single commodity index requires an annually fixed-ratio relation in positions. Next, we describe the Masters algorithm and highlight the fixed-ratio relations it is based upon. Then, the fixed-ratio relations are formally tested using index positions from the SCOT and IID reports. Last, we explain the spurious relationship between index positions based on mapping algorithms and crude oil prices.

10.4.1 Replication of commodity indices

The two most popular commodity indices are the Standard and Poor's Goldman Sachs Commodity

Table 10.7. Coefficient estimates for 1-week excess return forecasting for WTI crude oil with index positions from alternative sources, June 13 2006–January 12 2010.[a]

	RSP	*REM*	*REPO*	*IIP*13	*MMS*13	*OI*13	*AVB*	R_{lag}	R^2_{adj}
Panel A: Index positions from the iShares S&P GSCI Commodity-Indexed Trust (GSG)									
*ER*1	0.936	−0.884	−1.253	0.568	1.243	0.291	−1.950	−0.555	0.25
	(2.08)	(−1.19)	(−2.57)	(1.27)	(2.97)	(0.47)	(−9.86)	(−1.85)	
*ER*3	0.942	−0.728	−1.186	0.512	1.048	0.403	−1.781	−0.687	0.24
	(2.01)	(−1.08)	(−2.48)	(1.22)	(3.00)	(0.62)	(−9.50)	(−2.15)	
*ER*6	0.987	−0.828	−1.080	0.463	0.906	0.373	−1.493	−0.486	0.23
	(2.28)	(−1.27)	(−2.34)	(1.19)	(2.86)	(0.60)	(−7.81)	(−1.55)	
*ER*12	1.080	−0.967	−0.982	0.397	0.734	0.357	−1.033	−0.321	0.20
	(2.62)	(−1.47)	(−2.15)	(1.13)	(2.53)	(0.63)	(−5.20)	(−0.97)	
*ER*24	0.997	−1.073	−0.814	0.296	0.581	0.313	−0.611	−0.139	0.16
	(2.53)	(−1.80)	(−1.95)	(0.97)	(2.32)	(0.65)	(−2.86)	(−0.40)	
Panel B: Index positions from the Fund									
*ER*1	1.003	−1.012	−1.449	0.452	1.270	0.299	−2.076	−0.537	0.26
	(2.23)	(−1.36)	(−2.60)	(0.68)	(2.89)	(0.43)	(−8.50)	(−1.42)	
*ER*3	1.020	−0.850	−1.360	0.370	1.073	0.413	−1.883	−0.677	0.25
	(1.95)	(−1.18)	(−2.38)	(0.62)	(2.97)	(0.54)	(−8.14)	(−1.90)	
*ER*6	1.050	−1.012	−1.266	0.358	0.935	0.362	−1.554	−0.388	0.24
	(2.14)	(−1.46)	(−2.31)	(0.69)	(2.85)	(0.49)	(−6.90)	(−1.08)	
*ER*12	1.130	−1.205	−1.187	0.400	0.756	0.336	−1.045	−0.171	0.22
	(2.31)	(−1.71)	(−2.24)	(0.87)	(2.51)	(0.51)	(−4.66)	(−0.44)	
*ER*24	1.039	−1.325	−1.030	0.418	0.592	0.300	−0.595	0.004	0.19
	(2.16)	(−2.11)	(−2.19)	(1.06)	(2.25)	(0.55)	(−2.56)	(0.01)	

[a]Independent variables include: *RSP* and *REM*, the 1-week returns (in percent) on the S&P 500 Index and the MSCI Emerging Asia Index; *REPO*, the 1-week change in overnight repo positions on Treasury bonds held by primary dealers in trillion dollars; *IIP*13, the 13-week change in index positions (million contracts) from the GSG or the Fund; *MMS*13 and *OI*13, the 13-week changes in managed-money spread positions and total open interest (million contracts); *AVB*, the 1-week change in average basis; R_{lag}, lagged excess return. Dependent variable *ERm* is the 1-week realized excess return (in percent) on WTI crude oil futures contract that expires in *m* months. Independent variables are standardized by dividing by standard deviations. Standard errors are adjusted by Newey-West's (1994) approach and *t*-statistics are shown in parentheses.

Index (S&P GSCI) and the Bloomberg Commodity Index (BCI),[17] which provide broad exposure to commodity futures via index-based instruments such as exchange-traded products (ETPs) and swaps. Providers of these instruments have to hedge their risk by replicating the index using underlying futures contracts. The way the indices are constructed implies annually fixed ratios in positions that are needed to replicate the index. Since the two commodity indices are constructed in similar ways, we focus on the S&P GSCI in the following analysis.

The S&P GSCI, first launched by Goldman Sachs in 1991, is the first major investable commodity index and serves as a benchmark for investment in the commodity markets. The index currently comprises 24 commodities from all commodity sectors, providing diversification across sectors and within each sector. The S&P GSCI holds long positions in the nearby contract, which usually is the most liquid contract, and rolls positions forward from contracts that are close to maturity to contracts that have later maturity dates. Roll over takes place within a 5 day window from the fifth to the ninth business day each month, and on each day, an equal amount (one-fifth) of the positions are rolled. The S&P GSCI is a production-weighted index with the weights based on the quantity of world production of each commodity over the last 5 years of available data. The adjustment of production happens once a year over the January roll period. We use the same concepts and notations as in the manual of S&P Dow Jones Indices (2015) throughout the paper.

We show in Appendix A that in order to replicate the S&P GSCI Total Return (TR) Index the required positions X_d^{c1} and X_d^{c2} for commodity $c1$ and $c2$ on day d must satisfy

$$\frac{X_d^{c1}}{X_d^{c2}} = \frac{CPW^{c1} / CS^{c1}}{CPW^{c2} / CS^{c2}} \tag{10.3}$$

where *CPW* is the contract production weight, which reflects the relative significance of each of the commodities included in the index, and *CS* is the contract size. Equation 10.3 shows that the position ratio is equal to the ratio of contract production weights scaled by contract size. Note that there is no time subscript in CPW^c, reflecting the fact that production data are only updated on an annual basis. Therefore, the ratio of futures positions required to track the S&P GSCI is annually fixed because both *CPW* and *CS* are constant during a calendar year except for the January roll period when *CPW*s are updated (see details in Appendix A at the end of the chapter).

10.4.2 The Masters algorithm

The Masters algorithm is a method of imputing commodity index positions for individual futures markets that are not available based on the weekly agricultural index positions from the SCOT report.[18] The basic idea is to take a SCOT market that is unique to one of the two major indices – S&P GSCI and BCI – and use the notional value of index trader positions in that market to estimate total investment in the particular index. The uniqueness is key because it allows all of the SCOT positions for the unique market to be attributed to investment in a particular index. Any non-SCOT market (e.g. WTI crude oil) is simply assigned a notional value according to its weight in the index and positions can be computed from the notional value in a straightforward manner.

These procedures can be stated formally by taking WTI crude oil as an example. Let $X_{d,SPGSCI}^{CL}$ and $X_{d,SPGSCI}^{FC}$ be positions for WTI crude oil (CL) and feeder cattle (FC) on day d that are tied to the S&P GSCI. Since feeder cattle is only included in the S&P GSCI and its positions are known from the SCOT report, the Masters algorithm implies

$$X_{d,SPGSCI}^{CL} = \frac{X_{d,SPGSCI}^{FC} * CS^{FC} * DCRP_d^{FC}}{w_d^{FC}} \frac{w_d^{CL}}{CS^{CL} * DCRP_d^{CL}} \tag{10.4}$$

where w_d^{CL} and w_d^{FC} are dollar weights of WTI crude oil and feeder cattle in the S&P GSCI on day d, *CS* is contract size, and $DCRP_d$ is daily contract reference price. $X_{d,SPGSCI}^{FC} * CS^{FC} * DCRP_d^{FC}$ is the dollar value of feeder cattle positions, which is divided by its own dollar weights to obtain the total dollar value of investment in the S&P GSCI. The crude oil position is then the product of total dollar value of index investment and dollar weights of WTI crude oil divided by the value of WTI contracts. A second position estimate for WTI crude oil can be generated from Kansas wheat that is also unique to the S&P GSCI, and Michael W. Masters (2008) suggests taking the average of imputed WTI positions based on Kansas wheat and feeder cattle. Similarly, we can generate the crude oil index positions that are tied to the BCI and the total crude oil index positions are simply the sum of the two implied from each index.[19]

We next show that the Masters algorithm is essentially based on an annually fixed-ratio relation. By definition, the dollar weight w_d^c for any commodity c is defined as (S&P Dow Jones Indices, 2015, p. 13),

$$w_d^c = \frac{CPW^c * DCRP_d^c}{\sum_c CPW^c * DCRP_d^c} \tag{10.5}$$

Substituting Eqn 10.5 into Eqn 10.4 and simplifying,

$$X_{d,SPGSCI}^{CL} = X_{d,SPGSCI}^{FC} \frac{CPW^{CL} / CS^{CL}}{CPW^{FC} / CS^{FC}} \tag{10.6}$$

which is exactly the same as the annually fixed ratio result shown in Eqn 10.3. Under the Masters assumption that all index investment tracks the two commodity indices, Eqn 10.6 is an exact relation because $X_{d,S\&PGSCI}^{FC}$ is only tied to a single index. This relation collapses for commodities that are tracked by more than one index. For example, consider crude oil and corn (C), both of which are included in the two indices. We have shown that the components tracking either index maintain annually fixed ratios, i.e. $\frac{X_{d,S\&PGSCI}^{CL}}{X_{d,S\&PGSCI}^{C}} = k_1$ and $\frac{X_{d,BCI}^{CL}}{X_{d,BCI}^{C}} = k_2$, where k_1 and k_2 are constants. However, the ratio of aggregate index positions $\frac{X_{d,S\&PGSCI}^{CL} + X_{d,BCI}^{CL}}{X_{d,S\&PGSCI}^{C} + X_{d,BCI}^{C}}$ is not constant but varies between k_1 and k_2. Consequently, the Masters assumption that all index investment tracks only two indices ensures the uniqueness in certain markets.[20]

10.4.3 Testing fixed-ratio relations

Given that the Masters algorithm implicitly assumes fixed-ratio relations in index positions, we formally test these relations using index positions from the SCOT and IID reports. Consider the following regression:

$$\begin{aligned} ratio_t = \alpha + \sum_{i=1}^{m-1} \alpha_i D_{it} + \sum_{i=1}^{m} \beta_i D_{it} trend \\ + \sum_{i=1}^{m} \gamma_i D_{it} trend^2 + \varepsilon_t \end{aligned} \tag{10.7}$$

where m is the number of years, D_{it} is equal to 1 if t falls into year i and 0 otherwise, and *trend* and *trend*^{2} are the linear and quadratic time trends. The α_i's reflect the inter-year differences while β_i's and γ_i's capture the intra-year variation. The null hypothesis of interest is $H_0 : \beta_1 = \beta_2 = \cdots = \beta_m = \gamma_1 = \gamma_2 = \cdots = \gamma_m = 0$, which can be tested using an ordinary F-test. Our test is conservative in the sense that an absence of linear and quadratic time trends does not imply annually constant index position ratios since there might be high-order variations, while rejection of H_0 suggests that index position ratios are not constant at least within a calendar year. Observations for the January roll periods are excluded because they are different from those in the rest of the year due to updates in production weights.

We first use index positions of commodities that are unique to the S&P GSCI to test the fixed ratio relation underlying the Masters algorithm. If index investment only tracks two commodity indices, the ratio in index positions between those unique commodities should be annually fixed except for the January roll period. The unique commodities include Kansas wheat, feeder cattle, and cocoa, whose index positions are available from the SCOT report. For illustration, Fig. 10.5 shows gross long index position ratios between Kansas wheat and feeder cattle for 2006–2015. Clearly, there is a substantial intra-year variation in the ratios. Panel A of Table 10.8 presents p-values of the F-test on H_0 for 2006–2015 and individual years. With a few exceptions, the tests suggest rejecting the fixed-ratio assumption at a significance level of 5%. The most likely reason is that these commodities are not truly unique so that only a portion of their index positions are tied to the S&P GSCI. Since the fixed-ratio relation does not hold between unique commodities, it is even more unlikely to hold between commodities that are tracked by more than one index. We confirm this by testing index position ratios calculated from all 66 pairs of the 12 agricultural commodities. Panel B of Table 10.8 presents the number and percentage of cases in which the fixed-ratio assumption is not rejected. Not surprisingly, no pair of commodities has an annually fixed ratio in index positions for 2006–2015 with a few exceptions for individual years. Results are similar when ratios are constructed based on net long positions, which are omitted to save space.

Next, we test the fixed-ratio relation in index positions between crude oil and all agricultural

Fig. 10.5. Index position ratios between Kansas wheat and feeder cattle (KW/FC) based on the weekly *Supplemental Commitments of Traders* (SCOT) report for January 3 2006–December 29 2015.

commodities. We use index positions from the IID report for this test because it is the only CFTC report containing WTI crude oil positions for index investors. The IID report is available from December 2007 on a quarterly basis and from June 2010 on a monthly basis. The ratios are calculated based on gross long positions and the results (not reported) are similar based on net long positions. Panel C of Table 10.8 presents the *p*-values of the *F*-test on H_0 using quarterly (Q) and monthly (M) positions. The first 12 columns present results for ratios between WTI crude oil and individual agricultural commodities and the last column is for ratios between WTI crude oil and the weighted average of all 12 SCOT markets.[21] In all but a few cases, we reject the null hypothesis, indicating a lack of constant relations in index position ratios between WTI crude oil and individual agricultural commodities or their aggregate. These results suggest that it can be problematic to generate crude oil index positions from index positions of individual agricultural commodities or their aggregate.[22]

Overall, the test results provide strong evidence against fixed-ratio relations between index commodities, consistent with the argument that the fixed-ratio relation does not hold for aggregate positions. In fact, there are many other commodity indices or sub-indices in commodity markets.[23] The fixed-ratio relations fail because index investment in all commodity markets track multiple indices. In addition, the absolute error can be greatly amplified through mapping if WTI crude oil has a larger weight than agricultural commodities. For example, the average multiplier between WTI crude oil and feeder cattle is 48 for 2006–2010, which means that a 1000-contracts increase in positions of feeder cattle raises imputed positions of WTI crude oil by 48,000 contracts. This further undermines the accuracy of mapping algorithms. In sum, mapping algorithms may produce inaccurate and misleading measures of index positions for WTI crude oil due to the failure of the fixed-position ratio assumption.

10.4.4 Decomposition of the mapping algorithm results

We compare index positions in WTI crude oil from the Masters algorithm with IID position data – the most precise and reliable measure available – in order to better understand the

Table 10.8. The *p*-values for testing the fixed-ratio relation using different sources of index positions.[a]

	2006–2015	2006	2007	2008	2009	2010	2011	2012	2013	2014	2015		
Panel A: Markets that are unique to the S&P GSCI													
KW/FC	0.00	0.04	0.00	0.00	0.00	0.00	0.62	0.00	0.00	0.04	0.00		
KW/CC	0.00	0.00	0.00	0.01	0.00	0.00	0.11	0.00	0.00	0.00	0.05		
FC/CC	0.00	0.00	0.00	0.04	0.00	0.00	0.82	0.00	0.00	0.00	0.00		
Panel B: 12 agricultural commodity markets from the SCOT report													
Number	0	2	6	12	1	8	8	5	7	9	9		
Percentage	0.0	3.0	9.1	18.2	1.5	12.1	12.1	7.6	10.6	13.6	13.6		
# of pairs	660	66	66	66	66	66	66	66	66	66	66		
Panel C: WTI crude oil and agricultural commodities from the IID report													
	CC	KC	C	CT	FC	LH	LC	BO	S	SB	W	KW	AVE
CL (Q)	0.24	0.07	0.01	0.00	0.00	0.00	0.00	0.06	0.03	0.13	0.02	0.03	0.00
CL (M)	0.00	0.00	0.00	0.19	0.01	0.00	0.00	0.08	0.00	0.00	0.00	0.00	0.00

[a] Panel A presents the *p*-values of *F*-test on the null hypothesis that index position ratios are annually fixed between commodities that are unique to the S&P GSCI using the SCOT index positions for January 3 2006–December 29 2015. Panel B presents the number and percentage of cases in which the fixed-ratio assumption is not rejected for the 12 agricultural commodity markets included in the SCOT report. Panel C presents the *p*-values of *F*-test on the null hypothesis that index position ratios between WTI crude oil (CL) and each of the agricultural commodities are annually fixed using index positions from the IID report. Q represents quarterly data from December 2007 to September 2015 and M represents monthly data from June 2010 to October 2015. Agricultural commodities include cocoa (CC), coffee (KC), corn (C), cotton (CT), feeder cattle (FC), lean hogs (LH), live cattle (LC), soybean oil (BO), soybeans (S), sugar (SB), Chicago wheat (W), and Kansas wheat (KW). AVE is the weighted-average position measure of all 12 agricultural markets (see Online Supplemental Appendix for details of it construction). The January roll period is excluded in all tests.

source of the inaccuracy from these algorithms. Figure 10.6 shows the two index position measures for December 2007 through September 2015. The mapping algorithm overestimates index positions for WTI crude oil most of the time in this period. One of the largest deviations occurred in 2008. The Masters index positions increased from 527,000 contracts to 687,000 contracts in the first half of 2008 and dropped back to 447,000 contracts at the end of the year. In contrast, the IID positions fell by 47,000 contracts over the same period. Another dramatic mismeasurement occurred in late 2014, when the IID positions increased substantially from 463,000 contracts to 639,000 contracts but the measure from mapping algorithms showed no sign of increase. Consequently, the Masters algorithm provides a poor estimate of the magnitude and direction of WTI crude oil index positions, especially during 2008, in which a seemingly significant price impact was identified.[24]

We decompose index positions from the mapping algorithm to more precisely determine the source of mismeasurement. Figure 10.7(a) shows crude oil index positions from the Masters algorithm and the two components tied to the S&P GSCI and BCI. The pattern of total index positions is mainly determined by its S&P GSCI component, consistent with the fact that the S&P GSCI is heavily concentrated in energy markets. By construction (Eqn 10.4), the pattern of the S&P GSCI component is driven by changes in index positions of feeder cattle and Kansas wheat (Fig. 10.7b, c). Index positions in these two markets display very different patterns, especially during 2008 when a spike occurred in feeder cattle but not in Kansas wheat. In particular, the gross long index positions of feeder cattle increased from 9000 contracts to 11,600 contracts in the first half of 2008 and dropped back to 7000 contracts at the end of 2008. In other words, it appears that the spike of index positions in feeder cattle (Fig. 10.7b) causes a spike in the Masters estimate of crude oil index positions (Fig. 10.7a). If this is true, index positions in feeder cattle should have a similar impact on crude oil prices as the original Masters algorithm estimates. Figure 10.8 shows nearby futures prices of WTI crude oil and 13-week index position changes in Kansas wheat, feeder cattle, and soybean oil for the SNG's sample period. A common decline in prices and index positions is

Fig. 10.6. WTI crude oil index positions from the *Index Investment Data* (IID) report and the Masters algorithm, quarterly, December 2007–September 2015.

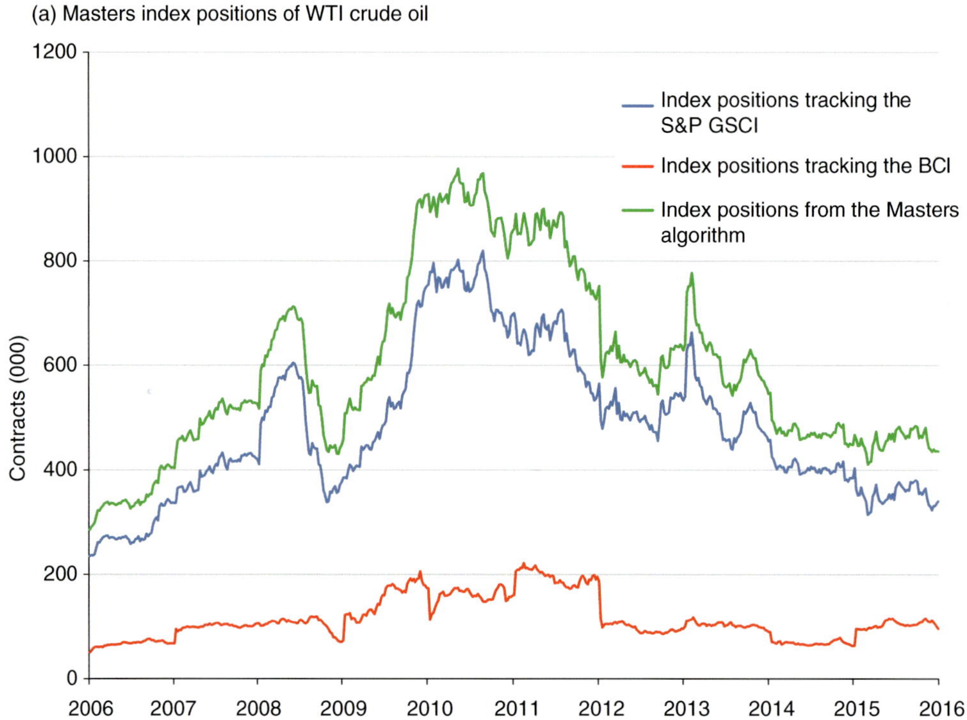

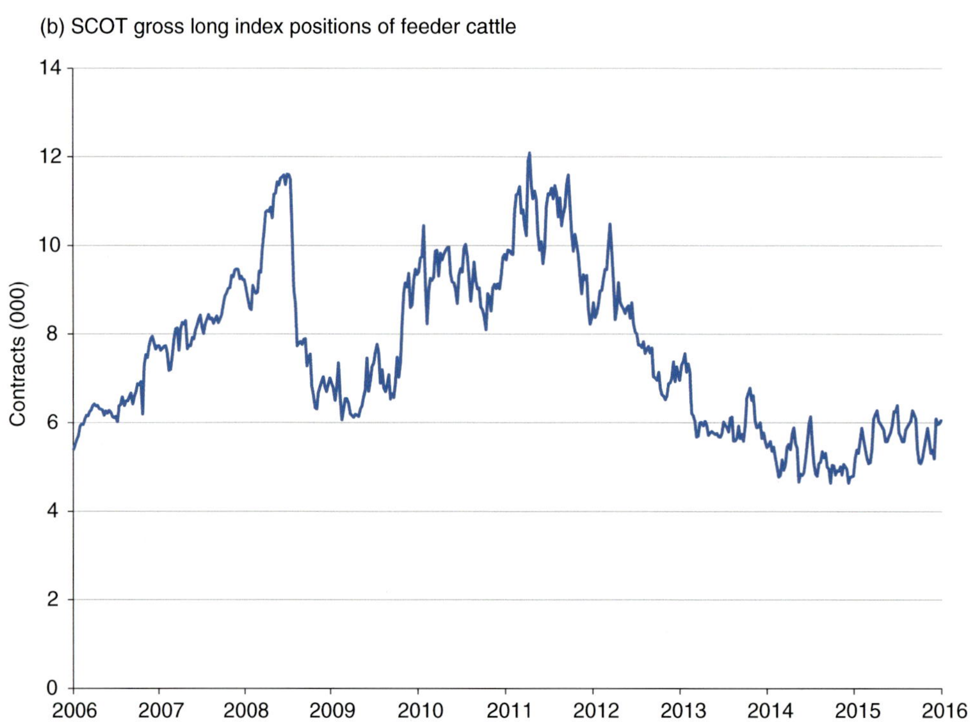

Fig. 10.7. Decomposition of index positions from the Masters algorithm, weekly, January 3 2006–December 29 2015.

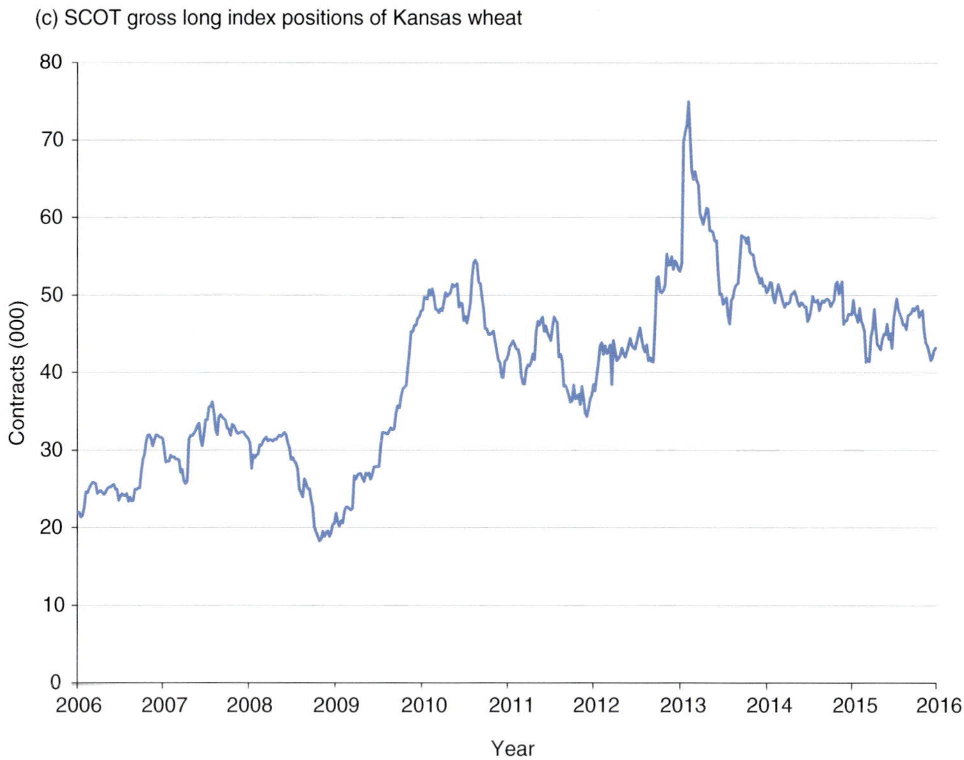

Fig. 10.7. Continued

observed for feeder cattle and soybean oil during 2008. For Kansas wheat, the index positions dropped much earlier than prices. To formally test this relationship, we re-estimate model Eqn 10.1, in which $IIP13_t$ represents 13-week changes in index positions of Kansas wheat, feeder cattle, or soybean oil, and all other predictor variables remain the same. The estimation results in Table 10.9 show that index positions of feeder cattle have an economically large and statistically significant impact on crude oil prices and Kansas wheat does not. Although index positions in soybean oil are also significant, the magnitude is much smaller since the BCI component is relatively small in generating the imputed index positions based on the Masters algorithm.

For additional perspective, Fig. 10.9 shows gross long index positions for the 12 SCOT agricultural commodities and the weighted-average measure (in bold) of WTI crude oil positions used in some studies.[25] Most markets display a notable spike in positions with different degrees of magnitude during 2007–2008 (cocoa, coffee, corn, cotton, feeder cattle, lean hogs, live cattle, soybeans, sugar), while a few others do so to a lesser degree or not at all (Chicago wheat, Kansas wheat, soybean oil). The weighted average shows a similar spike to the Masters measure during 2007–2008, making it a poor estimate of crude oil index positions relative to the IID measure. Regression results (Online Supplemental Appendix) for nearby futures show that SCOT index positions in 11 of the 12 SCOT agricultural markets have significant impacts on crude oil prices.[26] In ten of the SCOT markets, the estimated coefficient on $IIP13_t$ is between 1.2 and 1.6, close to the 1.792 coefficient estimate for WTI crude oil using nearby futures and the Masters algorithm (see Table 10.3). This shows that using most of the 12 SCOT markets would have generated similar regression results to the Masters algorithm in WTI crude oil.[27]

The common spike in agricultural index positions in 2007–2008 is likely due to the rise of specialized agricultural index funds during this period. Figure 10.10 shows the weighted-average position measure and total assets of the PowerShares DB Agriculture Fund (DBA), collected from Bloomberg. The DBA fund tracks an index based on prices for 11 of the 12

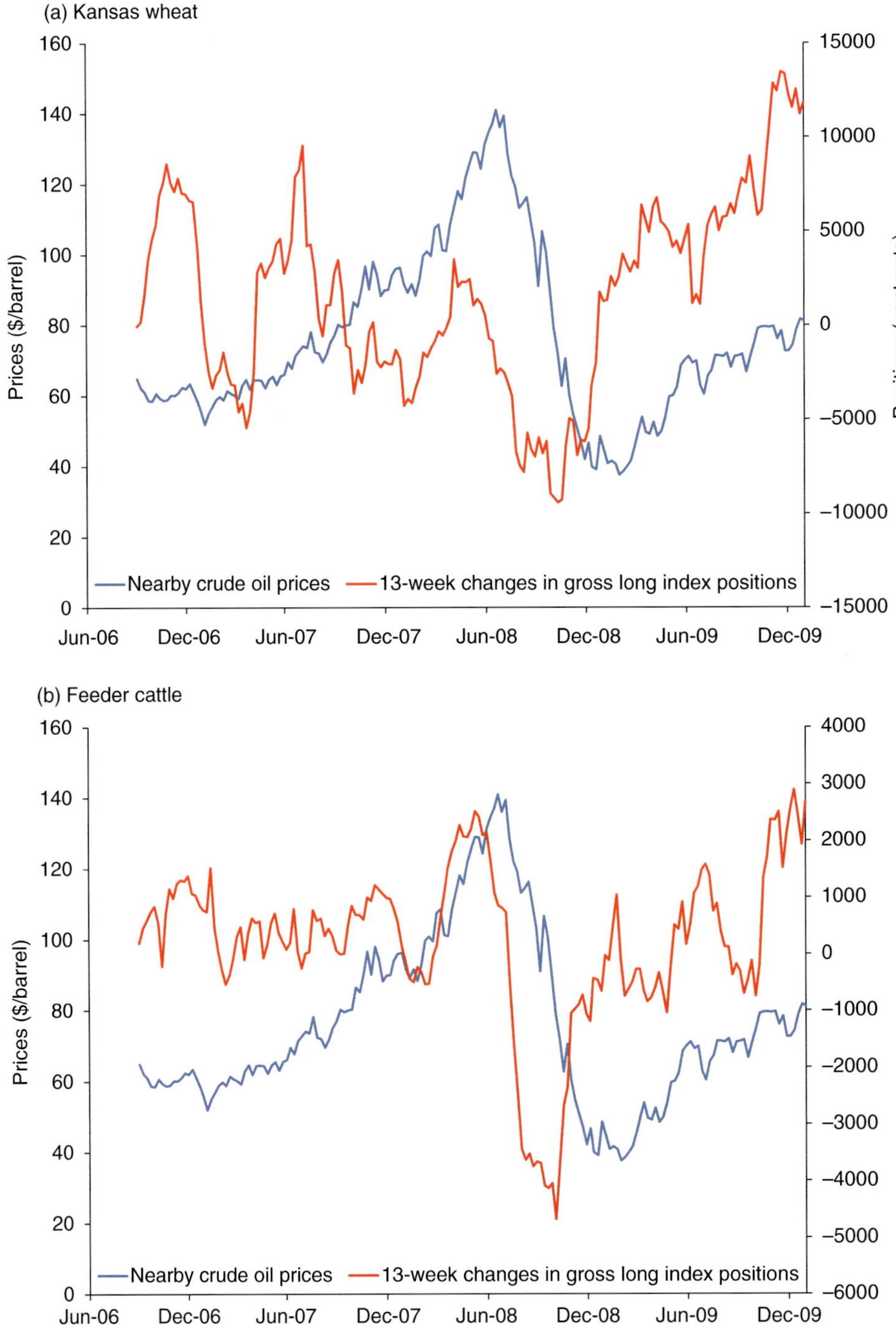

Fig. 10.8. Nearby prices of WTI crude oil and 13-week changes in gross long index positions of Kansas wheat, feeder cattle, and soybean oil, January 3 2006–January 12 2010.

Fig. 10.8. Continued

SCOT agricultural futures markets (soybean oil is excluded). The close parallel in movements in the weighted-average measure and total assets of DBA during 2007–2008 is striking. One possible explanation is that DBA total assets were more affected by rising agricultural prices than new investment flows, which would create the appearance of a larger boom in positions than actually took place. However, the total net assets of DBA increased almost six times from its launch in January 2007 through mid-2008, peaking at just under $3 billion. The sixfold increase in assets was several times larger than the increase in agricultural futures prices during the same period. This indicates the surge in DBA total assets during 2007–2008 was mainly due to new investment flows, consistent with the spike in SCOT positions and the weighted average in Fig. 10.9. Since the weighted-average measure is weighted by fixed futures prices, this measure is not affected by the spike in agricultural prices during 2007–2008. So, the evidence in Fig. 10.10 clearly points to a surge in agricultural index investment and associated market positions during 2007–2008.[28] The lack of a pronounced spike in the IID, GSG, and Fund WTI position measures indicates that the surge in agricultural index positions was not accompanied by a similar spike in crude oil index positions.

Overall, the pattern of index positions from mapping algorithms is determined by index position changes in agricultural commodity markets. There was a surge in agricultural index positions during 2007–2008, most likely due to the rise of specialized agricultural index funds like DBA. Illustrating the changing mix of index investment, the surge in agricultural index positions during 2007–2008 was not matched by a surge in crude oil index positions. As a result, the coincidental timing of the spike in agricultural index positions and the spike in crude oil prices during 2007–2008 creates the spurious impact of index positions on crude oil futures prices found in earlier studies that employ mapping algorithms (e.g. Mayer, 2012; Singleton, 2014; Cheng *et al.*, 2015).

10.5 Summary and Conclusions

This article investigates how index positions based on mapping algorithms lead to a spurious

Table 10.9. Coefficients estimates for 1-week excess return forecasting for WTI crude oil with index positions in Kansas wheat, feeder cattle, and soybean oil, June 13 2006–January 12 2010.[a]

	RSP	*REM*	*REPO*	*IIP*13	*MMS*13	*OI*13	*AVB*	R_{lag}	R^2_{adj}
Kansas wheat									
*ER*1	0.859	−0.947	−1.189	0.885	1.270	−0.199	−1.891	−0.495	0.25
	(2.45)	(−1.46)	(−3.11)	(1.29)	(2.99)	(−0.63)	(−13.69)	(−1.92)	
*ER*3	0.874	−0.790	−1.124	0.657	1.063	0.016	−1.721	−0.573	0.24
	(2.33)	(−1.27)	(−2.37)	(1.13)	(3.01)	(0.03)	(−10.85)	(−1.81)	
*ER*6	0.928	−0.888	−1.022	0.561	0.918	0.041	−1.442	−0.372	0.22
	(2.34)	(−1.38)	(−2.12)	(1.09)	(2.83)	(0.07)	(−8.37)	(−1.25)	
*ER*12	1.031	−1.024	−0.931	0.484	0.749	0.069	−1.000	−0.226	0.19
	(2.52)	(−1.54)	(−2.05)	(1.08)	(2.55)	(0.13)	(−4.99)	(−0.70)	
*ER*24	0.963	−1.126	−0.772	0.431	0.593	0.075	−0.588	−0.074	0.16
	(2.33)	(−1.86)	(−1.98)	(1.19)	(2.31)	(0.18)	(−2.78)	(−0.22)	
Feeder cattle									
*ER*1	0.870	−0.895	−1.172	1.191	1.179	−0.280	−1.964	−0.554	0.27
	(2.91)	(−1.75)	(−3.19)	(3.21)	(4.15)	(−1.28)	(−13.17)	(−1.64)	
*ER*3	0.871	−0.757	−1.099	1.249	0.982	−0.183	−1.793	−0.680	0.27
	(4.42)	(−2.43)	(−4.31)	(5.07)	(5.90)	(−1.06)	(−19.51)	(−3.65)	
*ER*6	0.924	−0.855	−0.995	1.254	0.841	−0.201	−1.514	−0.512	0.26
	(3.89)	(−2.12)	(−2.90)	(3.71)	(4.36)	(−0.91)	(−11.40)	(−2.06)	
*ER*12	1.027	−0.990	−0.905	1.216	0.675	−0.191	−1.068	−0.386	0.24
	(3.21)	(−1.94)	(−2.12)	(3.07)	(2.92)	(−0.66)	(−6.09)	(−1.62)	
*ER*24	0.954	−1.098	−0.749	1.091	0.523	−0.159	−0.644	−0.213	0.20
	(2.98)	(−2.03)	(−1.62)	(3.33)	(2.34)	(−0.54)	(−3.59)	(−0.87)	
Soybean oil									
*ER*1	0.871	−0.794	−1.267	0.645	1.234	0.165	−1.918	−0.513	0.25
	(2.17)	(−1.25)	(−2.47)	(2.03)	(3.38)	(0.23)	(−10.40)	(−1.65)	
*ER*3	0.889	−0.679	−1.178	0.425	1.035	0.288	−1.741	−0.583	0.23
	(2.28)	(−1.19)	(−2.37)	(1.85)	(3.26)	(0.40)	(−10.35)	(−1.80)	
*ER*6	0.944	−0.792	−1.066	0.346	0.894	0.274	−1.461	−0.385	0.22
	(2.39)	(−1.34)	(−2.15)	(1.72)	(3.00)	(0.39)	(−8.31)	(−1.26)	
*ER*12	1.042	−0.934	−0.974	0.354	0.729	0.269	−1.017	−0.251	0.19

Continued

Table 10.9. Continued.

	RSP	*REM*	*REPO*	*IIP*13	*MMS*13	*OI*13	*AVB*	R_{lag}	R^2_{adj}
	(2.52)	(−1.52)	(−2.07)	(1.73)	(2.58)	(0.43)	(−5.09)	(−0.74)	
*ER*24	0.966	−1.038	−0.816	0.386	0.575	0.252	−0.602	−0.112	0.16
	(2.31)	(−1.84)	(−2.09)	(1.91)	(2.27)	(0.52)	(−2.78)	(−0.32)	

[a]Independent variables include: *RSP* and *REM*, the 1-week returns (in percent) on the S&P 500 Index and the MSCI Emerging Asia Index; *REPO*, the 1-week change in overnight repo positions on Treasury bonds held by primary dealers in trillion dollars; *IIP*13, the 13-week change in index positions (million contracts) in Kansas wheat, feeder cattle, and soybean oil from the CFTC SCOT report; *MMS*13 and *OI*13, the 13-week changes in managed-money spread positions and total open interest (million contracts) in WTI crude oil; *AVB*, the 1-week change in average basis for WTI crude oil; R_{lag}, lagged excess return. Dependent variable *ERm* is the 1-week realized excess return (in percent) on WTI crude oil futures contract that expires in *m* months. Independent variables are standardized by dividing by standard deviations. Standard errors are adjusted by Newey-West's (1994) approach and *t*-statistics are shown in parentheses.

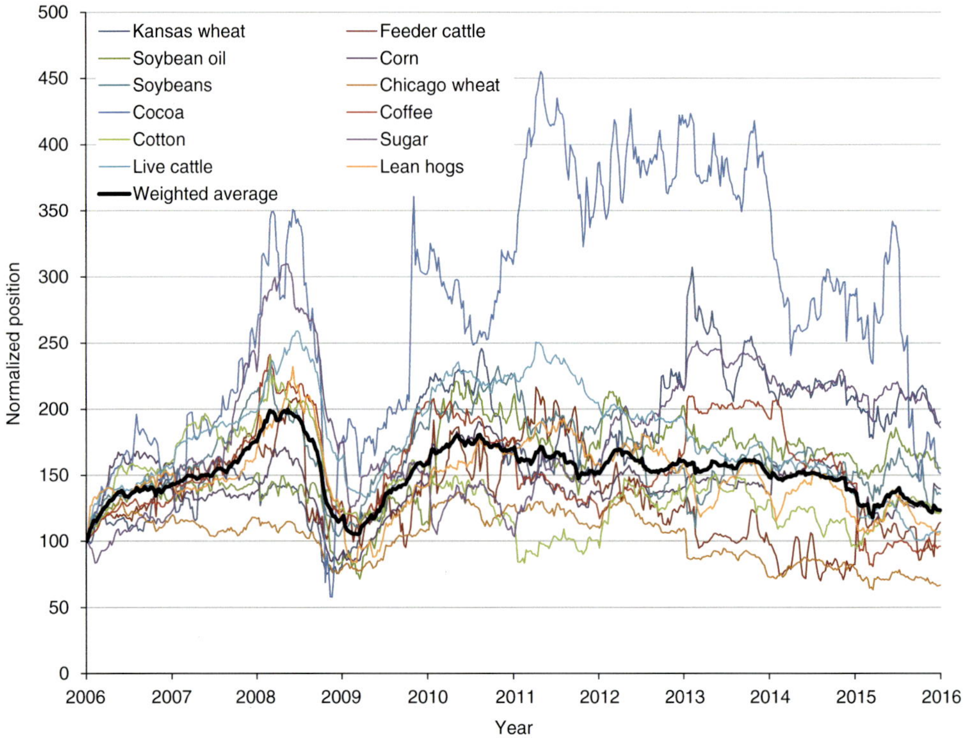

Fig. 10.9. Index position measures of WTI crude oil from the weighted-average algorithm and net long index positions of 12 agricultural commodities from the SCOT report, normalized by setting January 3 2006 = 100, weekly, January 3 2006–December 29 2015.

impact on crude oil prices. Using the same forecasting model as Singleton (2014), we confirm previous findings that index positions based on the best-known algorithm – the Masters algorithm – seem to have a large and statistically significant impact on excess returns for WTI crude oil futures. However, the forecasting power comes mainly from 2008, especially the second half of 2008, and the price impact is reversed in a post-sample period. Using two alternative index position measures we fail to find any significant impact of index positions on crude oil prices. In light of the sensitivity to market events in 2008 – the post-sample period – and alternative position measures, results from previous research based on mapping algorithms are highly likely to be spurious.

To evaluate the accuracy of mapping algorithms, we show theoretically that the mapping algorithms implicitly assume fixed-ratio relations in index positions between WTI crude oil and agricultural commodities. A formal test is implemented using position data from the US Commodity Futures Trading Commission (CFTC) *Supplemental Commitments of Traders* (SCOT) and *Index Investment Data* (IID) reports and this fixed-ratio assumption is rejected. Compared with the IID measure – the most accurate available – the Masters algorithm provides poor estimates of index positions for WTI crude oil in both direction and magnitude. In particular, index position measures based on the Masters algorithm shows a clear spike during 2007–2008 while the IID measure does not. Decomposition shows that the spike in index positions based on the Masters algorithm is largely driven by changes in positions of feeder cattle – one of the unique agricultural markets underlying this mapping algorithm. Within the same regression framework, we show that 13-week changes in index positions of feeder cattle have a significant impact on crude oil prices, which is obviously spurious. We also show that there was a surge in agricultural index investment during

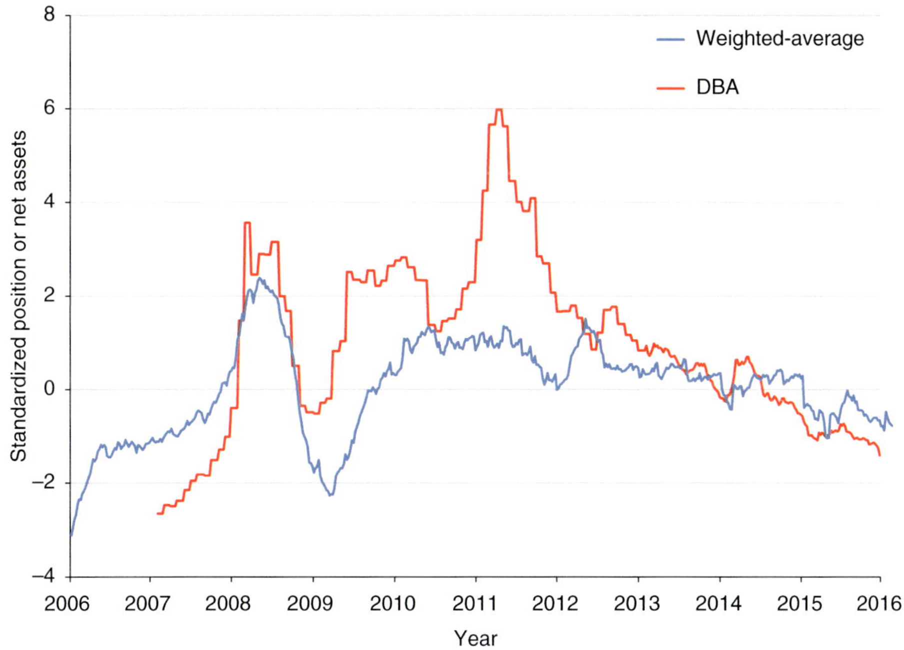

Fig. 10.10. The weighted-average index position measure and the total net assets of the PowerShares DB Agriculture Fund (DBA), standardized, January 3 2006–December 29 2015.

2007–2008, most likely due to the rise of specialized agricultural index funds, that was not matched by a surge in crude oil index investment. Hence, the idiosyncratic spike in agricultural index positions during 2007–2008, together with a coincidental spike in crude oil prices, causes the spurious relationship between index investment and crude oil prices found in previous studies that employ mapping algorithms (e.g. Mayer, 2012; Singleton, 2014; Cheng *et al.*, 2015).

Our results do not rule out a rational ('financialization') impact of index investment on commodity futures prices. For example, in the theoretical models of Acharya *et al.* (2013), Etula (2013), Brunetti and Reiffen (2014), and Hamilton and Wu (2014, 2015) index traders help reduce the risk premium that accrues to speculators by providing long positions to hedgers, although they may offer fewer long positions in times of financial stress (Cheng *et al.*, 2015). A recent paper by Basak and Pavlova (2016) explores theoretically how the financialization may affect commodity price dynamics in terms of price levels, volatilities, correlations, and inventories by comparing economies with and without institutional investors. Rational impacts along these lines, if evident, are likely to be economically small but long-lasting (e.g. Henderson *et al.*, 2015). In contrast, our study adds to the growing body of literature showing that buying pressure from index funds did not cause a massive bubble in commodity futures prices during 2007–2008. In particular, our results are consistent with previous studies (Irwin and Sanders, 2012; Sanders and Irwin, 2013, 2014; Hamilton and Wu, 2015), which question the accuracy of the Masters algorithm and regression results based upon it. Furthermore, we present direct evidence that index position measures based on mapping algorithms are misleading and lead to spurious estimates of the impact on crude oil prices.

Our results have important implications for researchers interested in the role of index funds in commodity futures markets. Clearly, mapping algorithms should not be used to infer index positions in non-agricultural futures markets from index positions in agricultural futures markets. Across all markets, the IID report provides the most accurate and preferred measure of total index positions. However, the CFTC discontinued the IID report in November 2015, and it was only available on a quarterly or monthly basis when it was released, which severely restricts the number of observations available for time-series statistical tests. Unfortunately, alternatives to the IID report are limited for non-agricultural futures markets such as crude oil and metals. The CFTC publishes the weekly DCOT report that includes swap dealer positions some have used to proxy index positions in energy and metals futures markets. Due to the very active non-index swap trade and internal netting of positions by swap dealers in the energy and metals markets, DCOT swap dealer positions are a poor proxy for total index positions in these markets (Irwin and Sanders, 2012). The best alternatives for energy and metals futures markets appear to be the index position measures based on the iShares S&P GSCI Commodity-Indexed Trust (GSG) and the private fund used in this article.

Our findings also have important implications for policy makers. According to the Commodity Exchange Act (CEA), position limits are 'necessary' if excessive speculation causes 'sudden or unreasonable fluctuations or unwarranted changes in the price of such commodity' (US House of Representatives, 2019). Our results undermine the necessity for position limits by providing evidence against the seemingly significant impact of index investment on crude oil prices found in several previous studies. For example, Singleton's (2014) study is included in the CFTC's list of 'preferred studies' in their latest proposed rule-making on position limits (CFTC, 2016); a highly debatable classification in the light of the results presented here. Our findings support the position of the CFTC Energy and Environmental Markets Advisory Committee (EEMAC) in February 2016, that 'there is little to no evidence that the CFTC's proposed rule mandating new federal position limits is sufficiently "necessary" to satisfy the explicit requirement under the CEA.'[29] New regulations on speculative positions may be expensive to market participants in terms of potential effects on trading liquidity and hedging effectiveness. Consequently, any policy move should be cautious given the lack of compelling evidence about a 'large' adverse impact on crude oil prices and the potential costs to market participants of additional regulations.

Appendix A. Derivation of Equation (10.3)

This Appendix provides a proof of Eqn 10.3 (i.e. the positions needed to replicate the S&P GSCI maintain fixed ratios within a calendar year except for the January roll period).

The S&P GSCI TR Index is constructed in a cumulative way (S&P Dow Jones Indices, 2015, p. 42),[30]

$$S_d = S_{d-1}\left(1 + CDR_d + TBR_d\right) \qquad (A10.1)$$

where S_d is the level of the S&P GSCI TR Index, $CDR_d \equiv \frac{TDW_d}{TDW_{d-1}} - 1$ is the contract daily return computed as percentage change of the total dollar weight (*TDW*), and TBR_d is the Treasury bill return on day *d*. Substituting CDR_d into Eqn A10.1,

$$S_d = S_{d-1}\frac{TDW_d}{TDW_{d-1}} + S_{d-1} * TBR_d \qquad (A10.2)$$

Depending on the rolling activity, we proceed with the derivation in three cases.

Case 1. Non-roll days

On non-roll days, the notional value of the index on day *d* is equal to the sum of notional values of positions held in each contract of commodity *c* plus the interest earned on the notional value on the previous day,

$$S_d = \sum_c \left(X_d^c * CS^c * DCRP_d^c\right) + S_{d-1} * TBR_d \qquad (A10.3)$$

where X_d^c is the number of contracts, CS^c is the contract size, and $DCRP_d^c$ is the daily contract reference price on day *d*. Combining Eqns A10.2 and A10.3,

$$S_{d-1}\frac{TDW_d}{TDW_{d-1}} = \sum_c \left(X_d^c * CS^c * DCRP_d^c\right) \qquad (A10.4)$$

On non-roll days, TDW_d is the sum of product of contract production weights and daily contract reference prices (S&P Dow Jones Indices, 2015, p. 33),

$$TDW_d = \sum_c \left(CPW^c * DCRP_d^c \right) \quad (A10.5)$$

Substituting Eqn A10.5 into Eqn A10.4,

$$\sum_c \left(\frac{S_{d-1}}{TDW_{d-1}} CPW^c * DCRP_d^c \right) = \sum_c \left(X_d^c * CS^c * DCRP_d^c \right) \quad (A10.6)$$

Note that $DCRP_d^c$ is the only unknown variable on day $d - 1$. To replicate the index, one has to specify X_d^c on day $d - 1$ to make sure that Eqn A10.6 holds for arbitrage $DCRP_d^c$. This implies that the coefficients of $DCRP_d^c$ on both sides must be equal,

$$X_d^c = \frac{S_{d-1}}{TDW_{d-1}} \frac{CPW^c}{CS^c} \quad (A10.7)$$

Equation A10.7 tells that the number of contracts that is needed to replicate the S&P GSCI TR Index consists of two parts. The first part $\left(\frac{S_{d-1}}{TDW_{d-1}}\right)$ is predetermined on day $d - 1$ and the same for all commodities, while the second part $\left(\frac{CPW^c}{CS^c}\right)$ is time invariant. The position ratio between any two commodities ($c1$ and $c2$) is then,

$$\frac{X_d^{c1}}{X_d^{c2}} = \frac{CPW^{c1} / CS^{c1}}{CPW^{c2} / CS^{c2}} \quad (A10.8)$$

which is constant and depends only on contract size and production weight.

Case 2. Non-January roll days

On roll days investors have to switch contracts over a 5-day window – the fifth to ninth business day every month. Let $X1_d^c$ and $X2_d^c$ be the positions of the old and new contracts. In this case, the notional value of the index on day d is equal to the sum of notional values of $X1_d^c$ and $X2_d^c$ plus the interest on the previous day,

$$S_d = \sum_c CS^c \left(X1_d^c * DCRP1_d^c + X2_d^c * DCRP2_d^c \right) + S_{d-1} * TBR_d \quad (A10.9)$$

where $DCRP1_d^c$ and $DCRP2_d^c$ are the daily contract reference prices of the old and new contracts. Combining Eqns A10.2 and A10.9,

$$S_{d-1} \frac{TDW_d}{TDW_{d-1}} = \sum_c CS^c \left(X1_d^c * DCRP1_d^c + X2_d^c * DCRP2_d^c \right) \quad (A10.10)$$

On non-January roll days, TDW_d is defined as follows (S&P Dow Jones Indices, 2015, p. 39),

$$TDW_d = \sum_c \left[CRW_d * DCRP1_d^c + (1 - CRW_d) * DCRP2_d^c \right] \quad (A10.11)$$

where CRW_d and $1 - CRW_d$ are the contract roll weights of the old and new contracts. The CRW_d is set to be 0.8, 0.6, 0.4, 0.2, and 0 on respective days over a 5-day roll period. Substituting Eqn A10.11 into Eqn A10.10,

$$\sum_c \frac{S_{d-1}}{TDW_{d-1}} CPW^c \left[CRW_d * DCRP1_d^c + (1 - CRW_d) * DCRP2_d^c \right] = \sum_c CS^c \left(X1_d^c * DCRP1_d^c + X2_d^c * DCRP2_d^c \right) \quad (A10.12)$$

To ensure that Eqn A10.12 holds for arbitrage prices, the coefficients of $DCRP1_d^c$ $DCRP1_d^c$ and $DCRP2_d^c$ on both sides must be equal:

$$X1_d^c = \frac{S_{d-1} * CRW_d}{TDW_{d-1}} \frac{CPW^c}{CS^c}, \quad X2_d^c = \frac{S_{d-1} * (1 - CRW_d)}{TDW_{d-1}} \frac{CPW^c}{CS^c} \quad (A10.13)$$

The total position is:

$$X1_d^c + X2_d^c = \frac{S_{d-1}}{TDW_{d-1}} \frac{CPW^c}{CS^c} \quad (A10.14)$$

and the ratio of total positions for any two commodities $c1$ and $c2$ is:

$$\frac{X1_d^{c1} + X2_d^{c1}}{X1_d^{c2} + X2_d^{c2}} = \frac{CPW^{c1} / CS^{c1}}{CPW^{c2} / CS^{c2}} \quad (A10.15)$$

which is exactly the same as Eqn A10.8 in Case 1.

Case 3. January roll days

On January roll days there are two changes for the index – switching contracts and updating production weights. Let $X1_d^c$ and $X2_d^c$ be the positions of the old and new contracts. We can obtain the same Eqn A10.10 as in Case 2. Here, the difference is TDW_d, which is defined as follows (S&P Dow Jones Indices, 2015, p. 40):

$$TDW_d = \sum_c + CPW_{new}^c \begin{bmatrix} \frac{NC_{new}}{NC_{old}} CPW_{old}^c * CRW_d * DCRP1_d^c \\ *\left(1 - CRW_d\right) * DCRP2_d^c \end{bmatrix} \tag{A10.16}$$

where CPW_{old} and CPW_{new} are the production weights of the previous and current year and NC is the normalizing constant, which is introduced to assure the continuity of the index and to allow comparison over time.[31] Substituting Eqn A10.16 into Eqn A9.10,

$$\sum_c \frac{S_{d-1}}{TDW_{d-1}} \begin{bmatrix} \frac{NC_{new}}{NC_{old}} CPW_{old}^c * CRW_d * DCRP1_d^c + CPW_{new}^c \\ *\left(1 - CRW_d\right) * DCRP2_d^c \end{bmatrix} = \sum_c CS^c \left(X1_d^c * DCRP1_d^c + X2_d^c * DCRP2_d^c\right). \tag{A10.17}$$

To ensure that Eqn A10.16 holds for arbitrary prices, the coefficients of $DCRP1_d^c$ and $DCRP2_d^c$ on both sides must be equal:

$$\begin{aligned} X1_d^c &= \frac{S_{d-1} * CRW_d}{TDW_{d-1}} \frac{NC_{new}}{NC_{old}} \frac{CPW_{old}^c}{CS^c}, \\ X2_d^c &= \frac{S_{d-1} * \left(1 - CRW_d\right)}{TDW_{d-1}} \frac{CPW_{new}^c}{CS^c} \end{aligned} \tag{A10.18}$$

The total position is:

$$X1_d^c + X2_d^c = \frac{S_{d-1}}{TDW_{d-1}} \frac{\frac{NC_{new}}{NC_{old}} CRW_d * CPW_{old}^c + \left(1 - CRW_d\right) * CPW_{new}^c}{CS^c} \tag{A10.19}$$

and the ratio of total positions for any two commodities $c1$ and $c2$ is:

$$\frac{X1_d^{c1} + X2_d^{c1}}{X1_d^{c2} + X2_d^{c2}} = \frac{\left[\frac{NC_{new}}{NC_{old}} CRW_d * CPW_{old}^{c1} + \left(1 - CRW_d\right) * CPW_{new}^{c1}\right] / CS^{c1}}{\left[\frac{NC_{new}}{NC_{old}} CRW_d * CPW_{old}^{c2} + \left(1 - CRW_d\right) * CPW_{new}^{c2}\right] / CS^{c2}} \tag{A10.20}$$

which is different from Cases 1 and 2. On January roll days, the position ratio varies over time because of the change of production weights.

Appendix B. Supplementary data

Supplementary data to this article can be found online at: https://doi.org/10.1016/j.eneco.2018.04.005 (accessed February 9 2022).

Acknowledgments

The authors thank James Hamilton, Aaron Smith, seminar participants at the 2016 Agricultural & Applied Economics Association Annual Meeting, 2017 Commodity Markets Winter Workshop in Norway, 2017 International Conference on Energy Finance in China, and anonymous referees for helpful suggestions and comments. James Hamilton and Cynthia Wu generously shared their estimates of WIT crude oil positions from the Masters algorithm for comparison to our own estimates.

Notes

1 Original citation: Yan, L., Irwin, S.H. and Sanders, D.R. (2018) Mapping algorithms, agricultural futures, and the relationship between commodity investment flows and crude oil futures prices. *Energy Economics* 72, 486–504. Reprinted by permission of Elsevier B.V. The Online Supplemental Appendix for the article is available at: https://ars.els-cdn.com/content/image/1-s2.0-S0140988318301282-mmc1.pdf (accessed February 9 2022).

2 Financial investors seek exposure to commodity futures markets through either exchange-traded products or over-the-counter swap contracts, whose returns are tied to an index of commodity prices (e.g.

the S&P GSCI). In the remainder of this paper, the term 'index investment' is used to generally refer to commodity index-based investment.

[3] See ESMA's reports available at: https://www.esma.europa.eu/policy-rules/mifid-ii-and-mifir (accessed February 9 2022).

[4] See Irwin and Sanders (2011), Fattouh *et al.* (2013), Cheng and Xiong (2014), and Haase *et al.* (2016) for thorough reviews.

[5] Instead of linking price changes to index positions, an alternative approach is to apply bubble tests directly to crude oil futures prices (e.g. Phillips and Yu, 2011; Shi and Arora, 2012; Harvey *et al.*, 2016; Tsvetanov *et al.*, 2016). In spite of mixed results, compelling evidence of large and long-lasting bubbles is limited.

[6] Staff report on commodity swap dealers & index traders with Commission recommendations, Commodity Futures Trading Commission, 2008. http://www.cftc.gov/idc/groups/public/@newsroom/documents/file/cftcstaffreportonswapdealers09.pdf (accessed February 9 2022).

[7] It is interesting to note that when Cheng *et al.* (2015) consider index positions for WTI crude oil drawn directly from large trader positions in WTI futures they find no significant impact of index investment flows on prices.

[8] Gilbert (2010, 2018) and Gilbert and Pfuderer (2014) also use a weighted-average algorithm to estimate a measure of index positions in commodity futures markets. These studies report significant effects of index investment in some markets pre-2012 including WTI crude oil. However, the authors do not interpret the findings as a direct relationship market by market, but, rather, as the influence of investor sentiment on the changing attractiveness of the commodity asset class as an investment.

[9] The price determination of crude oil has also been studied using structural models (e.g. Lombardi and Robays, 2012; Chevallier, 2013; Morana, 2013; Kilian and Murphy, 2014; Juvenal and Petrella, 2015; Knittel and Pindyck, 2016). Here, we follow SNG and use a reduced form model conditional on several macroeconomic and market specific variables to facilitate comparison.

[10] The Masters algorithm imputes index positions in crude oil from agricultural index positions. Details of the Masters algorithm will be presented in the following section.

[11] Forecasting regression Eqn 10.1 is a version of the long-horizon regression model frequently used to test the predictability of stock returns (e.g. Boudoukh and Richardson, 1994).

[12] *REPO* data are from the Federal Reserve Bank of New York, available at: http://www.newyorkfed.org/markets/gsds/search.html (accessed February 9 2022).

[13] SNG shows that the impact is even larger on futures returns over a 4-week horizon.

[14] Historical net assets of the GSG are collected from Bloomberg and dollar weights of WTI crude oil in the S&P GSCI index are provided by Standard & Poor's.

[15] We do not use IID positions as another measure of WTI index positions in the regression analysis because of sample limitations. As shown in Fig. 10.4, IID positions do not start until December 2007 and are only available quarterly through March 2010. This means only nine observations would be available for SNG's original sample period which ends in January 2010, too few for reliable estimation of the regression models. We also do not use DCOT positions because previous research (Irwin and Sanders, 2012) has shown that DCOT and IID positions in WTI crude oil are uncorrelated.

[16] It is interesting to note that Sanders and Irwin (2014) estimated a simplified version of SNG's regression model using Fund positions for WTI crude oil, RBOB gasoline, heating oil, and natural gas and, consistent with the findings here, found no evidence of a significant relationship in any of the four energy markets.

[17] The Bloomberg Commodity Index was originally launched in 1998 as the Dow Jones-AIG Commodity Index, renamed as the Dow Jones-UBS Commodity Index in 2009, and it received the current name (Bloomberg Commodity Index) on July 1 2014.

[18] While the Masters algorithm is widely credited to Michael W. Masters, apparently others developed similar algorithms around the same time. Masters (2008) contains this 'final note' in the Appendix: 'This method of calculating Index Speculators is almost identical to the methods used by Philip Verleger (www.pkverlegerllc.com), Steve Briese (www.commitmentsoftraders.org) and others. It is not clear who deserves the credit for developing it but it clearly is not us.'

[19] See Sanders and Irwin (2013) for numerical examples and Hamilton and Wu (2015) for a similar mathematical description.

[20] Hamilton and Wu (2015) show that the Masters algorithm can be generalized to infer index positions for non-SCOT markets from any SCOT market beyond the unique ones although the assumption that all commodity index investment tracks the S&P GSCI and BCI is required. The bottom line remains the same – inferring index positions from agricultural commodities. Here, such a generalization is not included because we focus on Masters' original procedure used in previous studies.

[21] See the Online Supplemental Appendix for details on the construction of the weighted average.
[22] We also tested the null hypothesis of fixed-position ratios between agricultural markets and the other US futures markets included in the IID report: RBOB gasoline, heating oil, natural gas, copper, gold, silver, and platinum. The null hypothesis of fixed ratios is rejected in every single case. These additional results are available in the Online Supplemental Appendix.
[23] The Global Commodity ETP Quarterly released by the *ETF securities* provide a list of commodity exchange-traded products (ETPs) and their target indices.
[24] A similar conclusion is reached if mapping algorithm positions are compared to GSG or private fund positions in WTI crude oil futures (see Fig. 10.3).
[25] See the Online Supplemental Appendix for details on the construction of the weighted-average measure and WTI return regression estimates using this measure. The return regression results are similar to the regression results shown earlier for the Masters algorithm measure.
[26] The nearby futures regression results in the Online Supplemental Appendix for Kansas wheat, feeder cattle, and soybean oil are the same as the *ER*1 regression results for these markets shown in Table 10.9.
[27] In the Online Supplemental Appendix we also report regressions of excess own returns for each of the 12 agricultural markets on their SCOT index positions and find that index positions are not significantly related to own returns in any of the 12 markets. This reinforces the conclusion that the reported relationship in some studies between index positions estimated via mapping algorithms and WTI crude oil futures returns is spurious.
[28] While there was a spike in agricultural index positions during 2007–2008, this was not generally the period of fastest growth of index positions in these markets. Using non-public daily position data from the CFTC's Large Trader Reporting System, Aulerich *et al.* (2013) show that the fastest period of growth in all but a few SCOT markets was 2004–2006. It is therefore more correct to say that the 2007–2008 spike in agricultural index positions occurred after index funds had built up a large position base in these markets.
[29] This report was released on February 25, 2016 and withdrawn two weeks later.
[30] The same results can be obtained for the S&P GSCI Excess Return Index.
[31] The normalizing constant ratio is computed as $\frac{NC_{new}}{NC_{old}} = \frac{\sum_c CPW_{new}^c * DCRP_d^c}{\sum_c CPW_{old}^c * DCRP_d^c}$ on the particular day preceding the January roll period, which is the same for all commodities.

References

Acharya, V.V., Lochstoer, L.A. and Ramadorai, T. (2013) Limits to arbitrage and hedging: evidence from commodity markets. *Journal of Financial Economics* 109, 441–465.

Aulerich, N.M., Irwin, S.H. and Garcia, P. (2013) Bubbles, food prices, and speculation: evidence from the CFTC's daily large trader data files. NBER Working Paper No. 19065. National Bureau of Economics Research (NBER). Available at: https://www.nber.org/system/files/working_papers/w19065/w19065.pdf (accessed Janaury 23 2022).

Basak, S. and Pavlova, A. (2016) A model of financialization of commodities. *Journal of Finance* 71, 1511–1556.

Boudoukh, J. and Richardson, M. (1994) The statistics of long-horizon regressions revisited. *Mathematical Finance* 4, 103–119.

Brunetti, C. and Reiffen, D. (2014) Commodity index trading and hedging costs. *Journal of Financial Markets* 21, 153–180.

Brunetti, C., Buyuksahin, B. and Harris, J.H. (2016) Speculators, prices and market volatility. *Journal of Financial and Quantitative Analysis* 51(5), 1545–1574.

Buyuksahin, B. and Harris, J.H. (2011) Do speculators drive crude oil futures prices? *Energy Journal* 32, 167–202.

Cheng, I.-H. and Xiong, W. (2014) The financialization of commodity markets. *Annual Review of Financial Economics* 6, 419–441.

Cheng, I.-H., Kirilenko, A. and Xiong, W. (2015) Convective risk flows in commodity futures markets. *Review of Finance* 19, 1733–1781.

Chevallier, J. (2013) Price relationships in crude oil futures: new evidence from CFTC disaggregated data. *Environmental Economics and Policy Studies* 15, 133–170.

CFTC (Commodity Futures Trading Commission) (2016) Position limits for derivatives. Available at: https://www.cftc.gov/LawRegulation/DoddFrankAct/Rulemakings/PositionLimitsforDerivatives/index.htm (accessed February 9 2022).

Etula, E. (2013) Broker-dealer risk appetite and commodity returns. *Journal of Financial Econometrics* 11, 486–521.

Fattouh, B., Lutz, K. and Mahadeva, L. (2013) Financial speculation in the oil markets and the determinants of the price of oil. *Energy Journal* 34, 7–33.

Gilbert, C.L. (2010) Speculative influences on commodity futures prices 2006–2008. United Nations Conference on Trade and Development Discussion Paper No.197. Available at: https://unctad.org/system/files/official-document/osgdp20101_en.pdf (accessed January 23 2022).

Gilbert, C.L. (2018) Investor sentiment and market fundamentals: the impact of index investment on energy and metals markets. *Mineral Economics* 31, 87–102.

Gilbert, C.L. and Pfuderer, S. (2014) The role of index trading in price formation in the grains and oilseeds markets. *Journal of Agricultural Economics* 65, 303–322.

Grossman, S.F. and Miller, M.H. (1988) Liquidity and market structure. *The Journal of Finance* 43, 617–633.

Haase, M., Zimmermann, Y.S. and Zimmermann, H. (2016) The impact of speculation on commodity futures markets: a review of the findings of 100 empirical studies. *Journal of Commodity Markets* 3, 1–15.

Hamilton, J.D. and Wu, J.C. (2014) Risk premia in crude oil futures prices. *Journal of International Money and Finance* 42, 9–37.

Hamilton, J.D. and Wu, J.C. (2015) Effects of index-fund investing on commodity futures prices. *International Economic Review* 56, 187–205.

Harvey, D., Leybourne, S.J., Sollis, R. and Taylor, R. (2016) Tests for explosive financial bubbles in the presence of non-stationary volatility. *Journal of Empirical Finance* 38, 548–574.

Henderson, B.J., Pearson, N.D. and Wang, L. (2015) New evidence on the financialization of commodity markets. *Review of Financial Studies* 28, 1285–1311.

Irwin, S.H. and Sanders, D.R. (2011) Index funds, financialization, and commodity futures markets. *Applied Economic Perspectives and Policy* 33, 1–31.

Irwin, S.H. and Sanders, D.R. (2012) Testing the Masters Hypothesis in commodity futures markets. *Energy Economics* 34, 256–269.

Juvenal, L. and Petrella, I. (2015) Speculation in the oil market. *Journal of Applied Econometrics* 30, 621–49.

Kilian, L. and Murphy, D. (2014) The role of inventories and speculative trading in the global market for crude oil. *Journal of Applied Econometrics* 29, 454–478.

Knittel, C.R. and Pindyck, R.S. (2016) The simple economics of commodity price speculation. *American Economic Journal: Macroeconomics* 8, 85–110.

Lombardi, M.J. and Robays, I.V. (2012) Do financial investors destabilize the oil price? Working paper. *Bank for International Settlements*. Available at: https://www.imf.org/external/np/seminars/eng/2011/tur/pdf/lombardi.pdf (accessed February 9 2022).

Masters, M.W. (2008) Testimony before the Committee on Homeland Security and Governmental Affairs, US Senate. May 20. Available at: http://hsgac.senate.gov/public/_files/052008Masters.pdf (accessed January 14 2022).

Masters, M.W. and White, A.K. (2008) The accidental Hunt brothers: how institutional investors are driving up food and energy prices. Available at: https://www.cftc.gov/sites/default/files/idc/groups/public/@swaps/documents/file/plstudy_31_ahb.pdf (accessed January 14 2022).

Mayer, J. (2012) The growing financialisation of commodity markets: divergences between index investors and money managers. *Journal of Development Studies* 48, 751–767.

Morana, C. (2013) Oil price dynamics, macro-finance interactions and the role of financial speculation. *Journal of Banking and Finance* 37, 206–226.

Newey, W.K. and West, K.D. (1994) Automatic lag selection in covariance matrix estimation. *Review of Economic Studies* 61, 631–653.

Phillips, P.C.B. and Yu, J. (2011) Dating the timeline of financial bubbles during the subprime crisis. *Quantitative Economics* 2, 455–491.

Sanders, D.R. and Irwin, S.H. (2011) The impact of index funds in commodity futures markets: a systems approach. *Journal of Alternative Investments* 14, 40–49.

Sanders, D.R. and Irwin, S.H. (2013) Measuring index investment in commodity futures markets. *Energy Journal* 34, 105–127.

Sanders, D.R. and Irwin, S.H. (2014) Energy futures prices and commodity index investment: new evidence from firm-level position data. *Energy Economics* 46, S57–S68.

Shi, S. and Arora, V. (2012) An application of models of speculative behaviour to oil prices. *Economics Letters* 115, 469–472.

Singleton, K.J. (2014) Investor flows and the 2008 boom/bust in oil prices. *Management Science* 60, 300–318.

Sockin, M. and Xiong, W. (2015) Informational frictions and commodity markets. *Journal of Finance* 70, 2063–2098.

S&P Dow Jones Indices (2015) S&P GSCI methodology. S&P Manual, S&P Global, New York. Available at: https://us.spindices.com/indices/commodities/sp-gsci (accessed February 9 2022).

Summers, L.H. (1986) Does the stock market rationally reflect fundamental values. *Journal of Finance* 41, 591–601.

Tsvetanov, D., Coakley, J. and Kellard, N. (2016) Bubbling over! The behaviour of oil futures along the yield curve. *Journal of Empirical Finance* 38, 516–533.

US House of Representatives (2019) 7 U.S.C., United States Code, Title 7 – Agriculture; Chapter 1 – Commodity Exchanges; Section 6a – Excessive speculation. Office of Law Revision Counsel (OLRC), US House of Representatives. Available at: https://www.govinfo.gov/content/pkg/USCODE-2019-title7/html/USCODE-2019-title7-chap1-sec6a.htm (accessed November 28 2022).

11

Sunshine versus Predatory Trading Effects in Commodity Futures Markets: New Evidence from Index Rebalancing[1]

New Author Foreword

It should be clear at this point that there is very little (if any) evidence of price impacts in commodity futures markets consistent with the Masters Hypothesis. Importantly, this does not rule out the possibility of smaller price impacts from commodity index funds. We have always been careful to point out this distinction, whether anyone was paying attention or not. For example, in one of our previous articles (see Chapter 6, this volume) we argued that financialization could have either irrational and harmful market impacts or rational and beneficial impacts. Large price bubbles under the Masters Hypothesis clearly fall on the irrational and harmful side of the ledger. However, it is important to remember that this is just one of the possible market impacts of financialization. Hence, the failure of the Masters Hypothesis does not mean that we should end the search for price impacts of financialization. Rather, the search should focus on smaller price impacts associated with more rational market dynamics. Price impacts of this type may be less interesting from a policy standpoint, but they are actually more interesting from a purely academic perspective because the smaller magnitude is more plausible.

This is the background for our interest in studying the price impact of the annual rebalancing of commodity indexes. The major commodity indexes are typically rebalanced across markets once a year to reflect changes in the weights of the underlying indexes. The most important rebalancing takes place in January for index funds that track the Standard and Poor's Goldman Sachs Commodity Index (S&P GSCI). For years we had been reading reports in the financial press about the large price impacts supposedly associated with the annual rebalancing in January. So, naturally, this attracted our interest.

There are some characteristics of annual rebalancing 'events' that are important to point out. First, the money flows in and out of the various commodity futures markets can indeed be quite large. Second, rebalancing money flows reflect a redistribution of existing index investment rather than new money flows into the investment class. Third, rebalancing money flows are predictable because the weight changes for major commodity indexes are pre-announced and occur during a predefined period each year. This set of characteristics means that rebalancing order (money) flows are not relevant to testing the Masters Hypothesis because the flows, while large, represent a predictable redistribution of index investment. Tests of the Masters Hypothesis require new and unpredictable order flows into commodity index instruments. But the same characteristics imply that annual rebalancing is a tailor-made event for testing different theories of the market impact of financialization.

Theoretical models suggest two opposite mechanisms through which predictable order flows from commodity index investors can affect market prices. The theory of sunshine trading of Admanti and Pfleiderer (1991) claims that predictable order flows attract additional liquidity suppliers, and hence, the flows have modest and temporary effects on market quality. In contrast, the theory of predatory trading proposed by Brunnermeier and Pederson (2005) contends that a trader who learns that another trader will transact a large quantity can profit by trading in the same direction prior to or simultaneous with the trader, which may cause prices to move away from the equilibrium and harm market quality.

DOI:10.1079/9781800622104.0011

Continuing our very productive partnership with Lei Yan, we used event study methods in this article to test the two competing theories about the market impact of S&P GSCI rebalancing in commodity futures markets. We found that the cumulative abnormal return during the S&P GSCI rebalancing period over 2004–2019 tends to have the same sign as the change in index weights. The price impact reaches a peak of 72 basis points in the middle of the week following the rebalancing period, but the impact is temporary as it declines to near zero within the next week. The findings showed that the impact of rebalancing order flows in commodity futures prices is modest and temporary, consistent with the prediction of sunshine trading theory.

This article is important because it is one of the first to document a price impact of financialization in commodity futures markets that unambiguously makes economic sense. Simply put, the rebalancing order flow of index investors consumes liquidity, and this comes at a cost in the form of higher execution prices in markets with increasing index weights, and vice versa. The key from a policy perspective is that the price impact of the rebalancing trades is both modest and temporary, rather than large and permanent, because the trades are both predictable and uninformed.

While we believe this article is an important and novel contribution to the literature on financialization in commodity futures markets, it will take some time to know if other researchers agree with us or not since it was just published in 2022. We are encouraged by citations to the article in several recently published papers in high-quality journals. As always, only time will tell.

Abstract

Annual rebalancing of the Standard and Poor's Goldman Sachs Commodity Index (S&P GSCI) provides a novel identification of the impact of predictable order flows from index investors in commodity futures markets. Using the 24 commodities included in the S&P GSCI for 2004–2019, we show that cumulative abnormal returns to a long-short strategy peaked at 72 basis points in the middle of the week following the rebalancing period, but the impact declines to near zero within the next week. The findings show that the impact of order flows from financial investors on commodity futures prices is modest and temporary, consistent with the prediction of sunshine trading theory.

Key words: commodity futures, financialization, index, order flow, rebalancing

JEL categories: G13, G14, G23

11.1 Introduction

Commodities as an alternative asset class began attracting a substantial amount of retail investment starting in the early 2000s and have become increasingly popular over time. According to data collected by Barclays and shown in Fig. 11.1, global commodity-linked investment rose from $55 billion in 2005 to a peak of nearly $450 billion in early 2012. Investment declined sharply in the years following the peak, but bounced back to about $300 billion in recent years, still a very large amount by historical standards. The increasing participation of financial investors in commodity markets is widely referred to as financialization.

Financial institutions and retail investors have gained access to commodities through a variety of instruments, such as exchange-traded funds (ETFs), exchange-traded notes (ETNs), and over-the-counter swaps whose returns are tied to prices of individual commodities or commodity indexes. The process of tracking an index in commodity futures markets results in large order flows. For example, commodity futures positions are typically rebalanced across markets annually to reflect changes in the weights of the underlying index. These rebalancing order flows are predictable because the weight changes for major commodity indexes are pre-announced and occur during a predefined period each year.

Theoretical models suggest two opposite mechanisms through which predictable order flows can affect market quality. The theory of sunshine trading of Admanti and Pfleiderer (1991) claims that predictable order flows attract additional liquidity suppliers, and hence, have modest and temporary price effects. Based on this theory, Bessembinder (2015) argues that predictable order flows should have a declining price impact over time as they become better understood by market participants. Bessembinder *et al.* (2016) extend Admanti and Pfleiderer's model by considering the role of market resilience

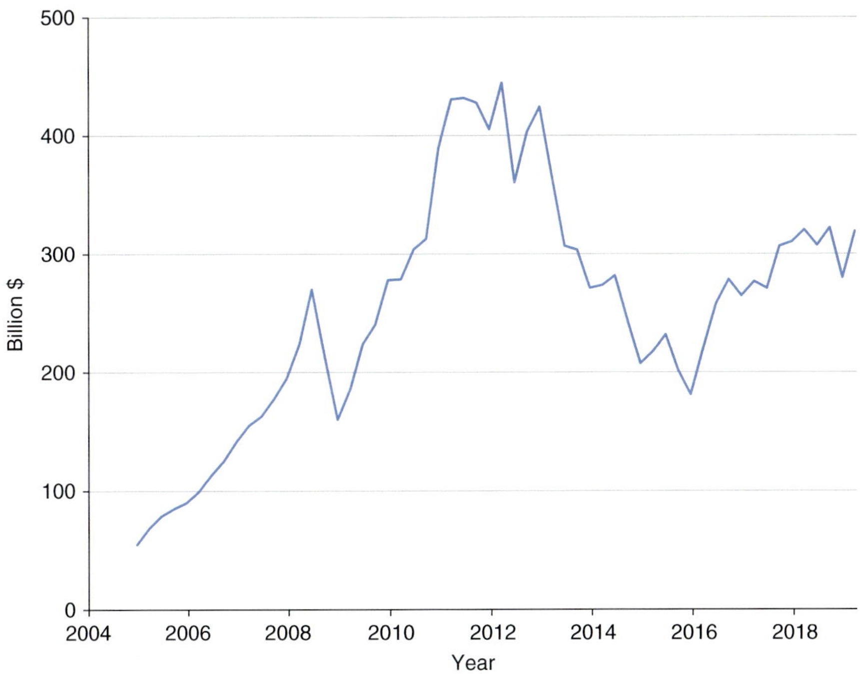

Notes: Three categories of commodity-linked investment are included in the the total: (I) broad Index swaps; (ii) exchange-traded products; and (iii) medium-term notes. The sample period is the fourth quarter of 2004 through the first quarter of 2019.

Fig. 11.1. Total global commodity-linked investment. (Data estimated by quarter from Barclays.)

and show that other traders in resilient markets supply liquidity to predictable order flows rather than exploiting them for predatory gains. In contrast, the theory of predatory trading proposed by Brunnermeier and Pederson (2005) contends that a trader who learns that another trader will transact a large quantity can profit by trading in the same direction prior to or simultaneous with the trader, which may cause prices to move away from the equilibrium and harm market quality.

Empirical evidence on the impact of predictable order flows in commodity futures markets is limited and focuses on the monthly 'roll' trades of financial index traders. Mou (2011) studies the roll trades for investors tracking the Standard and Poor's Goldman Sachs Commodity Index (S&P GSCI) and shows that a predatory strategy of selling expiring contracts and buying deferred contracts in advance of monthly roll dates earns significant profits. In contrast, Stoll and Whaley (2010), Aulerich *et al.* (2013), and Hamilton and Wu (2015) find that nearby calendar spreads in commodity futures markets either do not change or narrow as the amount of index trading increases around roll dates. Neuhierl and Thompson (2016) also focus on calendar spreads and design a trend-following strategy that exploits the momentum in spreads, finding that an increase in returns coincides with increased participation in commodity markets by financial investors. Bessembinder *et al.* (2016) study predictable roll trades undertaken by the US Oil Fund – the largest ETF in the crude oil futures market – and document narrower bid-ask spreads, greater order book depth, and improved resilience on roll dates. They conclude that other traders provide liquidity rather than exploit the predictable ETF roll trades in a predatory manner.

In this paper, we provide a novel identification strategy to test for sunshine versus predatory trading impacts in commodity futures markets. Our identification strategy is straightforward and built on an important exogenous event – the annual rebalancing of the S&P GSCI.

The S&P GSCI is widely acknowledged to be the leading benchmark for commodity investment. Market index weights for the S&P GSCI are based on historical production and rebalanced once a year. New weights are publicly announced in early November each year and implemented in the following January. Thus, the changes in index weights cause uninformed order flows into or out of individual commodity futures markets during the rebalancing period.[2] These order flows are economically significant as well, often exceeding several hundred million dollars of notional value. Our tests utilize 24 commodities included in the S&P GSCI and rebalancing weights for each year over 2004–2019.

There are several reasons why S&P GSCI rebalancing provides strong identification to estimate the price impact of predictable order flows from index investors in commodity futures contracts. First, the rebalancing is a public and predetermined event, with a gap between the announcement in early November and implementation from the fifth through the ninth business day of the following January. If S&P GSCI rebalancing contains new fundamental information, it should affect prices following the announcement date and not during the implementation period. Second, the rebalancing is a purely exogenous event. Weight changes due to rebalancing are fully determined by historical information about world production of the physical commodity and futures market liquidity, which often have a 1½–2-year lag due to data availability. The stale production and liquidity information conveyed by weight changes should be irrelevant to prices during the announcement or implementation periods associated with S&P GSCI rebalancing. This allows us to examine the impact of index order flows on commodity futures prices without concerns about reverse causality. Third, the changes in positions due to S&P GSCI rebalancing reflect a single trading motive, unlike aggregate position data that may reflect multiple and heterogeneous trading motives (Cheng *et al.*, 2015). Fourth, the rebalancing requires investors that track the S&P GSCI index to initiate trades by buying or selling additional futures contracts rather than passively reacting to orders of other traders in the market. This enables us to attribute potential price impacts during the rebalancing period to the activities of index traders, which is important in a zero-sum market.

We show that the cumulative abnormal returns (CARs) during the S&P GSCI rebalancing period over 2004–2019 tend to have the same signs as changes in index weights. A strategy that assigns an equal weight to each commodity and holds a long (short) position in commodities that experience positive (negative) index weight changes yields positive abnormal returns. The price impact reaches a peak of 72 basis points in the middle of the week following the rebalancing period, but the impact is temporary as it declines to near zero within the next week. Regression estimates show that the percentage changes in index weights, as a proxy for the size of the rebalancing flows, generally are not significantly related to the CARs for the front-running period but are significantly and positively related during the rebalancing period and the following week. The regressions also show that rebalancing not only impacts prices of the front-month futures contracts held directly by S&P GSCI investors, but also has a similar impact on prices for deferred contracts. Thus, regression analysis provides important new evidence that the impact of predictable order flows from index investors in commodity futures markets is modest and temporary.

Our finding of little if any price impact during front-running periods is inconsistent with the theory of predatory trading (Brunnermeier and Pederson, 2005). Instead, our results are consistent with the theory of sunshine trading (Admanti and Pfleiderer, 1991) because the predictable nature of rebalancing order flows attracts additional liquidity suppliers, and hence, the order flows have modest and temporary impacts. Finally, the weakening of price impacts in the second half of the sample period is consistent with Bessembinder's (2015) argument that predictable order flows should have a declining impact as they become better understood by market participants.

The article proceeds as follows. Section 11.2 describes the construction and rebalancing scheme of the S&P GSCI. A formal derivation is presented to link order flows due to the rebalancing to changes in weights of the index. Section 11.3 presents the event study framework used to examine price behavior around the announcement and rebalancing dates. Section 11.4 discusses the main results and presents robustness checks. Section 11.5 provides conclusions. We provide extra tables regarding the composition of the S&P GSCI in the Data Appendix (Appendix C in the current chapter).

11.2 S&P GSCI Construction and Rebalancing

In this section, we first describe the S&P GSCI and its rebalancing scheme. We then develop a relationship between order flows due to the rebalancing and changes in weights of the index. Last, we evaluate the economic significance of order flows due to the S&P GSCI rebalancing.

11.2.1 The S&P GSCI

The S&P GSCI, launched by Goldman Sachs in 1991, is the first major investable commodity index and serves as a benchmark for investment in commodity markets. The index comprises 24 commodities from all sectors (energy, metals, grains, softs, and livestock) and the composition has remained the same since 2002.[3] The wide range of constituent commodities provides diversification across sectors and within each sector. The S&P GSCI holds long positions in front-month futures contracts, which usually are the most liquid contracts. The index rolls positions from expiring contracts to later-month contracts to avoid physical delivery. The roll takes place within a 5-day window from the fifth through the ninth business day each month, and on each day, an equal amount (one-fifth) of the positions are rolled. Table C11.1 of the Data Appendix shows the futures contracts held at the beginning of the calendar month for each of the 24 commodities included in the S&P GSCI. Rolling occurs in the 5-day window for a given calendar month and commodity when the designated futures contract changes for the next calendar month in Table C11.1. Note that positions are not rolled every month in some commodity markets because futures contracts are not designated for every calendar month (e.g. corn: March, May, July, September, and December expirations only).

The S&P GSCI is a production-weighted index. The percentage dollar weight of commodity c on day d is,

$$w_d^c = \frac{CPW^c * DCRP_d^c}{\sum_c \left(CPW^c * DCRP_d^c\right)} \tag{11.1}$$

where $DCRP$ is the daily contract reference price, which is the settlement price of the futures contract held by the index and expressed in US dollars per physical unit, and CPW is the contract production weight, which measures the relative significance of the commodity and is expressed in physical units. Just as market capitalization is used to assign weights to components of equity indexes, the S&P GSCI assigns weight to each commodity in proportion to the amount of that commodity flowing through the economy. Note that CPW does not contain a time subscript because it remains the same during each calendar year except the rebalancing period. Although CPW is fixed within a year, the percentage dollar weight w_d^c varies from day to day as price changes.[4]

The CPW is constructed based on historical world production and futures trading volume (S&P Dow Jones Indices, 2017, p. 12). For commodity c,

$$CPW^c = Percentage\,TQT^c \,\frac{WPA^c}{10^6} \tag{11.2}$$

where *Percentage TQT* is the percentage total quantity traded and *WPA* is the world production average. *TQT* is measured by total trading volume of futures contracts during the relevant annual calculation period, which refers to the 12-month period ending on August 31 of the calendar year immediately preceding the year for which the composition of the S&P GSCI is being determined. *WPA* is the average quantity of world production of each commodity over the most recent 5 years for which complete world production data is available for all S&P GSCI commodities. The production data often has a lag of 2 years due to reporting delays. The same *WPA* is used for commodities that belong to any of the following groups: (i) Chicago wheat and Kansas wheat; (ii) feeder cattle and live cattle; and (iii) WTI crude oil, Brent crude oil, heating oil, RBOB gasoline, and gasoil. The *CPW*s are obtained by allocating *WPA* in proportion to *Percentage TQT*, which equals the *TQT* of one commodity divided by the total *TQT* of commodities within a given group. In this case, *CPW*s are jointly determined by world production and trading volume. For the rest of the commodities,

Percentage TQT equals one and *CPWs* are fully determined by world production. Equations 11.1 and 11.2 state that the constituent weights of the S&P GSCI depend on *CPWs*, which in turn rely on historical world production and futures market trading volume.

11.2.2 S&P GSCI rebalancing

The S&P GSCI adjusts the *CPWs* once a year to reflect changes to production and trading volume. Specifically, the index provider announces new *CPWs* in late October or early November each year, which will take effect in the following January from the fifth through the ninth business day. We refer to this 5-day window as the S&P GSCI rebalancing period. Implementation of new *CPWs* causes changes in weights, which in turn requires investors to adjust their positions tied to the S&P GSCI to minimize tracking error. This position adjustment is mechanical in the sense that investors must complete it regardless of changes in futures prices. The index transitions from old *CPWs* to new ones at an even pace, requiring investors to buy or sell an equal number of futures contracts on each of the rebalancing dates to achieve minimal tracking error.

The specific futures contracts used in rebalancing of the S&P GSCI are listed in the fourth column (labelled '2') of Table C11.1 in the Data Appendix. It turns out that rebalancing trades occur in only two futures contract months – March or April – for each of the 24 commodities in the S&P GSCI. In the remainder of this article we will refer to these as 'front-month' contracts, even though in some markets (e.g. WTI crude oil) contracts are still trading at the time of rebalancing that have a few days left until expiration.

It should be noted that the rebalancing period overlaps the January roll period for some commodities (14 of 24), which implies that investors need to not only roll their positions for these markets in January but also buy or sell additional contracts in response to weight changes due to the rebalancing.[5] The key difference is that the January roll in these markets simply closes positions in the expiring contract and re-establishes them in the front-month contract without generating new net flows, whereas rebalancing causes new net flows into or out of the market through trading additional contracts in the front month. We will discuss distinct effects of the rebalancing and the rolls on prices in the 'Results' section.

The weights of commodity *c* prior to and after the rebalancing are defined as

$$w_{old}^{c} = \frac{CPW_{old}^{c} * DCRP_{0}^{c}}{\sum_{c}\left(CPW_{old}^{c} * DCRP_{0}^{c}\right)} \tag{11.3}$$

and

$$w_{new}^{c} = \frac{CPW_{new}^{c} * DCRP_{0}^{c}}{\sum_{c}\left(CPW_{new}^{c} * DCRP_{0}^{c}\right)} \tag{11.4}$$

respectively, based on old and new *CPWs*. $DCRP_0$ denotes the settlement price of the contract held by the index on day 0, the last business day prior to the rebalancing period.[6] Since the same prices are used in Eqns 11.3 and 11.4, the changes in weight due to the rebalancing are driven only by *CPW* changes. An increase in *CPW* for a commodity raises its own weight but simultaneously reduces weights for all the other commodities. The sum of weight changes over commodities is equal to zero.

We show in Appendix B that order flows due to the S&P GSCI rebalancing are proportional to changes in weights:

$$Order\ flow^{c} = k * \left(w_{new}^{c} - w_{old}^{c}\right) \tag{11.5}$$

where *k* is a constant and can be interpreted as the notional value of total investment tied to the S&P GSCI. Equation 11.5 allows us to measure order flows due to the rebalancing by changes in weights, which are purely driven by *CPW* changes and reflect no new fundamental information about futures price.

The timing of rebalancing trades within the day could impact estimated price impacts (Ready and Ready, 2019). In particular, financial investors tracking the S&P GSCI could in theory decide when to trade within the day based on price trends. However, financial investors in commodity futures markets are most likely to trade during times of higher trading volume and lower trading costs. For example, Bessembinder *et al.* (2016) show that the US Oil Fund, the largest of the ETFs that track crude oil prices, routinely trades at the settlement price to complete its roll trades. Ready and Ready (2019) document that index traders in agricultural futures trade at or near the daily settlement. As rebalancing trades occur at the same time as roll trades, it is reasonable to assume that rebalancing trades occur at or very near the daily settlement price. This supports our use of daily settlement prices to measure the market impact of uninformed order flows in commodity futures markets due to the annual rebalancing of the S&P GSCI.

11.2.3 Economic significance of the S&P GSCI rebalancing

We evaluate the economic significance of the S&P GSCI rebalancing by estimating the magnitude of order flows in commodity futures markets. The pre- and post-rebalancing weights are constructed based on Eqns 11.3 and 11.4 using old and new *CPWs* and prices of the futures contract held by the index on the last business day prior to the rebalancing period (day 0). We assume that the notional value of total assets tied to the S&P GSCI is two-thirds of the total index investment in major US futures markets, as measured by the Commodity Futures Trading Committee (CFTC) *Index Investment Data* (IID) report for 2008–2015.[7,8] The fourth quarter or December report is used depending on the release frequency of the IID. Hence, order flows are obtained by multiplying the notional value of total investment in the S&P GSCI by weight changes. The number of futures contracts equals order flow divided by value of the contract. Note that the IID data are only used in this section to provide an estimation of the size of aggregated rebalancing order flows.

Figure 11.2 shows the estimated order flows due to the S&P GSCI rebalancing for WTI crude oil for 2008–2015. The number of contracts

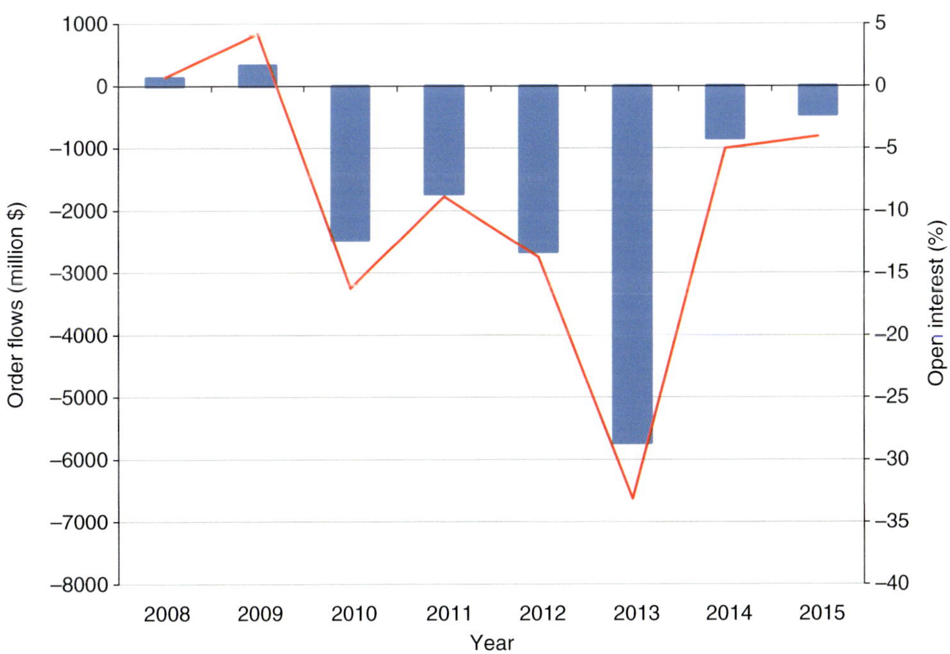

Notes: Order flows are measured in million dollars (left axis, bar) and percentages of open interest (right axis, line), where open interest is the average of daily open interest of the contract held by the index over a 2-week window prior to the rebalancing period. The notional value of total assets tied to the S&P GSCI is assumed to be two-thirds of the total index investment in major US futures markets based on the CFTC Index. The sample period is 2008–2015 since IID data are available only in this period.

Fig. 11.2. Order flows due to the S&P GSCI rebalancing for WTI crude oil. Based on data from the *Index Investment Data* (IID) report.

is expressed as a percentage of open interest, which is the average of daily open interest of the contract that the index rolls into over a 2-week window prior to the rebalancing period. The period from 2010 through 2013 saw substantial flows out of the WTI crude oil market over the rebalancing period, with a peak of $5731 million or 33% of daily open interest in 2013. These flows are still very large even if the trades occurred evenly on the five rebalancing dates. The large outflows during this period are due to the fact that the S&P GSCI continually reduced the weight of WTI crude oil and increased the weights of Brent crude oil and other energy products.

We calculate the order flows due to the S&P GSCI rebalancing for all 24 commodities and report them in Table 11.1. Because either positive or negative flows could affect price, we focus on the magnitude of weight changes and order flows by taking the average of their absolute values across years for each commodity. Order flows are measured in dollars and numbers of contracts. The number of contracts is also expressed as a percentage of volume and open interest, where volume (open interest) is the average of daily trading volume (open interest) of the contract that the index rolls into over a 2-week window prior to the S&P GSCI rebalancing period.[9] Table 11.1 shows a large variation in order flows

Table 11.1. Estimated order flows due to S&P GSCI rebalancing.[a]

		Order flow due to the rebalancing in absolute terms			
	Absolute weight change (%)	Dollar (million)	Number of contracts	Volume (%)	Open interest (%)
Chicago wheat	0.11	80.04	2,160	5.42	1.04
Kansas wheat	0.14	106.68	2,832	31.42	4.20
Corn	0.09	76.77	3,097	2.60	0.55
Soybeans	0.06	53.43	942	1.15	0.42
Coffee	0.01	11.01	184	1.59	0.24
Sugar	0.02	15.17	689	1.58	0.23
Cocoa	0.01	6.22	222	2.68	0.29
Cotton	0.03	26.87	605	5.16	0.56
Lean hogs	0.05	41.68	1,271	14.38	2.18
Live cattle	0.06	55.08	1,311	11.85	1.73
Feeder cattle	0.04	30.57	437	16.41	2.98
WTI crude oil	1.87	1,795.78	20,400	25.59	10.74
Heating oil	0.41	303.29	2,874	14.79	6.06
RBOB gasoline	0.60	376.08	3,909	24.77	8.14
Brent crude oil	1.39	1,208.65	12,878	31.67	16.96
Gasoil	0.56	513.13	7,034	23.61	12.28
Natural gas	0.09	70.83	1,404	4.37	0.95
Aluminum	0.06	51.69	233	0.39	0.07
Copper	0.04	31.32	88	0.21	0.06
Nickel	0.01	6.59	15	0.13	0.03
Lead	0.01	10.22	195	1.38	0.41
Zinc	0.01	10.19	798	2.37	0.63
Gold	0.08	69.71	2,313	25.24	4.48
Silver	0.00	2.74	55	0.15	0.06
Average	0.24	206.41	2,748	10.37	3.14

[a]The table reports the average weight changes and order flows due to rebalancing in absolute values for each of the 24 commodities included in the S&P GSCI for 2008–2015. Weight change is the difference of percentage dollar weights before and after the rebalancing. Order flows are measured in dollars, number of contracts, percentage of volume, and percentage of open interest, where volume (open interest) is the average of daily volumes (open interests) of the contract held by the index over a 2-week window prior to the rebalancing period. The notional value of total assets tied to the S&P GSCI is assumed to be two-thirds of the total index investment in major US futures markets, which is available from the CFTC *Index Investment Data* (IID) report for 2008–2015.

across commodities. The absolute change in weights due to the rebalancing is 1.87% for WTI crude oil, creating a flow of $1795.8 million in absolute terms. This means that investors need to buy or sell 20,400 additional March WTI crude oil contracts during the rebalancing period, accounting for 25.59% and 10.74% of daily volume and open interest, respectively. The flows are also large for wheat, cattle, and other energy products. The average flow across commodities is $206.4 million in total or $41.3 million/day. In comparison, Henderson *et al.* (2015) document average proceeds of $14.8 million related to Commodity Linked Note (CLN) issues and report that the induced hedge flows have a significant and permanent impact on futures price. Consequently, the order flows due to the S&P GSCI rebalancing are economically large enough to allow for a potential price impact.

The economic significance of the S&P GSCI rebalancing is further examined by calculating changes in positions held by index traders around the rebalancing period based on the CFTC *Supplemental Commitments of Traders* (SCOT) report. The SCOT report provides a breakdown of each Tuesday's open interest for 12 agricultural futures markets since January 3 2006. We examine the gross long positions held by index traders for 11 of the 12 SCOT agricultural futures (soybean oil is excluded) included in the S&P GSCI for four different periods: (i) the 2 weeks containing the rebalancing period; (ii) the 2 weeks prior to the rebalancing period; (iii) the 2 weeks after the rebalancing period; and (iv) other non-rebalancing weeks. Table 11.2 presents the average absolute percentage changes in positions held by index traders for those four periods. For nine of the 11 commodities (except cocoa and feeder cattle), the absolute changes in index positions for the 2 weeks containing the rebalancing period are two to four times larger than changes that precede or follow the rebalancing period.[10] The relatively large changes in positions for the 2 weeks containing the rebalancing period are likely driven by changes in weights, lending further support to the economic significance of order flows due to S&P GSCI rebalancing.

11.3 Methods

We follow standard event study methodology and use abnormal returns to measure the price impact of order flows due to the S&P GSCI rebalancing. The abnormal return is the difference between the actual return and the expected return,

Table 11.2. Changes (%) in agricultural index trader positions around the S&P GSCI rebalancing period.[a]

	Two weeks containing the rebalancing period	Two weeks prior to the rebalancing period	Two weeks after the rebalancing period	Other non-rebalancing weeks
Chicago wheat	3.99	1.00	2.13	1.85
Kansas wheat	4.60	1.75	2.50	2.65
Corn	3.90	1.46	1.46	1.76
Soybeans	3.17	2.15	1.94	1.96
Coffee	5.97	1.15	2.40	1.94
Sugar	4.55	0.97	2.47	1.65
Cocoa	2.71	2.10	5.00	3.46
Cotton	3.56	1.10	2.32	1.83
Lean hogs	4.62	1.24	1.96	1.85
Live cattle	3.80	0.91	1.66	1.38
Feeder cattle	3.19	2.18	3.46	3.15
Average	4.00	1.46	2.48	2.13

[a]The table presents the average absolute percentage changes in gross long positions held by index traders for four different windows: (i) the 2 weeks containing the rebalancing period; (ii) the 2 weeks prior to the rebalancing period; (iii) the 2 weeks after the rebalancing period; and (iv) other non-rebalancing weeks. Index positions are from the weekly CFTC *Supplemental Commitments of Traders* (SCOT) report for 2006–2019. The SCOT report provides a breakdown of each Tuesday's open interest for 12 agricultural commodity markets and our calculations are for 11 of the 12 commodities included in the S&P GSCI (soybean oil is excluded).

$$AR_{i,t} = R_{i,t} - E\left[R_{i,t} | X_t\right] \quad (11.6)$$

where $AR_{i,t}$, $R_{i,t}$, and $E[R_{i,t} | X_t]$ are the abnormal, actual, and expected returns, respectively, and X_t is the conditioning information. The actual return is the difference in log daily settlement prices of the front-month futures contract (March or April) held by investors in the S&P GSCI. The expected return refers to the return that would be expected if the rebalancing does not take place and is defined as the predicted value of $R_{i,t}$ conditional on X_t.

We estimate the expected return based on a multi-factor model, which assumes a linear relationship between futures returns and a group of economic factors (Z_t). Specifically,

$$R_{i,t} = \mu_i + \beta_i' Z_t + \varepsilon_{i,t} \quad (11.7)$$

where β_i is a vector of coefficients that measures dependence of $R_{i,t}$ on Z_t. We follow Henderson *et al.* (2015) and include the following factors in Z_t:

1. The return on the S&P 500 Index, which measures the impact of changes in expectations about US economic growth.
2. The return on the Morgan Stanley Capital International (MSCI) Emerging Markets Asia Index, which measures the impact of changes in expectations about emerging markets' economic growth. Its next-day return is also included to account for trade non-synchronicity between Asian and US markets.
3. The return on the JP Morgan Treasury Bond Index, which captures the linkage between commodity futures and interest rate markets.
4. The return on the Trade Weighted US Dollar Index, which reflects the fact that the US dollar is the most common settlement currency for commodity transactions.
5. The percentage change in the VIX Index, which measures the relationship between commodity prices and innovations to VIX found by Cheng *et al.* (2015).
6. The percentage change in the Baltic Dry Index, which is a widely used indicator of global economic conditions and measures changes to the cost of transporting raw materials by sea.
7. The change in the 10-year break-even inflation rate, representing changes in expected inflation.
8. The 1-day lagged dependent variable is included to control for potential autocorrelation in returns.

We obtain futures prices for the 24 commodities included in the S&P GSCI for 2004–2019 from Barchart, Inc. (available at: https://www.barchart.com/ (accessed January 14 2022)). The Baltic Dry Index is collected from Bloomberg and the rest of the factors are collected from the Federal Reserve Bank of St. Louis.

We also consider two alternative models for estimating abnormal returns. The first alternative is a constant-mean model,

$$R_{i,t} = \mu_i + \varepsilon_{i,t} \quad (11.8)$$

where μ_i is the mean return for commodity *i*. In this case, the expected return is equal to a constant that is estimated from historical returns. In the second alternative model, the constant is assumed to be zero, as implied by the long-standing view that historical average returns to most individual commodity futures do not differ from zero (e.g. Erb and Harvey, 2006). Hence, the expected return is zero and the abnormal return is identical to the actual return.

The multi-factor and constant-mean models are estimated over a 60-business-day window that ends 2 weeks preceding the first rebalancing period. The event window starts 2 weeks before the first rebalancing date and ends 3 weeks after the last rebalancing date to allow for any potential effects from front-running trades or delayed trades. Recall that the S&P GSCI rebalancing dates in January are represented by days 1–5. The event window is then from day –9 to day 20. For each day within the event window, the expected return is the product of the estimated parameters and the values of the factors on that day and the abnormal return is the actual return minus the expected return. For the multi-factor model, the R-squared has an average of 18.9% and is larger for energy and metals, which is not surprising given that energy and metal markets are integrated more closely with the global economy and financial markets. Detailed estimation results for the expected return are omitted to save space.

We define the cumulative abnormal return (CAR) as the sum of abnormal returns for a given interval [d_1, d_2], where d_1 and d_2 take values of −9–20. For instance, CAR[−9, 0] represents the cumulative abnormal return over the 2 weeks before the rebalancing period and CAR[1, 5] represents the cumulative abnormal return for the rebalancing period. If order flows due to the rebalancing impact the futures price, we expect the CARs to diverge over the rebalancing period – positive CARs for weight increases and negative CARs for weight decreases. However, the CARs can be biased due to cross-market dependence, which may arise from common shocks that occur on the rebalancing dates and affect all markets (e.g. quantitative easing by the Federal Reserve) or markets within a sector (e.g. an oil production cut by the Organization of the Petroleum Exporting Countries).[11,12] This is similar to issues encountered in cross-sectional equity returns in which stocks from the same industry tend to be correlated with each other. To mitigate the problem, we calculate the CARs to a long-short strategy each year, which assigns an equal weight to each commodity and holds a long position in commodities with positive weight changes and a short position in commodities with negative weight changes. The long-short strategy provides a more robust measure of the price impact of order flows due to the rebalancing since the effects of cross-market dependence on the long side should largely offset the effects on the short side. The CARs from this strategy are then averaged across 16 years. In the presence of a price impact, the average CARs should be positive for the rebalancing period and reverse afterwards if the impact is not permanent. This long-short strategy, while informative, uses only information on the direction of weight changes and ignores the magnitude of weight changes. Moreover, inference regarding the average CAR of the long-short strategy may lack statistical power since there is a total of 16 annual observations.

In order to improve statistical power, we employ panel regressions to link CARs to proxies for the magnitude of the order flows. In particular, we estimate the following regression model,

$$CAR[d_1, d_2]_{it} = \alpha_t + \beta(\Delta w / w)_{it}^{S\&P GSCI} + \varepsilon_{it} \quad (11.9)$$

where $CAR[d_1, d_2]_{it}$ is the cumulative abnormal return in percentage terms for commodity *i* from d_1 to d_2 in year *t*, α_t is the time-fixed effect that captures the common component in returns across commodities in year *t*, and $(\Delta w/w)^{S\&P\ GSCI}$ is the percentage change in weights. We consider various combinations of d_1 and d_2 to allow for the possibilities of front-running trades and delayed trades. Since the same weight change is unlikely to impact prices for corn to the same degree as for WTI crude oil, we use the percentage changes in weights to control for the effects of market size (Hau *et al.*, 2010).[13] In the presence of limits to arbitrage, we expect that CARs increase in percentage weight changes, i.e. $\beta > 0$. The regressions are estimated using a fixed-effects model. Time-clustered standard deviations are used to construct *t*-statistics for the estimated coefficients. We consider alternative model specifications (no fixed effects, firm fixed effects, time and firm fixed effects), various approaches to calculating standard errors (ordinary least squares (OLS), firm-clustered, time- and firm-clustered) and Fama-MacBeth (1973) estimation and the results are similar.[14] We present these alternative results in the Online Appendix.

11.4 Results

In this section, we measure the impact of order flows due to the S&P GSCI rebalancing on commodity futures prices. We first examine cumulative abnormal returns (CARs) following the announcement date of the rebalancing. Next, we estimate CARs around the rebalancing period. Last, we provide panel regressions to show the relationship between CARs and proxies for the magnitude of order flows due to the rebalancing, followed by a set of robustness checks.

11.4.1 CARs following the announcement date

The assertion that order flows due to the S&P GSCI rebalancing convey no new fundamental information about price is essential to our identification strategy. Recall that the changes in weights due to the rebalancing are driven by CPW changes that rely on historical production and trading volume information. Hence, the rebalancing order flows reflect no new information.

More importantly, new CPWs are announced in late October or early November each year, which is about 2 months prior to the rebalancing period.[15] If weight changes due to the rebalancing incorporate new information on fundamentals, we would expect to observe price reactions immediately following the announcement date, not during the actual rebalancing period. Even if the announcement of new CPWs acts as a shock to the market, the corresponding changes in weights should not be treated as relevant information for a rebalancing period that occurs 2 months later.

To examine whether commodity futures prices react to the announcement of new CPWs, we calculate CARs to a long-short strategy following the S&P GSCI announcement date each year. The long-short strategy assigns an equal weight to each commodity and takes a long position in commodities with positive weight changes and a short position in commodities with negative weight changes. Figure 11.3 shows the average CARs across years and their 95% confidence intervals. Abnormal returns are calculated from the multi-factor model. Results for abnormal returns from the constant-mean and zero-mean models are similar and reported in the Online Appendix. The CAR is normalized to be zero on day 0, the day immediately prior to the announcement date. The average CARs for the long-short strategy are 15 and 26 basis points on the announcement date (day 1) and the following day (day 2), respectively, and decline to nearly zero thereafter. The 95% confidence interval indicates that the average CARs are statistically significant only on day 1 and do not differ from zero on the rest of the days. These results suggest that the announcement of new weights has a significant but very short-lived impact on futures prices. The impact peaks on day 2 instead of day 1 because the announcements are often released after futures markets closed.[16]

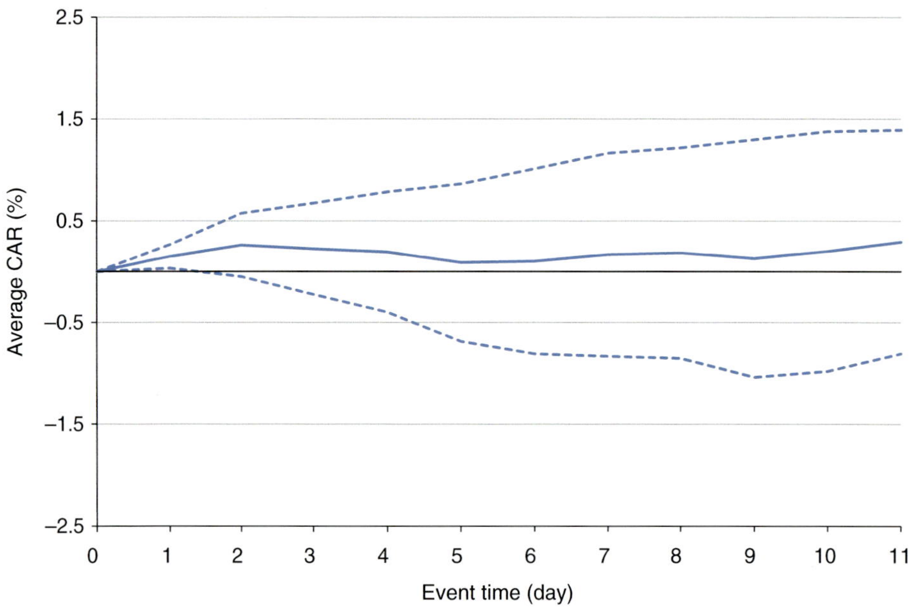

Notes: The figure shows the average CARs in percentage terms for a long-short strategy following the S&P GSCI rebalancing announcement date. The long-short strategy assigns an equal weight to each commodity and holds a long (short) position in commodities that experience positive (negative) weight changes. The CARs are normalized to be zero on day 0, the day immediately prior to the rebalancing announcement date. Day 1 is the announcement date. Abnormal returns are based on the multi-factor model. The sample consists of 24 commodities included in the S&P GSCI for 2004–2019. Dashed lines indicate 95% confidence intervals.

Fig. 11.3. Average cumulative abnormal returns (CARs) for a long-short strategy following the S&P GSCI rebalancing announcement date.

It is not clear why commodity futures markets react to the announcements given that new weights are highly unlikely to contain new information, as argued earlier. One possibility is that the order flow impacts during implementation of rebalancing trades 2 months later are large enough that they are anticipated by the market. Another possibility is an informational friction along the lines of Sockin and Xiong (2015), where some traders misperceive the announcement as representing changes in commodity demands. Finally, it is important to keep in mind that while a significant price impact is observed following CPW announcements, it is small and only lasts for a single day.

11.4.2 CARs around the rebalancing period

We calculate average CARs to the same long-short strategy for the rebalancing period and the following 10 business days, in the same way as we did for announcement dates, and present the results in Fig. 11.4. The CARs are normalized to be zero on day 0, the day immediately prior to the first rebalancing date, and expressed in percentage terms. Days 1–5 represent the rebalancing dates. Abnormal returns are based on the multi-factor return model. The average CARs show an upward trend for days [1, 7], reach a peak of 53 basis points on day 7, and decrease

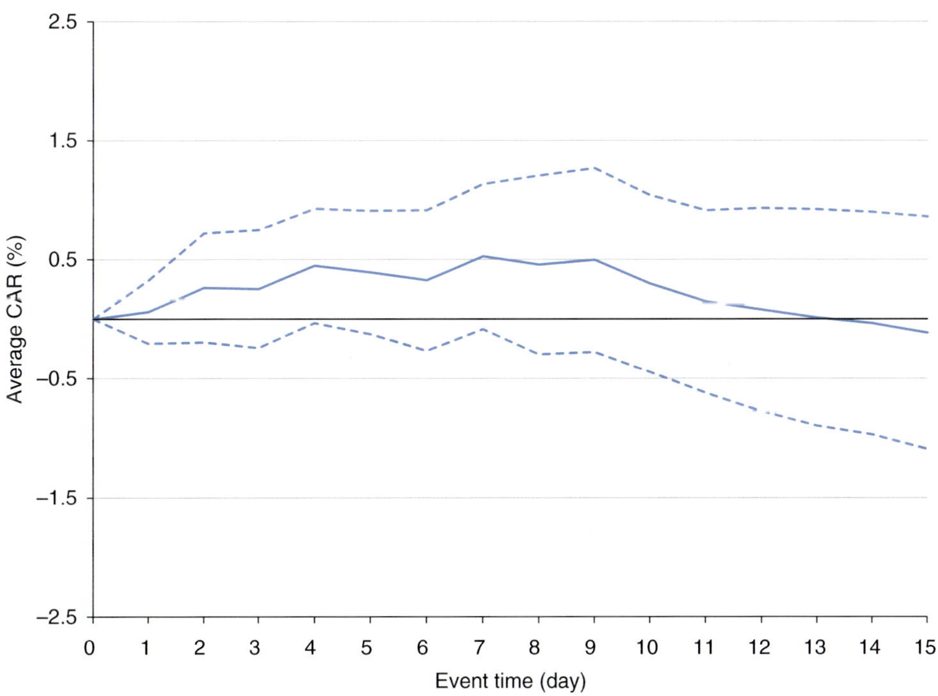

Notes: The figure shows the average CARs in percentage terms for a long-short strategy over the S&P GSCI rebalancing period and the following 2 weeks. The long-short strategy assigns an equal weight to each commodity and holds a long (short) position in commodities that experience positive (negative) weight changes. The CARs are normalized to be zero on day 0, the day immediately prior to the first rebalancing date. Days 1–5 denote the rebalancing dates. Abnormal returns are based on the multi-factor model. The sample consists of 24 commodities included in the S&P GSCI for 2004–2019. Dashed lines indicate 95% confidence intervals.

Fig. 11.4. Average cumulative abnormal returns (CARs) for a long-short strategy over the rebalancing period and the following 2 weeks.

continually to −11 basis points on day 15. However, the 95% confidence interval indicates that the average CARs do not differ from zero throughout the window, probably due to the small number of annual observations (16) used in the estimation. Despite the lack of statistical power, a clear pattern emerges that order flows due to the rebalancing increase futures return by up to 53 basis points, but the impact is completely reversed within a few days.

To allow for the possibility of front-running associated with predatory trading, we also calculate average CARs for the long-short strategy over a pre-rebalancing window. The front-running we consider entails buying in advance of rebalancing those commodities with positive weight changes and selling in advance those with negative weight changes. The advanced buying and selling is assumed to be placed in the front-month futures contracts (March and April) to match rebalancing trades. This is not the same type of front-running as considered by Mou (2011), who examined the monthly rolling of futures index positions from the expiring futures contract to the front-month contract. Monthly rolling of this type may put downward pressure on the expiring futures price and upward pressure on the front-month futures price, thereby increasing the calendar price spread between two contracts. Consequently, front-running the monthly index roll involves advanced selling of expiring contracts and advanced buying of front-month contracts. In contrast, front-running the annual rebalancing requires buying and selling only the front-month futures contract, with trading dependent on the sign of the weight change for individual commodities.

Average CARs for the long-short strategy during the front-running period of days −9–0 are shown in Fig. 11.5(a). The average CARs are small and statistically insignificant, providing little evidence of front-running annual rebalancing events. For completeness, we show in Fig. 11.5(b) the average CARs for the long-short strategy over the entire event window from day −9 to day 15, which combines the event windows previously used in Figs 11.4 and 11.5(a). The average CARs computed in this manner build slowly from near zero to reach a peak of 72 basis points on day 7, and then decrease steadily to 8 basis points on day 15. While the pattern in the average CARs is similar, with or without consideration of front-running, the peak in the average CAR does increase from 53 to 72 basis points with allowance for front-running. Overall, these results suggest notable but temporary price impacts associated with S&P GSCI rebalancing.[17]

Rebalancing versus roll effects

The rebalancing period overlaps with the January roll period for some commodities. The roll occurs because existing index positions must be moved from expiring contracts to next-month contracts. Investors tracking the S&P GSCI need to complete the rebalancing and roll trades during the same period. A natural question to ask is whether the identified rebalancing effects are due to roll trades that occur at the same time. To separate rebalancing effects from possible roll effects, we examine the returns for a group of commodities that do not have monthly rolls in January. Ten commodities satisfy this criterion, including cocoa, coffee, corn, cotton, feeder cattle, silver, soybeans, sugar, Chicago wheat, and Kansas wheat. Figure 11.6 shows the average CARs for the long-short strategy defined earlier using the ten non-roll commodities. The average CARs based on all commodities are included for comparison. While the average CARs based on non-roll commodities are somewhat less statistically significant because of a smaller number of observations, no major differences are found in the average level of CARs compared to those based on all commodities.

It is actually not surprising that rebalancing price impacts are largely unaffected by roll trades. First, rolling involves selling the expiring contract and buying the next-month contract, while rebalancing involves buying and selling the next-month contract only. Therefore, the only overlap between roll and rebalancing trades is when the next-month contract is bought. There is no overlap when rebalancing trades sell the next-month contract. Second, several studies test whether the spread between front-month contracts and the next month contracts increases during index roll periods, and the results are decidedly mixed. Four studies find that spreads in commodity futures markets are either unaffected or slightly narrow following index rolls (Stoll and Whaley, 2010; Aulerich *et al.*, 2013; Hamilton and Wu, 2015; Sanders and Irwin, 2016) and three studies (Mou, 2011; Brunetti and Reiffen, 2014; Bessembinder *et al.*, 2016) report evidence of expanded spreads after

(a) Front-running period

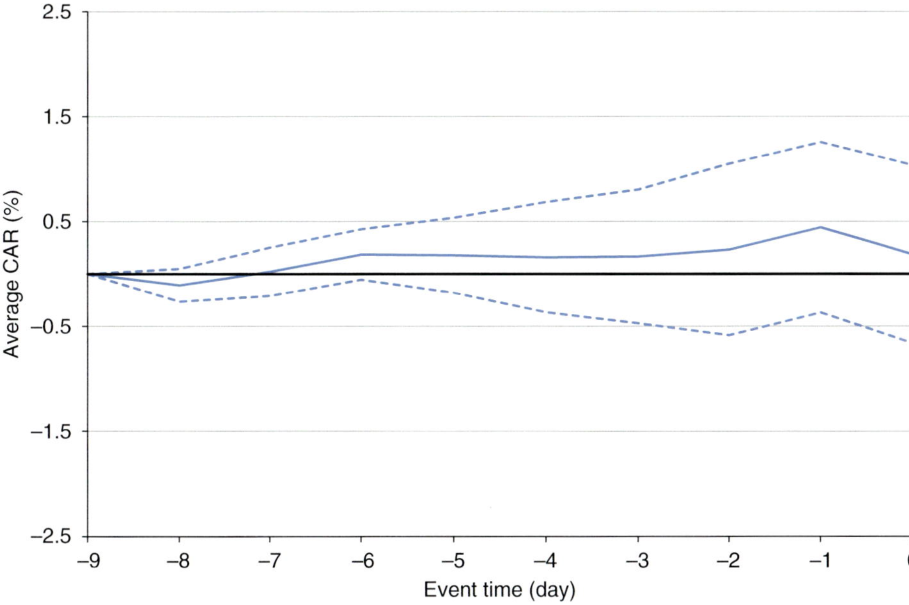

(b) Front-running and rebalancing period

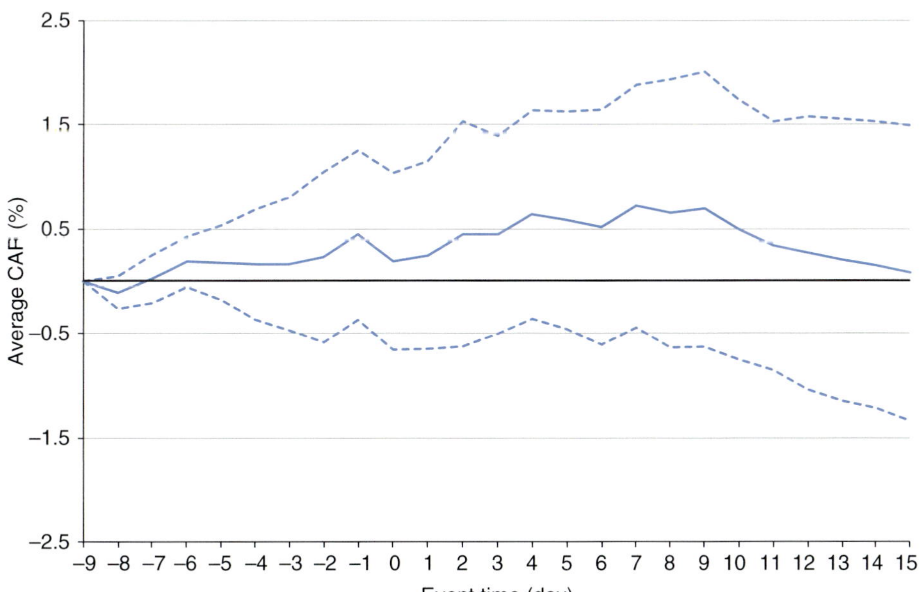

Notes: The figure shows the average CARs in percentage terms for a long-short strategy over the pre-rebalancing period (part (a) and an extended rebalancing window (part (b)). The long-short strategy assigns an equal weight to each commodity and holds a long (short) position in commodities that experience positive (negative) weight changes. The CARs are normalized to be zero on day −9, which precedes the first rebalancing date by 10 business days. Days 1–5 denote the rebalancing dates. Abnormal returns are based on the multi-factor model. The sample consists of 24 commodities included in the S&P GSCI for 2004–2019. Dashed lines indicate 95% confidence intervals.

Fig. 11.5. Average cumulative abnormal returns (CARs) for a long-short strategy over the pre-rebalancing period and an extended rebalancing period.

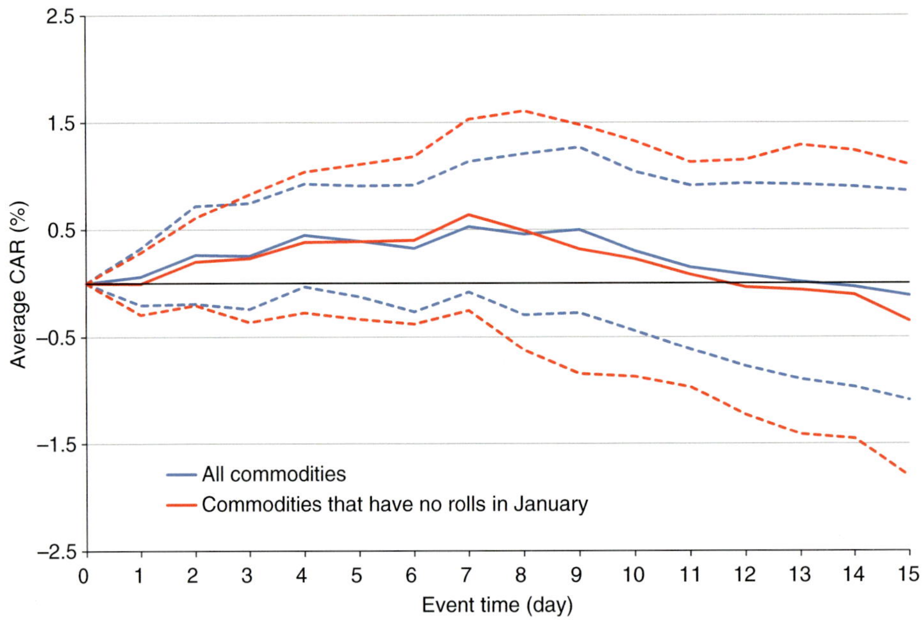

Notes: The figure shows the average CARs for a long-short strategy over the S&P GSCI rebalancing period and the following 2 weeks for all commodities (blue) and the ten commodities that have no rolls in January. The rebalancing-only commodities include corn, cocoa, coffee, cotton, feeder cattle, silver, soybeans, sugar, Chicago wheat, and Kansas wheat. The long-short strategy assigns an equal weight to each commodity and holds a long (short) position in commodities that experience positive (negative) weight changes. The CARs are normalized to be zero on day 0, the day immediately prior to the first rebalancing date. Days 1–5 denote the rebalancing dates. Abnormal returns are based on the multi-factor model. Dashed lines indicate 95% confidence intervals.

Fig. 11.6. Average cumulative abnormal returns (CARs) for a long-short strategy over the rebalancing period and the following 2 weeks for commodities that have no rolls.

index rolls. Altogether, previous studies and our own result (Fig. 11.6) suggest that the identified price impact is highly unlikely to be due to monthly January roll trades.

11.4.3 Regressions

Baseline model

We estimate regressions of CARs on percentage changes in weights and report the estimation results in Table 11.3. The CARs are calculated for nine different windows, including two front-running windows, four rebalancing windows, and three windows stretching to post-rebalancing days. The top panel of Table 11.3 presents results when abnormal returns are estimated based on the multi-factor model. The estimated coefficients on $\Delta w/w^{S\&P\,GSCI}$ are positive but not statistically significant when CARs are measured over the front-running period (days [–9, 0] and [–4, 0]). In contrast, the estimated coefficients are significantly positive at the 1% and 5% level when CARs are measured over the rebalancing period (days [1, 3], [1, 4], and [1, 5]). An estimated coefficient of 0.13 means that a 1% increase in weight increases the CAR by 13 basis points, or equivalently, a one-standard-deviation increase in $\Delta w/w^{S\&P\,GSCI}$ (4.41%) leads to a 57 basis points increase in the CAR for the rebalancing period (days [1, 5]). The price impact is larger when CARs are measured for days [1, 10], which mirrors the finding in Fig. 11.4 where the average CAR for the long-short strategy reaches a peak in the middle of the week following the rebalancing period (day 7). The estimated coefficients on $\Delta w/w^{S\&P\,GSCI}$ are smaller and insignificant when

Table 11.3. Regressions of cumulative abnormal returns (CARs) on percentage changes in weights due to rebalancing.[a]

	Dependent variables: CAR for days								
	[−9, 0]	[−4, 0]	[1, 2]	[1, 3]	[1, 4]	[1, 5]	[1, 10]	[1, 15]	[1, 20]
				Multi-factor model					
$\Delta w/w^{S\&P\ GSCI}$	0.09	0.07	0.06	0.10***	0.14***	0.13**	0.20***	0.11	0.09
	(1.41)	(1.32)	(1.35)	(2.56)	(2.83)	(2.27)	(2.55)	(1.07)	(0.53)
R^2 (%)	0.33	0.39	0.82	1.65	2.54	1.51	1.72	0.32	0.14
N	384	384	384	384	384	384	384	384	384
				Constant-mean model					
$\Delta w/w^{S\&P\ GSCI}$	0.10***	0.07*	0.04	0.07**	0.10***	0.09*	0.11	0.03	0.02
	(2.36)	(1.68)	(1.03)	(2.07)	(2.35)	(1.75)	(1.51)	(0.41)	(0.12)
R^2 (%)	0.52	0.50	0.48	1.14	1.87	1.02	0.77	0.04	0.01
N	384	384	384	384	384	384	384	384	384
				Zero-mean model					
$\Delta w/w^{S\&P\ GSCI}$	0.13**	0.08*	0.05	0.08***	0.12***	0.10**	0.14**	0.08*	0.08
	(2.22)	(1.86)	(1.26)	(2.34)	(2.63)	(2.31)	(2.08)	(1.76)	(0.91)
R^2 (%)	1.19	0.93	0.66	1.50	2.43	1.46	1.32	0.32	0.23
N	384	384	384	384	384	384	384	384	384

[a]The table presents the estimation results of regressions of CARs on percentage changes in weights due to the S&P GSCI rebalancing. The dependent variable is the CAR for various windows around the rebalancing period. Days 1–5 are the S&P GSCI rebalancing dates. Abnormal returns are calculated based on the multi-factor, zero-mean, and constant-mean models, respectively, using contracts held by the index at the beginning of February. Regressions are estimated using a fixed-effects model with time effect. The *t*-statistics are based on time-clustered standard errors and reported in parentheses. *, **, and *** indicate significance at levels of 10%, 5%, and 1%, respectively. R^2 (%) is the R-squared from the projected model where individual fixed effects are not included. N is the number of observations, which consists of 24 commodities and 16 years (2004–2019).

CARs are measured for days [1, 15] or [1, 20], suggesting that the price impact disappears over a longer period. The R-squared ranges between 0.14% and 2.54% and is larger when CARs are measured for days [1, 3] through [1, 10], suggesting that percentage weight changes due to the rebalancing help explain the variation of CARs for the rebalancing period and the following week.

Table 11.3 also reports regression results when abnormal returns are calculated from the constant-mean model (middle panel) and the zero-mean model (bottom panel). Compared to estimates in the top panel, the estimated coefficients on $\Delta w/w^{S\&P\ GSCI}$ become larger and more statistically significant when CARs are measured for days [−9, 0] and [−4, 0]. This provides some evidence that weight changes due to the rebalancing are associated with front-running trades, but the results in the top panel show that the relationship weakens when abnormal returns are conditioned on a group of control variables. When CARs are measured for the rebalancing period and the following week, the estimated coefficients on $\Delta w/w^{S\&P\ GSCI}$ are slightly smaller but remain positive and statistically significant. The results of the three panels suggest that percentage changes in weights, as a proxy for the rebalancing order flows, are significantly related to cumulative abnormal returns for the rebalancing period and the following week, no matter which model is used for estimating abnormal returns. As such, we focus on abnormal returns from the multi-factor model in subsequent analyses and save the results from the constant-mean and zero-mean models to the Online Appendix. Finally, there is at best weak evidence of front-running associated with predatory trading strategies.

Rebalancing effects over time

Under the theory of sunshine trading, the impact of predictable order flows will be mitigated as they become better understood by the market (Bessembinder, 2015). This implies that the price impact of order flows due to S&P GSCI rebalancing will decline over time. To test this hypothesis, we estimate the same regression model for the first and second half of the sample period, respectively. A comparison of the estimated coefficients on percentage weight changes allows us to examine whether rebalancing impact is the same between sub-sample periods.[18] The estimation results are reported in Table 11.4. The estimated coefficients on $\Delta w/w^{S\&P\ GSCI}$ are generally larger and more statistically significant in the first half of the sample, suggesting that rebalancing impacts are likely smaller for 2012–2019. Moreover, the estimates remain large (though not statistically significant) over the post-rebalancing period (days [1, 15] and [1, 20]) for 2004–2011, while they are reduced to close to or below zero for 2012–2019. This implies that the market impact of rebalancing is diminished in recent years. None of the coefficients for the front-running periods ([–9, 0], [–4, 0]) are statistically significant in either sub-period. These results are largely consistent with Bessembinder's (2015) prediction that predictable trades such as the rebalancing-based order flows will have a reduced market impact as they become better known to the marketplace.

It should also be noted that the point estimates in Table 11.4 for days [1, 3], [1, 4], [1, 5] and [1, 10] are reasonably close across the two sub-samples, suggesting that the main rebalancing impact has persisted to some degree over the sample. This is possibly due to the inherent risk in arbitraging against this order flow as it entails positions with both low correlations across markets and through time (Gromb and Vayanos, 2010). Furthermore, the confidence bands presented in Figs 11.4 and 11.5 indicate that the returns to the long-short strategy are quite volatile from year to year relative to average returns, and hence, the low risk-return trade-offs may not attract sufficient counterparties to completely eliminate the order flow impact of annual rebalancing trades (Bessembinder, 2015).

Rebalancing effects along the futures curve

We have shown that the uninformed order flows due to the S&P GSCI rebalancing temporarily impact futures prices of front-month contracts. An interesting question is how the uninformed order flows impact the entire futures pricing curve. For example, Henderson *et al.* (2015) report that hedge flows of CLN issues have an equal impact across the futures pricing curve. This is consistent with the prediction of the rational theory of storage that futures prices for storable commodities are tightly integrated. To examine whether the identified rebalancing effects emerge across the futures pricing curve, we consider contracts with longer maturities. Here we calculate CARs for contracts that have

Table 11.4. Regressions of CARs on percentage changes in weights due to rebalancing: sub-sample analysis.[a]

	Dependent variables: CAR for days								
	[–9, 0]	[–4, 0]	[1, 2]	[1, 3]	[1, 4]	[1, 5]	[1, 10]	[1, 15]	[1, 20]
					2004–2011				
$\Delta w/w^{S\&P\ GSCI}$	0.12	0.06	0.03	0.11***	0.16***	0.12*	0.21***	0.17	0.2
	(1.33)	(0.60)	(1.34)	(2.55)	(2.45)	(1.68)	(2.33)	(1.38)	(1.35)
R^2 (%)	0.48	0.17	0.17	1.91	3.00	1.25	1.92	0.77	0.76
N	192	192	192	192	192	192	192	192	192
					2012–2019				
$\Delta w/w^{S\&P\ GSCI}$	0.06	0.09	0.09	0.09	0.12	0.14	0.19	0.05	–0.02
	(0.63)	(1.29)	(1.08)	(1.34)	(1.54)	(1.49)	(1.41)	(0.29)	(–0.08)
R^2 (%)	0.18	0.85	1.97	1.41	2.11	1.79	1.52	0.07	0.01
N	192	192	192	192	192	192	192	192	192

[a]The table presents the estimation results of regressions of cumulative abnormal returns (CARs) on percentage changes in weights due to the S&P GSCI rebalancing for sub-sample periods. The dependent variable is the CAR for various windows around the rebalancing period. Days 1–5 are the S&P GSCI rebalancing dates. Abnormal returns are calculated based on the multi-factor model using contracts held by the index at the beginning of February. Regressions are estimated using a fixed-effects model with time effect. The *t*-statistics are based on time-clustered standard errors and reported in parentheses. *, **, and *** indicate significance at levels of 10%, 5%, and 1%, respectively. R^2 (%) is the R-squared from the projected model where individual fixed effects are not included. N is the number of observations, which consists of 24 commodities and 8 years in each sub-sample.

at least 4, 6, and 9 months to maturity, respectively, and link them to weight changes using the regression model.[19] Contracts with maturity longer than 9 months are not considered because these contracts are much less frequently traded in many markets. The estimation results are reported in Table 11.5. In each panel, the estimated coefficients on percentage weight changes are positive and statistically significant for the rebalancing period and the following week, showing that price impacts of rebalancing prevail across the entire futures term structure. As an example, the coefficients for days [1, 10] are 0.20, 0.17, and 0.18, respectively, for contracts with at least 4, 6, and 9 months to maturity. These estimates are equal to or only slightly smaller than the 0.20 coefficient for front-month contracts (Table 11.3), which indicates that rebalancing causes a parallel shift in the futures pricing curve. Since coefficients on percentage weight changes become smaller and insignificant for days [1, 15] and [1, 20], the entire futures curve shifts back after the rebalancing period. In sum, uninformed order flows due to rebalancing not only impact prices of the futures contracts held directly by S&P GSCI investors but also have a similar impact on prices for deferred contracts.

Examining rebalancing impacts along the futures pricing curve also helps distinguish between rebalancing and roll effects. Recall that the S&P GSCI rebalancing period overlaps exactly with the January roll period for some commodities (14 out of 24), where index positions are rolled from the expiring contract to the front-month contract. Although rebalancing trades and roll trades happen in some markets during the same period in January, they should have markedly different impacts on the futures pricing curve. Since roll trades involve no net order flow into or out of a given commodity futures market, the impact of rolling results in simultaneous downward pressure on the expiring contract and upward pressure on front-month contracts. In other words, only the calendar price spread between the expiring and front-month contracts is impacted. Since prices for longer maturity contracts are not impacted by

Table 11.5. Regressions of CARs on percentage changes in weights due to rebalancing: effects along the futures pricing curve.[a]

	Dependent variables: CAR for days								
	[−9, 0]	[−4, 0]	[1, 2]	[1, 3]	[1, 4]	[1, 5]	[1, 10]	[1, 15]	[1, 20]
	Contracts that have at least 4 months to maturity								
$\Delta w/w^{S\&P\ GSCI}$	0.08	0.07	0.05	0.08**	0.12***	0.12**	0.20***	0.11	0.09
	(1.63)	(1.46)	(1.12)	(2.30)	(2.52)	(2.07)	(2.69)	(1.16)	(0.55)
R^2 (%)	0.40	0.46	0.60	1.39	2.30	1.46	1.95	0.37	0.16
N	384	384	384	384	384	384	384	384	384
	Contracts that have at least 6 months to maturity								
$\Delta w/w^{S\&P\ GSCI}$	0.06	0.05	0.04	0.07*	0.10**	0.10*	0.17***	0.09	0.03
	(1.00)	(0.94)	(0.81)	(1.88)	(1.96)	(1.72)	(2.33)	(0.81)	(0.20)
R^2 (%)	0.20	0.21	0.39	0.99	1.47	1.08	1.45	0.24	0.03
N	360	360	360	360	360	360	360	360	360
	Contracts that have at least 9 months to maturity								
$\Delta w/w^{S\&P\ GSCI}$	0.04	0.05	0.03	0.06	0.10**	0.11*	0.18***	0.10	0.06
	(0.77)	(1.11)	(0.67)	(1.63)	(1.99)	(1.76)	(2.59)	(0.91)	(0.36)
R^2 (%)	0.09	0.29	0.32	0.85	1.58	1.23	1.78	0.37	0.10
N	357	357	357	357	357	357	357	357	357

[a]The table presents the estimation results of regressions of cumulative abnormal returns (CARs) on percentage changes in weights due to the S&P GSCI rebalancing. The dependent variable is the CAR for various windows around the rebalancing period. Days 1–5 are the S&P GSCI rebalancing dates. Abnormal returns are calculated based on the multi-factor model using contracts that have at least 4, 6, and 9 months to maturity, respectively. Regressions are estimated using a fixed-effects model with time effect. The t-statistics are based on time-clustered standard errors and reported in parentheses. *, **, and *** indicate significance at levels of 10%, 5%, and 1%, respectively. R^2 (%) is the R-squared from the projected model where individual fixed effects are not included. N is the number of observations, which consists of 24 commodities and 16 years (2004–2019).

rolling, the more forward parts of the futures pricing curve are unaffected. In contrast, rebalancing trades create additional demands for futures contracts and reflect new order flows into or out of the market, which can increase or decrease the price level of the front-month futures contract. Arbitrage in rational storage markets will transmit the front month price impact to deferred futures prices, resulting in a parallel shift in the entire futures pricing curve. Since we find that the market impact of rebalancing causes a parallel shift in the entire futures pricing curve, it is highly unlikely that the identified price impacts are due to the order flow of roll trades.

Rebalancing and the Great Recession

The Great Recession of 2008–2009 witnessed large price drops in many commodity markets simultaneously. Institutional investors were also subject to serious funding constraints during the recession period. Acharya *et al.* (2013) and Etula (2013) emphasize the role of financial intermediaries as arbitrageurs in determining prices in the commodity futures market. The implication is that the price impacts should increase in the size of the rebalancing order flows because of limits to arbitrageurs' capacity to take the other side of the trades. Similarly, the price impact should be larger during the Great Recession when arbitrageurs confronted financial constraints.

We explore whether part of the rebalancing effects can be attributed to the Great Recession by excluding the January 2009 rebalancing, which falls within the recession interval specified by the National Bureau of Economic Research. Specifically, we include an interaction term of percentage weight changes and dummy variable D^{2009} in the regression model, where D^{2009} equals 1 for year 2009 and 0 otherwise. The results are presented in Table 11.6. The estimated coefficients on $\Delta w/w^{S\&P\,GSCI}$ are very similar to those in the baseline regression (Table 11.3) in terms of both size and significance. The estimated coefficients on $\Delta w/w^{S\&P\,GSCI}D^{2009}$ are positive for days [1, 3], [1, 4], [1, 5], and [1, 10], though not statistically significant. It is not surprising that the coefficients on $\Delta w/w^{S\&P\,GSCI}D^{2009}$ are estimated imprecisely as there is only one major recession in the sample period. Regarding magnitude, the coefficients on the interaction term for these days are 1.2–2.8 times larger than the coefficients on $\Delta w/w^{S\&P\,GSCI}$. This is consistent with expectation that the price impact of uninformed order flows is higher during the Great Recession of 2008–2009 due to financial constraints facing financial intermediaries.

Rebalancing and large commodity markets

Another interesting question is whether the identified price effects emerge only in the biggest

Table 11.6. Regressions of CARs on percentage changes in weights due to rebalancing: effects of the Great Recession.[a]

	Dependent variables: CAR for days								
	[−9, 0]	[−4, 0]	[1, 2]	[1, 3]	[1, 4]	[1, 5]	[1, 10]	[1, 15]	[1, 20]
$\Delta w/w^{S\&P\,GSCI}$	0.11	0.09	0.06*	0.09**	0.13***	0.11**	0.19***	0.11	0.07
	(1.29)	(1.42)	(1.73)	(2.28)	(2.75)	(2.04)	(2.34)	(1.02)	(0.54)
$\Delta w/w^{S\&P\,GSCI}D^{2009}$	−0.36	−0.32	−0.03	0.11	0.25	0.31	0.21	0.09	0.02
	(−0.96)	(−1.15)	(−0.19)	(0.63)	(1.21)	(1.24)	(0.57)	(0.20)	(0.74)
R^2 (%)	0.58	0.74	0.83	1.76	2.93	1.92	1.80	0.34	0.29
N	384	384	384	384	384	384	384	384	384

[a] The table presents the estimation results of regressions of cumulative abnormal returns (CARs) on percentage changes in weights due to the S&P GSCI rebalancing. The dependent variable is the CAR for various windows around the rebalancing period. Days 1–5 are the S&P GSCI rebalancing dates. Abnormal returns are calculated based on the multi-factor model using contracts held by the index at the beginning of February. D^{2009} is the dummy variable for the Great Recession, which equals 1 for 2009 and 0 for other years. Regressions are estimated using a fixed-effects model with time effect. The *t*-statistics are based on time-clustered standard errors and reported in parentheses. *, **, and *** indicate significance at levels of 10%, 5%, and 1%, respectively. R^2 (%) is the R-squared from the projected model where individual fixed effects are not included. N is the number of observations, which consists of 24 commodities and 16 years (2004–2019).

commodity markets. Recall that the S&P GSCI is concentrated in energy with a combined weight above 45% for WTI and Brent crude oil. These two oil markets also have the largest changes in weights (Table 11.1). To explore whether the rebalancing effects are concentrated in crude oil futures markets, we drop WTI and Brent crude oil and estimate the regressions for the remaining 22 commodities. Results in Table 11.7 show that the estimated coefficients on percentage weight changes are positive and statistically significant for days [1, 3], [1, 4], [1, 5], and [1, 10], supporting rebalancing price impacts in the 22 non-oil markets. The magnitude of the coefficient estimates is identical to or slightly smaller than that in the baseline regression (Table 11.3). This suggests that order flows due to the rebalancing have an almost equal impact on futures price when crude oil markets are excluded. Although crude oil has the largest order flows, the price impact is not necessarily the biggest thanks to their high levels of market depth. In contrast, the price impact may be more significant for small markets where market depth is lower. This likely explains the result in Table 11.7 that the rebalancing price effects are similar without crude oil. Thus, the identified price impacts of order flows due to the rebalancing are not specific to a few seemingly important markets such as crude oil.

Rebalancing and the Bloomberg Commodity Index

The S&P GSCI is not the only important commodity index that changes its weights. The Bloomberg Commodity Index (BCOM), another benchmark for investment in commodity futures markets, assigns weight to each commodity from the sixth through the tenth business day in January, 1 day lagged relative to the S&P GSCI rebalancing period.[20] We repeat the analysis by considering both S&P GSCI and BCOM rebalancing. The BCOM includes 22 commodities with 20 of them included in the S&P GSCI (soybean oil and soybean meal are excluded). The BCOM is constructed in a similar way to the S&P GSCI but differs by imposing caps on the weights of sectors and individual markets. Since the weights of the BCOM also depend on historical production and trading volume information, order flows due to the BCOM rebalancing are exogenous for the same reasons we argued earlier for the S&P GSCI. We evaluate the joint rebalancing effect of the two commodity indexes by including their percentage weight changes in the regression simultaneously and present the results in Table 11.8.

The estimated coefficients on $\Delta w/w^{S\,\&\,P\,GSCI}$ are positive and statistically significant for days [1, 3], [1, 4], [1, 5], and [1, 10]. More importantly, the magnitude of the coefficients is almost identical to those reported in Table 11.3. The estimated coefficients on $\Delta w/w^{BCOM}$ are overall smaller and less significant. The coefficient on $\Delta w/w^{BCOM}$ has a similar size as the coefficient on $\Delta w/w^{S\,\&\,P\,GSCI}$ for days [1, 2] but declines over the rebalancing period and its following week, suggesting that order flows due to the BCOM rebalancing are small and have little impact on futures price. Not surprisingly, the R-squared

Table 11.7. Regressions of CARs on percentage changes in weights due to rebalancing: role of large markets.[a]

	Dependent variables: CAR for days								
	[–9, 0]	[–4, 0]	[1, 2]	[1, 3]	[1, 4]	[1, 5]	[1, 10]	[1, 15]	[1, 20]
$\Delta w/w^{S\&P\ GSCI}$	0.07	0.07	0.06	0.11***	0.14***	0.12**	0.15**	0.08	0.07
	(1.01)	(1.21)	(1.25)	(2.39)	(2.66)	(2.02)	(2.08)	(0.76)	(0.40)
R^2 (%)	0.19	0.35	0.83	1.79	2.21	1.18	0.89	0.16	0.09
N	352	352	352	352	352	352	352	352	352

[a]The table presents the estimation results of regressions of cumulative abnormal returns (CARs) on percentage changes in weights due to the S&P GSCI rebalancing for commodities with WTI and Brent crude oil excluded. The dependent variable is the CAR for various windows around the rebalancing period. Days 1–5 are the S&P GSCI rebalancing dates. Abnormal returns are calculated based on the multi-factor model using contracts held by the index at the beginning of February. Regressions are estimated using a fixed-effects model with time effect. The *t*-statistics are based on time-clustered standard errors and report in parentheses. *, **, and *** indicate significance at levels of 10%, 5%, and 1%, respectively. R^2 (%) is the R-squared from the projected model where individual fixed effects are not included. *N* is the number of observations, which consists of 22 commodities and 16 years (2004–2019).

Table 11.8. Regressions of CARs on percentage changes in weights due to rebalancing: the Bloomberg Commodity Index (BCOM) rebalancing.[a]

	Dependent variables: CAR for days								
	[−9, 0]	[−4, 0]	[1, 2]	[1, 3]	[1, 4]	[1, 5]	[1, 10]	[1, 15]	[1, 20]
$\Delta w/w^{S\&P\ GSCI}$	0.09	0.07	0.06	0.10***	0.14***	0.13**	0.20***	0.11	0.09
	(1.40)	(1.38)	(1.31)	(2.50)	(2.78)	(2.24)	(2.53)	(1.05)	(0.51)
$\Delta w/w^{BCOM}$	0.08*	0.06	0.05**	0.06	0.03	0.03	0.04	0.06	0.04
	(1.82)	(1.53)	(2.30)	(1.46)	(0.70)	(0.56)	(0.48)	(0.64)	(0.42)
R^2 (%)	0.95	0.92	2.20	2.77	2.80	1.69	1.86	0.51	0.20
N	384	384	384	384	384	384	384	384	384

[a] The table presents the estimation results of regressions of cumulative abnormal returns (CARs) on percentage changes in weights due to both the S&P GSCI and BCOM rebalancing. The dependent variable is the CAR for various windows around the rebalancing period. Days 1–5 are the S&P GSCI rebalancing dates. Days 2–6 are the BCOM rebalancing dates. Abnormal returns are calculated based on the multi-factor model using contracts held by the S&P GSCI at the beginning of February. Regressions are estimated using a fixed-effects model with time effect. The *t*-statistics are based on time-clustered standard errors and reported in parentheses. *, **, and *** indicate significance at levels of 10%, 5%, and 1%, respectively. R^2 (%) is the R-squared from the projected model where individual fixed effects are not included. N is the number of observations, which consists of 24 commodities and 16 years (2004–2019).

does not improve compared to values in Table 11.3 given the limited impact of BCOM rebalancing. Once again, there is at best weak evidence of front-running associated with predatory trading strategies. Overall, the BCOM rebalancing is less important and its associated order flows have limited effect on commodity futures prices.

11.5 Conclusions

The rebalancing of the S&P GSCI provides a novel identification to estimate the impact of predictable order flows from financial investors in commodity futures markets. Each year, the S&P GSCI reassigns weights to each commodity in the index during the January rebalancing period and the changes in weights generate large financial index order flows into and out of the market. Using the 24 commodities included in the S&P GSCI for 2004–2019, we show that the cumulative abnormal returns (CARs) for the rebalancing period tend to have the same sign as changes in index weights. A strategy that assigns an equal weight to each commodity and holds a long (short) position in commodities that experience positive (negative) index weight changes yields positive abnormal returns. The price impact reaches a peak of 72 basis points in the middle of the week following the rebalancing period, but the impact is temporary as it declines to near zero within the next week.

Regression estimates show that the percentage changes in index weights, as a proxy for the size of the rebalancing flows, generally are not significantly related to the CARs for the front-running period but are significantly and positively related during the rebalancing period and the following week. The regressions also show that rebalancing not only impacts prices of the front-month futures contracts held directly by S&P GSCI investors but also have a similar impact on prices for deferred contracts. Since the market impact of rebalancing causes a parallel shift in the entire futures pricing curve, it is highly unlikely that the identified price impacts are due to the order flow of roll trades. Thus, regression analysis provides important new evidence that the impact of predictable order flows from index investors in commodity futures markets is modest and temporary.

Our finding of little if any price impact during front-running periods is inconsistent with the theory of predatory trading (Brunnermeier and Pederson, 2005). Instead, our results are consistent with the theory of sunshine trading (Admanti and Pfleiderer, 1991) because the predictable nature of rebalancing order flows attracts additional liquidity suppliers, and hence, the order flows have modest and temporary impacts. Finally, the weakening of price impacts in the second half of the sample period is consistent with Bessembinder's (2015) argument that predictable order flows should have a declining impact as they become better understood by market participants.

Appendix A. Supplementary Data

Supplementary data to this article can be found online at: https://doi.org/10.1016/j.jcomm.2021.100195 (accessed September 2 2022).

Appendix B. Mathematical Derivation of Eqn 11.5

This appendix provides a derivation of Eqn 11.5, which shows that order flows due to the S&P GSCI rebalancing are proportional to weight changes.

The S&P GSCI Excess Return Index is constructed in a cumulative manner with a base value of 100 on January 2 1970 (S&P Dow Jones Indices, 2017, p. 23).[21] That is,

$$S_d = S_{d-1}\left(1+CDR_d\right) \tag{B11.1}$$

where S_d is the value of the index on day d and CDR_d is the contract daily return computed as a percentage change in the total dollar weight (TDW), i.e. $CDR_d = \frac{TDW_d}{TDW_{d-1}} - 1$. The TDW_d on rebalancing dates is defined as (S&P Dow Jones Indices, 2017, p. 21),

$$TDW_d = \sum_c \begin{bmatrix} \frac{NC_{new}}{NC_{old}} CPW_{old}^c * CRW1_d * DCRP1_d^c + \\ CPW_{new}^c * CRW2_d * DCRP2_d^c \end{bmatrix} \tag{B11.2}$$

where CPW_{old}^c and CPW_{new}^c are the contract production weights of commodity c prior to and after the rebalancing, $CRW1_d$ and $CRW2_d$ are the contract roll weights of the first and second nearby futures contracts that take 0.8/0.2, 0.6/0.4, 0.4/0.6, 0.2/0.8, and 0/1 on the fifth through the ninth business days, $DCRP1_d^c$ and $DCRP2_d^c$ are the daily contract reference prices of the first and second nearby futures contracts, and NC_{old} and NC_{new} are the normalizing constants prior to and after the rebalancing, respectively.

For investors who track the S&P GSCI Excess Return Index, replication requires that the notional value of the index on day d must be equal to the sum of notional values of the positions held in each commodity market. That is,

$$S_d = \sum_c \begin{bmatrix} X1_d^c * CS^c * DCRP1_d^c + X2_d^c * \\ CS^c * DCRP2_d^c \end{bmatrix} \tag{B11.3}$$

where CS^c is the contract size of commodity futures, and $X1_d^c$ and $X2_d^c$ are the positions of the first and second nearby futures contracts, respectively. Substituting Eqns B11.2 and B11.3 into Eqn B11.1 gives

$$\sum_c \frac{S_{d-1}}{TDW_{d-1}} \begin{bmatrix} \frac{NC_{new}}{NC_{old}} CPW_{old}^c * CRW1_d * DCRP1_d^c + \\ CPW_{new}^c * CRW2_d * DCRP2_d^c \end{bmatrix} = \sum_c \begin{bmatrix} X1_d^c * CS^c * DCRP1_d^c + X2_d^c * \\ CS^c * DCRP2_d^c \end{bmatrix}. \tag{B11.4}$$

Note that Eqn B11.4 holds for any values of $DCRP1_d^c$ and $DCRP2_d^c$, implying

$$X1_d^c = \frac{S_{d-1}}{TDW_{d-1}} \frac{NC_{new}}{NC_{old}} \frac{CPW_{old}^c * CRW1_d}{CS^c} \tag{B11.5}$$

$$X2_d^c = \frac{S_{d-1}}{TDW_{d-1}} \frac{CPW_{new}^c * CRW2_d}{CS^c} \tag{B11.6}$$

The total futures position in commodity c is given by

$$X1_d^c + X2_d^c = \frac{S_{d-1}}{TDW_{d-1}} \frac{1}{CS^c} \begin{bmatrix} \frac{NC_{new}}{NC_{old}} CPW_{old}^c * CRW1_d + \\ CPW_{new}^c * CRW2_d \end{bmatrix} \tag{B11.7}$$

Recall that the S&P GSCI rebalancing occurs within a 5-day window from the fifth to the ninth business day in January. Let $d = \{1, 2, 3, 4, 5\}$ be the rebalancing dates and X_d^c be the total futures position. We have $CRW1_0 = 1$ and $CRW2_0 = 0$ for the day immediately prior to the first rebalancing date ($d = 0$) and $CRW1_5 = 0$ and $CRW2_5 = 1$ for the last rebalancing day ($d = 5$). The total futures positions on those 2 days are

$$X_0^c = \frac{S_0}{TDW_0} \frac{1}{CS^c} \frac{NC_{new}}{NC_{old}} CPW_{old}^c \text{ and} \quad X_5^c = \frac{S_5}{TDW_5} \frac{1}{CS^c} CPW_{new}^c \tag{B11.8}$$

respectively. Since Eqn B11.1 implies $\frac{S_0}{TDW_0}=\frac{S_5}{TDW_5}$, the change in position is

$$X_5^c - X_0^c = \frac{S_0}{TDW_0}\frac{1}{CS^c}\left(CPW_{new}^c - \frac{NC_{new}}{NC_{old}}CPW_{old}^c\right) \quad \text{(B11.9)}$$

The ratio of normalizing constant is defined as (S&P Dow Jones Indices, 2017, p. 17),

$$\frac{NC_{new}}{NC_{old}} = \frac{\sum_c \left(CPW_{new}^c * DCRP_0^c\right)}{\sum_c \left(CPW_{old}^c * DCRP_0^c\right)} \quad \text{(B11.10)}$$

where $DCRP_0^c$ is the price of the first nearby futures contract of commodity c on the day immediately prior to the first rebalancing date. Substituting Eqn B11.10 into Eqn B11.9 and rearranging the terms gives

$$\left(X_5^c - X_0^c\right) * DCRP_0^c * CS^c = k * \left(w_{new}^c - w_{old}^c\right) \quad \text{(B11.11)}$$

where $k = S_0\frac{NC_{new}}{NC_{old}}$, $w_{new}^c = \frac{CPW_{new}^c * DCRP_0^c}{\sum_c \left(CPW_{new}^c * DCRP_0^c\right)}$, and $w_{old}^c = \frac{CPW_{old}^c * DCRP_0^c}{\sum_c \left(CPW_{old}^c * DCRP_0^c\right)}$.

The left-hand side of Eqn B11.11 represents the notional value of investment flow into or out of the market. k is a constant and can be interpreted as the total investment tied to the index. The change in weight, $w_{new}^c - w_{old}^c$, is driven only by CPW change given that the same prices are used for calculating weights prior to and after the rebalancing period. Therefore, order flows due to the S&P GSCI rebalancing are proportional to changes in weights.

Appendix C. Data Appendix

Table C11.1. Commodity futures included in the S&P GSCI index.[a]

		Contract expiration at beginning of calendar month											
Commodity	Trading facility	1	2	3	4	5	6	7	8	9	10	11	12
Chicago wheat	CBT	H	H	K	K	N	N	U	U	Z	Z	Z	H
Kansas wheat	KBT	H	H	K	K	N	N	U	U	Z	Z	Z	H
Corn	CBT	H	H	K	K	N	N	U	U	Z	Z	Z	H
Soybeans	CBT	H	H	K	K	N	N	X	X	X	X	F	F
Coffee	ICE	H	H	K	K	N	N	U	U	Z	Z	Z	H
Sugar	ICE	H	H	K	K	N	N	V	V	V	H	H	H
Cocoa	ICE	H	H	K	K	N	N	U	U	Z	Z	Z	H
Cotton	ICE	H	H	K	K	N	N	Z	Z	Z	Z	Z	H
Lean hogs	CME	G	J	J	M	M	N	Q	V	V	Z	Z	G
Live cattle	CME	G	J	J	M	M	Q	Q	V	V	Z	Z	G
Feeder cattle	CME	H	H	J	K	Q	Q	Q	U	V	X	F	F
WTI crude oil	NYM	G	H	J	K	M	N	Q	U	V	X	Z	F
Heating oil	NYM	G	H	J	K	M	N	Q	U	V	X	Z	F
RBOB gasoline	NYM	G	H	J	K	M	N	Q	U	V	X	Z	F
Brent crude oil	ICE	H	J	K	M	N	Q	U	V	X	Z	F	G
Gasoil	ICE	G	H	J	K	M	N	Q	U	V	X	Z	F
Natural gas	NYM	G	H	J	K	M	N	Q	U	V	X	Z	F
Aluminum	LME	G	H	J	K	M	N	Q	U	V	X	Z	F
Copper	LME	G	H	J	K	M	N	Q	U	V	X	Z	F
Nickel	LME	G	H	J	K	M	N	Q	U	V	X	Z	F
Lead	LME	G	H	J	K	M	N	Q	U	V	X	Z	F
Zinc	LME	G	H	J	K	M	N	Q	U	V	X	Z	F
Gold	CMX	G	J	J	M	M	Q	Q	Z	Z	Z	Z	G
Silver	CMX	H	H	K	K	N	N	U	U	Z	Z	Z	H

[a]The table lists futures contracts of the 24 commodities included in the S&P GSCI. CBT represents the Chicago Board of Trade. KBT represents the Kansas City Board of Trade. ICE represents the Intercontinental Exchange. CME represents the Chicago Mercantile Exchange. NYM represents the New York Mercantile Exchange. LME represents the London Metal Exchange. CMX represents COMEX Exchange. The table also lists maturities of the futures contracts held by the index at the beginning of each calendar month. Futures month codes are: January (F), February (G), March (H), April (J), May (K), June (M), July (N), August (Q), September (U), October (V), November (X), and December (Z).

Table C11.2. Reference percentage dollar weights of the components in the S&P GSCI.[a]

Commodity	Year																
	2004	2005	2006	2007	2008	2009	2010	2011	2012	2013	2014	2015	2016	2017	2018	2019	Avg.
Chicago wheat	3.98	3.81	2.36	2.18	3.39	3.35	3.99	2.97	3.16	3.19	3.42	2.93	3.53	3.87	2.98	2.74	3.27
Kansas wheat	1.75	1.40	0.88	0.92	0.80	0.30	0.83	0.69	0.98	0.67	0.78	0.75	0.87	1.07	1.10	1.13	0.96
Corn	4.23	4.07	2.33	2.09	3.28	3.31	4.03	3.34	4.61	4.71	4.99	3.39	4.19	5.48	4.92	4.31	3.96
Soybeans	2.45	3.09	1.71	1.43	1.84	2.21	2.92	2.37	2.55	2.62	2.91	2.77	2.95	3.81	3.65	3.12	2.65
Coffee	0.68	0.67	0.77	0.65	0.69	0.55	0.78	0.76	1.02	0.81	0.57	0.67	0.93	1.03	1.01	0.72	0.77
Sugar	1.60	1.24	1.25	1.77	1.24	0.91	1.92	2.27	2.33	1.86	1.46	1.39	1.58	2.47	2.48	1.53	1.71
Cocoa	0.44	0.30	0.22	0.18	0.22	0.22	0.38	0.39	0.30	0.23	0.23	0.30	0.45	0.59	0.37	0.32	0.32
Cotton	1.64	1.72	0.96	0.86	0.89	0.78	0.97	1.25	1.88	1.07	1.01	1.07	1.19	1.54	1.61	1.41	1.24
Lean hogs	2.06	2.38	1.96	1.49	1.51	1.00	1.55	1.57	1.48	1.59	1.69	2.11	2.33	2.67	2.25	1.88	1.84
Live cattle	4.13	3.88	2.85	2.62	2.72	2.03	3.06	2.60	2.39	2.59	2.72	3.16	4.87	5.19	4.18	3.52	3.28
Feeder cattle	0.81	0.93	0.77	0.66	0.54	0.39	0.51	0.44	0.41	0.51	0.50	0.72	1.57	1.53	1.26	1.26	0.80
WTI crude oil	25.34	25.75	31.05	37.11	35.60	39.36	35.05	34.86	30.28	24.66	23.64	24.40	23.03	22.91	24.73	26.41	29.04
Heating oil	7.41	7.10	8.40	5.99	4.65	4.35	4.58	4.63	4.73	6.16	6.02	5.88	5.29	4.03	3.86	4.45	5.50
RBOB gasoline	7.99	8.02	7.65	1.46	4.64	4.72	4.33	4.66	4.80	5.99	6.03	5.74	5.36	4.64	4.64	4.51	5.32
Brent crude oil	11.85	11.82	14.20	14.94	13.02	13.71	14.00	15.17	17.38	22.43	23.22	24.69	20.20	16.40	16.72	18.71	16.78
Gasoil	3.76	3.89	4.58	5.25	4.56	5.25	5.79	6.29	7.48	8.58	8.29	7.36	5.82	4.87	4.63	5.56	5.75
Natural gas	11.36	10.09	9.59	9.87	7.17	6.32	5.18	4.10	2.83	1.96	2.55	3.18	3.28	3.20	3.93	3.14	5.49
Aluminum	3.12	3.31	2.78	3.08	3.52	2.45	2.51	2.73	2.55	2.14	2.02	2.00	2.88	3.27	3.64	3.89	2.87
Copper	1.76	2.39	2.24	3.39	4.03	3.02	2.85	3.68	3.74	3.27	3.21	3.11	3.84	4.08	4.44	4.45	3.34
Nickel	0.64	0.93	0.79	0.80	1.63	0.79	0.66	0.83	0.79	0.58	0.53	0.55	0.70	0.66	0.69	0.77	0.77
Lead	0.22	0.31	0.28	0.28	0.52	0.46	0.41	0.51	0.47	0.40	0.45	0.47	0.60	0.74	0.87	0.78	0.48
Zinc	0.51	0.57	0.53	0.93	1.30	0.57	0.57	0.72	0.61	0.51	0.53	0.59	0.88	1.01	1.31	1.28	0.78
Gold	2.10	2.11	1.67	1.82	1.95	1.70	2.82	2.81	2.67	2.99	2.80	2.41	3.25	4.40	4.20	3.71	2.71
Silver	0.19	0.23	0.19	0.23	0.29	0.24	0.31	0.36	0.54	0.49	0.44	0.34	0.41	0.55	0.52	0.42	0.36

[a] The table presents the reference percentage dollar weights for the 24 commodities included in the S&P GSCI for 2004–2019. Reference percentage dollar weights are calculated based on Eqn 11.1, where reference price is the average of the prices of the contract held by the index on the last business day of each calendar month over the annual calculation period. Avg. is the average across years.

Credit Author Statement

Lei Yan: conceptualization; data curation; formal analysis; methodology; software; visualization; roles/writing – original draft; writing – review and editing; Scott H. Irwin: conceptualization; methodology; visualization; roles/writing – original draft; writing – review and editing; Dwight R. Sanders: conceptualization; methodology; visualization; roles/writing – original draft; writing – review and editing.

Notes

[1] Original citation: Yan, L., Irwin, S.H. and Sanders, D.R. (2022) Sunshine vs. predatory trading effects in commodity futures markets: new evidence from index rebalancing. *Journal of Commodity Markets* 26, 100195. Reprinted by permission of Elsevier B.V. The Online Supplemental Appendix for the article is found here: https://view.officeapps.live.com/op/view.aspx?src=https%3A%2F%2Fars.els-cdn.com%2Fcontent%2Fimage%2F1-s2.0-S2405851321000295-mmc1.docx&wdOrigin=BROWSELINK (accessed September 2 2022).
[2] The total investment across all commodity futures markets included in the index does not change during the annual rebalancing period.
[3] The S&P GSCI transitioned from including unleaded gasoline to RBOB gasoline in 2007.
[4] To illustrate index composition, we calculate percentage dollar weights for each commodity and each year based on average prices for each year and report weights in Table C11.2 of the Data Appendix.
[5] The 14 markets that have both rebalancing and roll trades in January include: (i) lean hogs; (ii) live cattle; (iii) WTI crude oil; (iv) heating oil; (v) RBOB gasoline; (vi) Brent crude oil; (vii) gasoil; (viii) natural gas; (ix) aluminum; (x) copper; (xi) nickel; (xii) lead; (xiii) zinc; and (xiv) gold. See Table C11.1 in the Data Appendix.
[6] The use of day 0 price is due to the definition of the normalizing constant of the index (see Eqn B11.10 in Appendix B).
[7] The IID report, initially compiled under a special call issued by the CFTC to swap dealers and index traders in June 2008, provides the most accurate measure of total index investment in major US commodity futures markets. The CFTC discontinued release of the IID report in November 2015.
[8] Masters (2008) estimates the value of total assets invested in the S&P GSCI and the Bloomberg Commodity Index and argues that the former is nearly twice as large as the latter, which is consistent with our assumption.
[9] The reported volume and open interest for London Metal Exchange (LME) contracts cannot be directly compared with those for US contracts because of different contract specifications and trading times. We follow the Bloomberg Commodity Index methodology and use one-third of the reported volume and open interest for LME contracts for a fair comparison.
[10] The percentage changes in positions held by index traders are large for cocoa and feeder cattle during post-rebalancing and non-rebalancing weeks, which may be driven by idiosyncratic behaviors given that the total position held by index traders in these two markets is small.
[11] Another factor that contributes to cross-market dependence is the long-run upward trend in prices during the 2000s commodities boom.
[12] Estimating abnormal returns based on the multi-factor or constant-mean model does not address the cross-market dependence issue because the estimation window is 2 weeks ahead of the S&P GSCI rebalancing period.
[13] We also use the change in weight divided by open interest as the independent variable. By using open interest to proxy for market size, we obtain similar results that the weight changes scaled by open interests are positively and significantly related to CARs. Detailed results are provided in the Online Appendix.
[14] See Petersen (2009) for detailed discussions about different approaches for estimating standard errors in panel data sets.
[15] The announcement dates of the S&P GSCI rebalancing were: October 29 2003, November 3 2004, October 24 2005, November 6 2006, November 1 2007, November 3 2008, November 3 2009, November 4 2010, November 3 2011, November 5 2012, November 7 2013, November 11 2014, November 5 2015, November 10 2016, November 10 2017, and November 1 2018.
[16] Traders with good field knowledge may form accurate expectations for new weights before they are announced and enter positions accordingly. If this happens, the impacts on futures prices, if any, may be observed prior to the announcement date. We extend the event window by 10 days before the announcement

to allow for this possibility. There are no substantial changes compared to the results reported in the text. See the Online Appendix for details.
[17] We extend the front-running window by an additional 10 days to check whether this type of trading may have occurred even earlier. There are no substantial changes compared to the results reported in the text. See the Online Appendix for details.
[18] We do not include an interaction dummy for the Great Recession in the first half of the sample since we later show in section headed 'Rebalancing and the Great Recession' that the 2008–2009 recession is not associated with the identified rebalancing price impacts.
[19] The main analysis uses March or April contracts (see Table C11.1 in the Data Appendix), which have 2–3 months to maturity.
[20] The Bloomberg Commodity Index was launched in 1998 as the Dow Jones-AIG Commodity Index, renamed as the Dow Jones-UBS Commodity Index in 2009, and it received the current name (Bloomberg Commodity Index) on July 1 2014.
[21] A derivation based on the S&P GSCI Total Return Index leads to the same result.

References

Acharya, V.V., Lochstoer, L.A. and Ramadorai, T. (2013) Limits to arbitrage and hedging: evidence from commodity markets. *Journal of Financial Economics* 109, 441–465.

Admanti, A.R. and Pfleiderer, P. (1991) Sunshine trading and financial market equilibrium. *Review of Financial Studies* 4, 443–481.

Aulerich, N.M., Irwin, S.H. and Garcia, P. (2013) Bubbles, food prices, and speculation: evidence from the CFTC's daily large trader data files. NBER Working Paper No. 19065. National Bureau of Economic Research (NBER), Cambridge, Massachusetts. Available at: https://www.nber.org/system/files/working_papers/w19065/w19065.pdf (accessed September 2 2022).

Bessembinder, H. (2015) Predictable ETF order flows and market quality. *Journal of Trading* 10, 17–23.

Bessembinder, H., Carrion, A., Tuttle, L. and Venkataraman, K. (2016) Liquidity, resiliency and market quality around predictable trades: theory and evidence. *Journal of Financial Economics* 121, 142–166.

Brunetti, C. and Reiffen, D. (2014) Commodity index trading and hedging costs. *Journal of Financial Markets* 21, 153–180.

Brunnermeier, M.K. and Pedersen, L.H. (2005) Predatory trading. *Journal of Finance* 60, 1825–1863.

Cheng, I.H., Kirilenko, A. and Xiong, W. (2015) Convective risk flows in commodity futures markets. *Review of Finance* 19, 1733–1781.

Erb, C.B. and Harvey, C.R. (2006) The strategic and tactical value of commodity futures. *Financial Analysts Journal* 62, 69–97.

Etula, E. (2013) Broker-dealer risk appetite and commodity returns. *Journal of Financial Econometrics* 11, 486–521.

Fama, E.F. and MacBeth, J.D. (1973) Risk, return and equilibrium: empirical tests. *Journal of Political Economy* 81, 607–636.

Gromb, D. and Vayanos, D. (2010) Limits of arbitrage. *Annual Review of Financial Economics* 2, 251–275.

Hamilton, J.D. and Wu, J.C. (2015) Effects of index-fund investing on commodity futures prices. *International Economic Review* 56, 187–204.

Hau, H., Massa, M. and Peress, J. (2010) Do demand curves for currencies slope down? Evidence from the MSCI global index change. *Review of Financial Studies* 23, 1681–1717.

Henderson, B.J., Pearson, N.D. and Wang, L. (2015) New evidence on the financialization of commodity markets. *Review of Financial Studies* 28, 1285–1311.

Masters, M.W. (2008) Testimony before the Committee on Homeland Security and Government Affairs, US Senate. May 20. Available at: http://hsgac.senate.gov/public/_files/052008Masters.pdf (accessed September 2 2022).

Mou, Y. (2011) Limits to arbitrage and commodity index investment: front-running the Goldman roll. Working paper. Available at: http://www.cftc.gov/idc/groups/public/@swaps/documents/file/plstudy_33_yu.pdf (accessed September 2 2022).

Neuhierl, A. and Thompson, A.J. (2016) Liquidity timing in commodity markets and the impact of financialization. Working paper. Available at: https://papers.ssrn.com/sol3/papers.cfm?abstract_id=2682698 (accessed September 2 2022).

Petersen, M.A. (2009) Estimating standard errors in finance panel data sets: comparing approaches. *Review of Financial Studies* 22, 435–480.

Ready, M. and Ready, R.C. (2019) Order flows and financial investor impacts in commodity futures markets. Working paper. Available at: https://papers.ssrn.com/sol3/papers.cfm?abstract_id=3164757 (accessed September 2 2022).

Sanders, D.R. and Irwin, S.H. (2016) The 'necessity' of new position limits in agricultural futures markets: the verdict from daily firm level position data. *Applied Economic Perspectives and Policy* 38, 292–317.

Sockin, M. and Xiong, W. (2015) Informational frictions and commodity markets. *Journal of Finance* 70, 2063–2098.

S&P Dow Jones Indices (2017) S&P GSCI methodology. April, 2017. S&P Manual, S&P Global, New York. Available at: http://us.spindices.com/documents/methodologies/methodology-sp-gsci.pdf (accessed September 2 2022).

Stoll, H.R. and Whaley, R.E. (2010) Commodity index investing and commodity futures prices. *Journal of Applied Finance* 20, 7–46.

12

The Order Flow Cost of Index Rolling in Commodity Futures Markets[1]

New Author Foreword

Very early on in the commodity price boom, probably in 2005 or 2006, we began to hear about something called the 'Goldman roll.' Unlike stocks and bonds, futures contracts expire on a regular basis, requiring index investors to roll all their positions forward from the expiring contract to a deferred-month contract to avoid physical delivery. Most notably, the Standard and Poor's Goldman Sachs Commodity Index (S&P GSCI) rolls positions forward from the nearby contract to the next deferred contract over a fixed 5-day window from the fifth to the ninth business day of every month. This is the now famous Goldman roll.

The behavior of commodity futures prices, and in particular, calendar spreads, during the Goldman roll quickly became a hot topic. This is not really surprising because the entire notional value of index investment has to be rolled nearly every month. Critics of index investing argued that index rolling artificially decreased the price spread between the nearby and first deferred contracts (spread computed as nearby minus deferred). This is due to the simultaneous selling of nearby contracts and buying of deferred contracts during the Goldman roll, which pushes nearby prices down and deferred prices up, thereby decreasing 'calendar' spreads. As we discussed in Chapter 8 (this volume), this is alleged to cause a permanent decrease in calendar spreads as the impact of index rolling cumulated through time.

The first paper to take a serious look at the issue was by a PhD student at the Columbia Business School named Yiqun Mou. His paper initially appeared in 2010 and it had the attention-grabbing title, 'Limits to Arbitrage and Commodity Index Investment: Front-Running the Goldman Roll.' As an economist, it's hard not to be intrigued by something dealing with front-running and Goldman Sachs. Front-running is actually as old as futures markets. It refers to the practice in pit trading days of floor brokers taking advantage of their knowledge of customer orders in their 'book.' In simplest terms, the brokers would take the same position as their customers but before customer orders were submitted. This 'front-running' based on inside knowledge of customer order flow was obviously unfair and anticompetitive. Mou borrowed this concept for his paper in a clever way, arguing that it was possible for traders to legally front-run the Goldman roll and make large returns. Most notably, he estimated that front-running cost commodity index investors the princely sum of 4% per year.

Mou's paper caused something of a stir when it first appeared among both academic researchers and regulators. To date, the working paper versions (latest in 2011) have been cited more than 100 times, which makes it all the more surprising that the paper was never published in a finance journal. We always thought it was one of the more interesting papers in the literature on financialization impacts in commodity futures markets. It would be interesting to know the back story of why Mou never pushed the paper through to publication. It certainly was a lost opportunity in our eyes.

As time passed and it became clear that Mou's paper was never going to be published, his lost opportunity became our opportunity, and we could not pass it up. We also had just finished our rebalancing paper (see Chapter 11, this volume) and the same conceptual framework for evaluating order flow impacts from index

DOI:10.1079/9781800622104.0012

rebalancing trades could be applied to index roll trades. So we went to work, once again, benefiting greatly from our collaboration with Lei Yan.

The first issue we took a look at was whether traders were really front-running the Goldman roll. If all, or nearly all, index roll activity actually occurred during the 5-day window for the Goldman roll, then Mou's conclusion that massive front-running was occurring made sense. But we had access to data on the actual roll behavior of index investors that Mou did not, and it showed that index investors had expanded the roll window over time, presumably in an effort to try to minimize the market impact of roll trades. So we concluded that what Mou attributed to front-running was in actuality the market impact of index traders using a wider roll window.

The second issue we examined was what happened after the Goldman roll window. If index roll trades have a V-shape centered on the Goldman roll, then order flow impacts should have the same V-shape. Mou stopped his analysis of order flow impacts at the end of the Goldman roll window, which left open the question of whether calendar spreads snapped back as roll trades diminished. The answer was very important from a policy perspective. We found that there was in fact a very distinctive V-shape to the order flow impact of roll trades, which meant that the order flow impact of index rolling was temporary rather than permanent. This actually makes complete sense when one considers that index roll trades are predictable and uninformed. Simply put, index roll trades consume liquidity in the form of temporarily pushing nearby futures prices down and first deferred prices up. The greater the volume of index roll trades, the greater the price pressure, and vice versa. This explains the match between the V-shaped pattern in index roll trades and order flow impacts.

The third issue we wanted to examine was the order flow cost of index rolling over time. We first sliced the data into four different periods of financialization based on trends in overall index investment in commodity futures markets. We estimated that commodity index investors paid a total of $29 billion in order flow costs during monthly rolls over 1991–2019 and this was heavily concentrated in the growth period of financialization over 2004–2011. During this growth phase, order flow costs totaled an astounding $23 billion, or an average of $2.9 billion/year. Clearly, costs of this magnitude represented a major drag on the performance of commodity index funds during this period. But then the big surprise. Average annual order flow costs declined over 80% to $474 million during the post-financialization years from 2012 to 2019. What on earth could have happened that led to such enormous cost savings?

A careful examination of the yearly estimates revealed that order flow costs nosedived after 2006. This coincided almost perfectly with the transition to electronic trading in commodity futures markets. We considered other potential explanations and some additional factors may have contributed, but their role was marginal at best. The evidence was simply too obvious to ignore, and we concluded that the transition to electronic trading led to a dramatic increase in the supply of liquidity in commodity futures markets. This drove down order flow costs for all traders, including index investors.

In retrospect, this article may be our favorite because it is a perfect example of the joy and excitement that can be found in research. We started out investigating one thing and ending up finding something else that was completely unexpected. As far as we know, this article provides the best evidence to date of the profound impact that electronic trading has had on the cost of transacting in commodity futures markets. In 2012 (see Chapter 6, this volume), we wrote that electronic trading was the biggest structural change in commodity futures markets of the last 150 years. Little did we know that a decade later we would provide such compelling evidence of the impact of electronic trading in a study motivated by, of all things, the Goldman roll.

Abstract

Commodity index rolling is treated as a natural experiment and an event study of order flow costs in a wide array of futures markets is conducted. The spread between nearby and deferred futures prices decreases significantly in the early and growth phases of financialization (1991–2011), with the spreads reversing back after rolling is completed. Spread impacts disappear in the post-financialization period (2012–2019). We argue that a dramatic increase in the supply of liquidity brought on by the transition to electronic trading in commodity futures markets is primarily responsible for the decline of roll order flow costs.

Key words: commodity, electronic trading, financialization, futures, index, order flow, roll, spread

JEL categories: G12, G13, G14, Q02

12.1 Introduction

Financial investors, such as pension funds and other institutional investors, have substantially increased their participation in commodity futures markets since the early 2000s. In contrast to traditional hedgers and speculators, financial entities treat commodities as a new class of assets and seek diversification benefits and protection against inflation. The financial entities get access to commodities through a variety of investment tools such as exchange-traded funds (ETFs) and swaps, whose returns are often tied to prices of a group of commodities or a commodity index. Global commodity-linked investment as estimated by Barclays was about $300 billion at the end of 2019, with a peak of $450 billion in early 2012.[2] The rapid growth of commodity-linked investment is often labeled the 'financialization' of commodity futures markets (Tang and Xiong, 2012).

Investing in a commodity index is typically accomplished by taking long positions in the commodity futures contracts that make up the index. Unlike stocks and bonds, futures contracts expire on a regular basis, requiring investors to roll all their positions forward from the expiring contract to a deferred month contract to avoid physical delivery. For a given commodity index, all roll information, such as timing of the roll and futures contracts included in the index, is predetermined and publicly available. As an example, the Standard and Poor's Goldman Sachs Commodity Index (S&P GSCI) rolls positions forward from the nearby contract to the next deferred contract over a fixed 5-day window from the fifth to the ninth business day of every month. This is widely referred to as the Goldman roll (GR). Rolling of index positions is a major event in commodity futures markets because the entirety of index investment is rolled from one contract maturity to the next, and this can surpass a billion dollars of notional value in individual commodity futures markets (Stoll and Whaley, 2010).

To minimize tracking errors, commodity investors should complete roll trades during the roll window of the specified index. Hence, rolling activity should be predictable in terms of timing. Moreover, the roll-based order flows should not contain new information about contemporary or subsequent changes in commodity futures prices because the flows are fully determined by known rules. Therefore, the order flows associated with rolling for a specific index, such as the S&P GSCI, should be both predictable and uninformed. In practice, there may be some deviation from this expectation to the degree that investors track different commodity indexes and/or do not precisely follow pre-set rolling rules. Deviations like this may make the timing and volume of roll trades somewhat difficult to anticipate by other traders.

It is important to recognize that the process of rolling positions is as old as commodity futures markets. Nearly all market participants have to roll out of or exit the nearby contract prior to expiration. However, to the extent that both long and short positions need to roll, there is no reason to expect a systematic price impact because buyer-initiated and seller-initiated orders are roughly balanced. In contrast, we expect to observe order flow impacts from commodity index rolling because this causes an imbalance in the desire of long versus short positions to roll. In particular, because index roll trades involve seller-initiated trades of nearby contracts and buyer-initiated trades of next deferred contracts, nearby prices should be pushed lower than they otherwise would, and next deferred contract prices should be pushed higher than they otherwise would. This results in a decrease in the price spread between nearby and deferred futures contracts. The decrease in the spread represents the order flow cost of roll trades for commodity index investors. Note that the order flow impact of index rolling is expected to be temporary as the spread should revert to its starting level once index rolling is completed.[3]

To the best of our knowledge, only one previous study estimates the size of order flow costs due to commodity index rolling. Mou (2011) reports the cost to be almost 4% per year.[4] Order flow costs of this magnitude are several times larger than the direct cost of many index investments and would represent a substantial drag on investor returns. Irwin *et al.* (2020) and Moran *et al.* (2020) argue that large order flow costs are key to explaining the disappointing return of commodity investments since the early 2000s. While this may be the case, the available evidence on the size of order flow costs and how it varies over time is quite limited. For example,

Mou's sample period ends in 2010, over a decade ago. There is clearly a need for additional research on order flow costs of index rolling in commodity futures markets given the importance of such costs to the performance of index investments.

Here we treat index rolling as a natural experiment and conduct an event study of order flow costs in a wide array of commodity futures markets. Specifically, we examine 19 energy, metal, grain, soft, and livestock futures markets included in the S&P GSCI over the 40-year period from 1980 to 2019. A major advantage of the event study approach is that it does not place any constraints on the pattern of order flow impacts during the roll window. We find that the spread between nearby and deferred futures prices decreases significantly in the early (1991–2003) and growth (2004–2011) phases of financialization, with the spreads reversing back over the post-GR period. The maximum decline is roughly 30–40 basis points during the GR window. The reversal shows that the order flow impact of index roll trades is temporary. Spread impacts decline substantially in the post-financialization period (2012–2019).

It is estimated that commodity index investors paid a total of $29 billion in order flow costs during monthly rolls over 1991–2019. This was heavily concentrated in the growth period of financialization over 2004–2011 when order flow costs totaled an astounding $23 billion, or an average of $2.9 billion/year. By any reasonable standard, these costs represented a major drag on the performance of commodity index investments during this period. Average annual order flow costs declined to $474 million during the post-financialization years from 2012 to 2019. The more than 80% decline in average dollar order flow costs from index rolling is indicative of a dramatic increase in the supply of liquidity in commodity futures markets driven primarily by the transition to electronic trading.

12.2 The S&P GSCI Roll

The S&P GSCI is the first major investable commodity index and widely recognized as a benchmark for investment in commodity markets. The S&P GSCI was launched in 1991 and has historical data available back to 1970. The index currently comprises 24 commodities from all sectors (energy, industrial metals, precious metals, grains, softs, and livestock). The index composition changed over time, with only four commodities (corn, soybeans, Chicago wheat, and live cattle) included in 1970. There were 17 commodities in 1991 when the index was first published, and the universe expanded to 24 commodities in 2002 and has remained the same since. The S&P GSCI is a world-production-weighted index, with the weights primarily determined by the average quantity of global production of each commodity over the last 5 years of available data. Since the weight assigned to each commodity is in proportion to the amount of that commodity flowing through the economy, the index is concentrated in energy products.

Table 12.1 lists 19 out of the 24 commodity futures markets that are included in the S&P GSCI and traded in the USA.[5] These 19 markets are referred to as *index* commodities in the remainder of this paper. On average, the aggregate weight of the 19 markets in the S&P GSCI was 94.9% over 2004–2019. The five industrial metals (aluminum, copper, lead, nickel, and zinc) listed on the London Metal Exchange (LME) are excluded because of uncertainty about the maturity structure used for these markets in the S&P GSCI.[6] We have no reason to expect that the exclusion of the LME markets biases the results in any way. Because the year a commodity enters the index differs from one commodity to the next (see entry years in Table 12.1), we include all index commodities available for analysis prior to 1991 (the year when the index was created). For years after 1991, a commodity will not be used for analysis until it enters the index.

The S&P GSCI is designed to be investable by including the most liquid futures contract, which is typically the nearby contract. The positions are periodically rolled forward from the nearby contract to the next deferred-month contract to avoid physical delivery. The roll occurs within a 5-day window from the fifth through the ninth business day of each month, and on each date one-fifth of the positions are rolled. Note that the roll occurs every month for energy products but less frequently for other commodities because futures contracts are not designated for every calendar month. Whenever a roll happens, we refer to the contract that the index

Table 12.1. Commodity futures in and out of the S&P GSCI.[a]

	Commodities in the S&P GSCI		Commodities out of the S&P GSCI	
	Commodity	Entry year	Commodity	List year
Energy	Crude oil (WTI)	1987	Electricity, PJM	2003
	Heating oil #2	1983	Ethanol	2005
	Gasoline	1988	Propane	1987
	Crude oil (Brent)	1999		
	Gasoil	1999		
	Natural gas	1994		
Metals	Gold	1978	Copper	1959
	Silver	1973	Palladium	1977
			Platinum	1968
Grains	Corn	1970	Oats	1959
	Soybeans	1970	Rough rice	1986
	Wheat (Chicago)	1970	Soybean meal	1959
	Wheat (Kansas)	1999	Soybean oil	1959
			Wheat, spring	1970
Softs	Cocoa	1984	Lumber	1969
	Coffee 'C'	1981	Orange juice	1967
	Cotton #2	1977	Sugar, white	1990
	Sugar #11	1973		
Livestock	Live cattle	1970	Butter	1996
	Feeder cattle	2002	Milk, Class III	1996
	Lean hogs	1976	Pork bellies	1966

[a]This table lists 19 commodities included in the S&P GSCI and 17 commodities out of the index. Entry year refers to the year that a commodity entered the index. List year refers to the year that futures contracts of a commodity were listed. Gasoline includes unleaded gasoline and RBOB gasoline; unleaded gasoline futures were replaced by RBOB gasoline futures in October 2005 and the S&P GSCI transitioned from unleaded to RBOB in 2007. Lean hog futures replaced live hog futures in February 1997. Pork bellies futures were delisted in July 2011. Propane futures were delisted in October 2010.

rolls from as the nearby contract and the contract that the index rolls into as the first deferred contract.

All information about the S&P GSCI roll is publicly available; therefore, the resulting order flows are highly predictable in theory. Index investors may deviate in practice from the published rules and execute roll trades prior to or after the roll window. At the same time, the volume of roll trades should be reasonably predictable for two reasons. First, the size of index positions can be deduced from a number of Commodity Futures Trading Commission (CFTC) position reports (such as the *Index Investment Data* (IID) report). Second, since the roll is repeated each month for energy markets and essentially every other month for most other markets, the trader can quickly learn the typical size of roll volume and timing based on recent history. For these reasons we are confident that event study procedures can be applied to the S&P GSCI roll with a moderately wide event window.

To form a comparison group, we consider commodities that are not included in the S&P GSCI, referred to as *out-of-index* commodities. Table 12.1 lists 17 commodities out of the index and the year in which their futures contracts were listed. We apply a similar roll scheme as the S&P GSCI by matching sector and maturity structure. Out-of-index commodities are used as a control group because they are not included in the index and their prices are closely related to index commodities within the same sector.

While there is uncertainty about the exact timing of roll trades within the trading day, the available evidence suggest that roll trades occur at or near the daily settlement. For example, Ready and Ready (2022) document that index traders in agricultural futures trade at or near the daily settlement. In addition, Bessembinder *et al.* (2016) show that the US Oil Fund (USO), the largest of the ETFs that track crude oil futures prices, routinely trades at the settlement price to complete its roll trades. This evidence supports our use of daily futures settlement prices. We obtain the daily settlement prices for January 1980 through December 2019 from Barchart, Inc. (available at: https://www.barchart.com/ (accessed January 14 2022)).

12.3 Index Rolling and the Event Window

As noted earlier, a 5-day window for commodity index rolling is specified in the construction of the S&P GSCI. However, data on the actual rolling behavior of commodity index traders indicates that this may exclude a notable amount of roll activity. Important evidence regarding actual rolling behavior is presented in Fig. 12.1. Figure 12.1(a) shows the changes in net long positions held in aggregate by commodity index traders in the December 2004 and December 2008 corn futures contracts.[7] The rolling activity reaches a peak on day 2 of the GR but has a pronounced V-shape, with approximately 55% of rolling activity occurring within the GR window. Aulerich *et al.* (2014) also report the annual average percentage of commodity index traders' rolling activity that occurs during the GR period across all 12 agricultural futures markets included in the *Supplemental Commitments of Traders* (SCOT) report from the US Commodity Futures Trading Commission (CFTC). The amount of rolling activity in the GR window falls marginally from an average of 62% in 2004 to 56% in 2009. In Fig. 12.1(b), we show the average changes in positions for all markets and WTI crude oil for a large private index fund over 2007–2012.[8] The fund rolled about 1000 futures contracts/day during the pre- and post-GR windows, but the number reaches around 3000 contracts/day during the GR window. Overall, a similar V-shaped pattern of rolling activity is reported for the private index fund as for commodity index traders (Fig. 12.1a), with an average of 40% of fund rolling activity occurring in the GR window.

The data presented in Fig. 12.1 indicate that the actual rolling of positions by index traders in commodity futures markets centers on the GR due to the dominant role played by the S&P GSCI but rolling in aggregate occurs over a wider window and this has been the case at least since the early 2000s. An implication is that any price impact before the 5-day GR is not necessarily due to front-running by predatory traders, as suggested by Mou (2011), but instead may simply reflect the order flow impact of rolling across a wider window by index traders. Consequently, we consider a 25-business-day event window, including 10 pre-GR days, 5 GR dates, and 10 post-GR days, represented by days −10 to −1, 0 to 4, and 5 to 14, respectively.[9] Pre-GR days are included to capture the possibility of early execution of roll trades by index investors. Post-GR days capture rolling activity after the GR window and allow tests of whether the impact of roll order flows is permanent or temporary. Our event window, starting from 10 business days prior to the GR to 10 business days after the GR, should be wide enough to capture the vast majority of rolling activity by commodity index traders.

Bessembinder *et al.* (2016) show that the price impacts of predictable roll trades around roll dates for the USO ETF are temporary, reflecting that roll trades are information-less and the WTI crude oil futures market is resilient. We argue that the trades associated with commodity index rolling are uninformed and predictable, and therefore, should have temporary price impacts. There is the possibility that some roll trades are timed opportunistically in an effort to increase investment returns. This could imbue roll trades with an informational component. The available evidence on the relationship between lagged and contemporaneous commodity futures returns and changes in commodity index trader positions provides little support for this possibility. Aulerich *et al.* (2014) find a significant but small impact of past returns on daily index trader positions in agricultural markets, but this disappears when the analysis is limited to roll windows. Lehecka (2015) analyzes weekly index trader positions in agricultural futures markets and reports that past returns do not significantly impact index positions. On balance, there is very little evidence that index investors time trades opportunistically during roll windows, which makes sense given the passive nature of commodity index investments.

12.4 Growth of Order Flow Demand

The order flow demand associated with commodity index rolling has grown over time as index

(a) Changes in net long positions by commodity index traders.

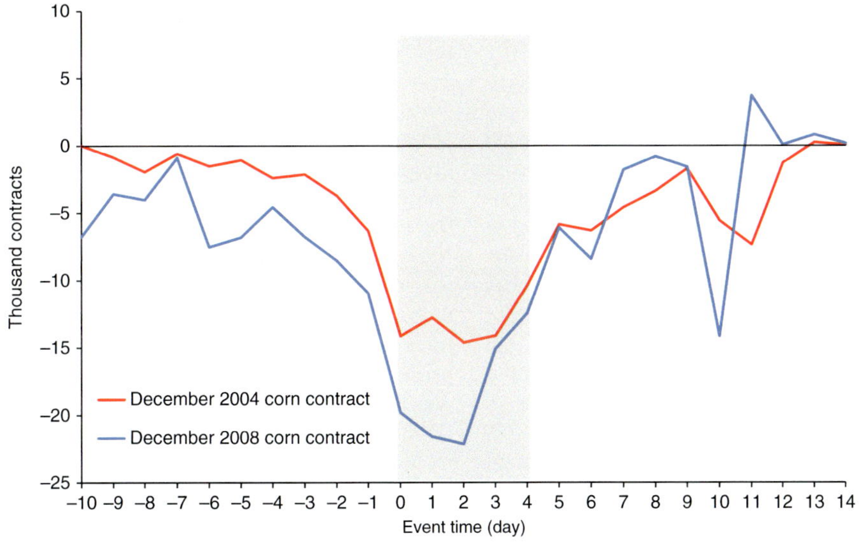

(b) Average changes in positions by a private index fund for 2007–2012.

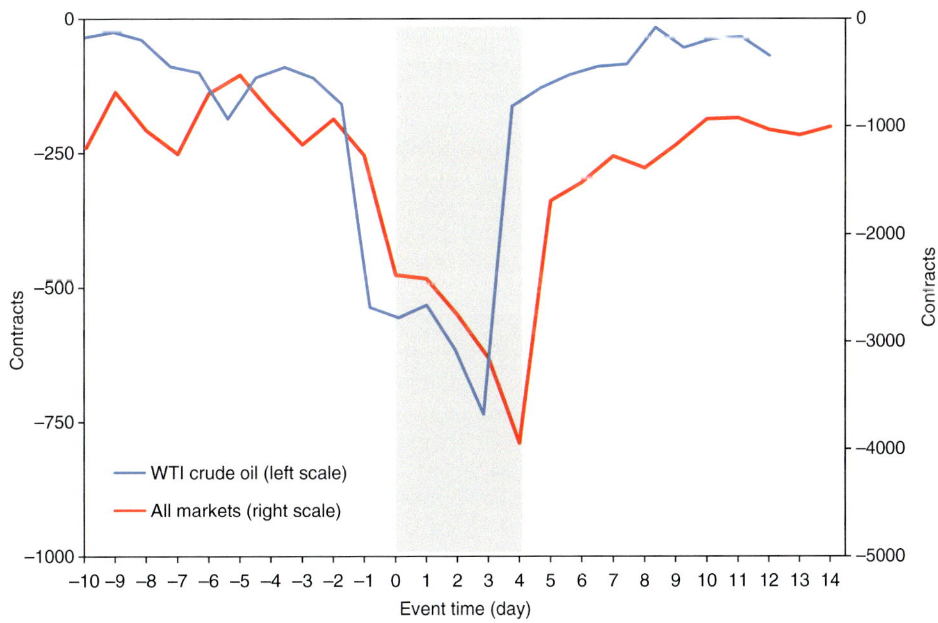

Notes: Days 0 to 4 represent the GR window, ranging from the fifth through the ninth business day of each month (shaded area). Days −10 to −1 represent the 10 business days prior to the GR window and days 5–14 represent the 10 business days after the GR window.

Fig. 12.1. Changes in index positions around the Goldman roll (GR). (a) Daily changes in net long positions held by index traders in December 2004 and December 2008 corn futures contracts. (b) Average changes in positions by a large private index fund in WTI crude oil and all commodities it tracks for 2007–2012. (Data from (a) Aulerich *et al*. (2014); and (b) Sanders and Irwin (2014))

investment has increased. Figure 12.2(a) shows the dollar value of broad index swaps in major US futures markets for each year over 1991–2017 as estimated by Barclays. In Fig. 12.2(b), the notional index investment is calculated based on positions from the weekly *Supplemental Commitments of Traders* (SCOT) report released by the US Commodity Futures Trading Commission (CFTC). The SCOT report covers 12 agricultural futures markets for 2000–2019.[10] Another source of commodity index investment is the CFTC *Index Investment Data* (IID), which includes the notional value of index positions in all major US futures markets. The IID data provide the most accurate measure of index investment but are only available for the relatively short period from December 2007 to October 2015. The IID-based measures of commodity index investment are included for comparison.

Figure 12.2 shows that both the Barclays and the SCOT-based measures of index investment move closely with the IID-based measure during their overlapping windows, helping to verify accuracy. Figure 12.2(a) shows that total index investment stayed at a low level from 1991 through 2003, began a period of rapid growth in 2004, peaked in 2011 near $200 billion, declined through 2015, and then recovered back to around $150 billion. A similar pattern is observed in agricultural index investment (Fig. 12.2b) with somewhat less of a recovery toward the end of the sample period. Because the entire notional value of index investment has to be rolled nearly every month, the growth in roll order flow demand from index investors since the early 1990s has been enormous. From 2004 to 2011 alone, the roll order flow demand from index investors increased nearly fourfold.

The data presented in Fig. 12.2 also suggest a clear division of the sample into three subperiods – 1991–2003, 2004–2011, and 2012–2019. We argue that these three periods represent different stages in the 'financialization' of commodity futures markets, and therefore, in the order flow demand from index investors. The 1980–1990 period is also included and referred to as the pre-financialization period since it pre-dates the development of major commodity indexes. While it was theoretically possible for individual investors to create their own commodity indexes at any point in the past, we argue that such efforts would have been of negligible size and market impact. Hence, the 1980–1990 period serves as our control period and we do not expect changes in spreads during this period.

The major commodity indexes were created during 1991–2003. In particular, the creation of the S&P GSCI in 1991 attracted considerable interest and facilitated the creation of other index-linked products. We refer to this period as the early stage of financialization. While the total amount of index investment specifically associated with the S&P GSCI during this period is unknown, there is evidence that the index attracted a notable amount of investment in this period. For example, the Oppenheimer Real Asset Fund, the first fund that invested in the Goldman Sachs Commodity Index, was launched on March 31 1997 and attracted $62.1 million in assets in the first 5 months (Kahn, 1997). Aulerich *et al.* (2014) report that index traders had an average of 8% of long open interest in 12 agricultural futures markets by 2000. Hence, the order flow demand from index rolling during this period was not negligible, but not yet large in absolute size.

We refer to the period from 2004 to 2011 as the growth stage of financialization, as index investment in commodity futures markets exploded. The popularity of commodities as a new asset class received a boost by the seminal works of Gorton and Rouwenhorst (2006) and Gorton *et al.* (2012), who concluded that investing in commodities could yield 'equity-like' returns and diversification benefits. Using 2004 as the starting date of the growth stage of financialization is consistent with previous studies of financialization impacts in commodity futures markets (e.g. Tang and Xiong, 2012; Hamilton and Wu, 2014). As noted earlier, the rapid growth and magnitude of index investment caused a large increase in the order flow demand from index rolling during this period.

The last period, 2012–2019, is referred to as the post-financialization period because global commodity-linked investment leveled out during this period, albeit at a high level from a historical perspective. Therefore, the order flow demand from index rolling remained at a stable but elevated level.

In sum, there is a clear pattern in commodity index investment that leads to splitting the sample into four 'financialization' stages. These

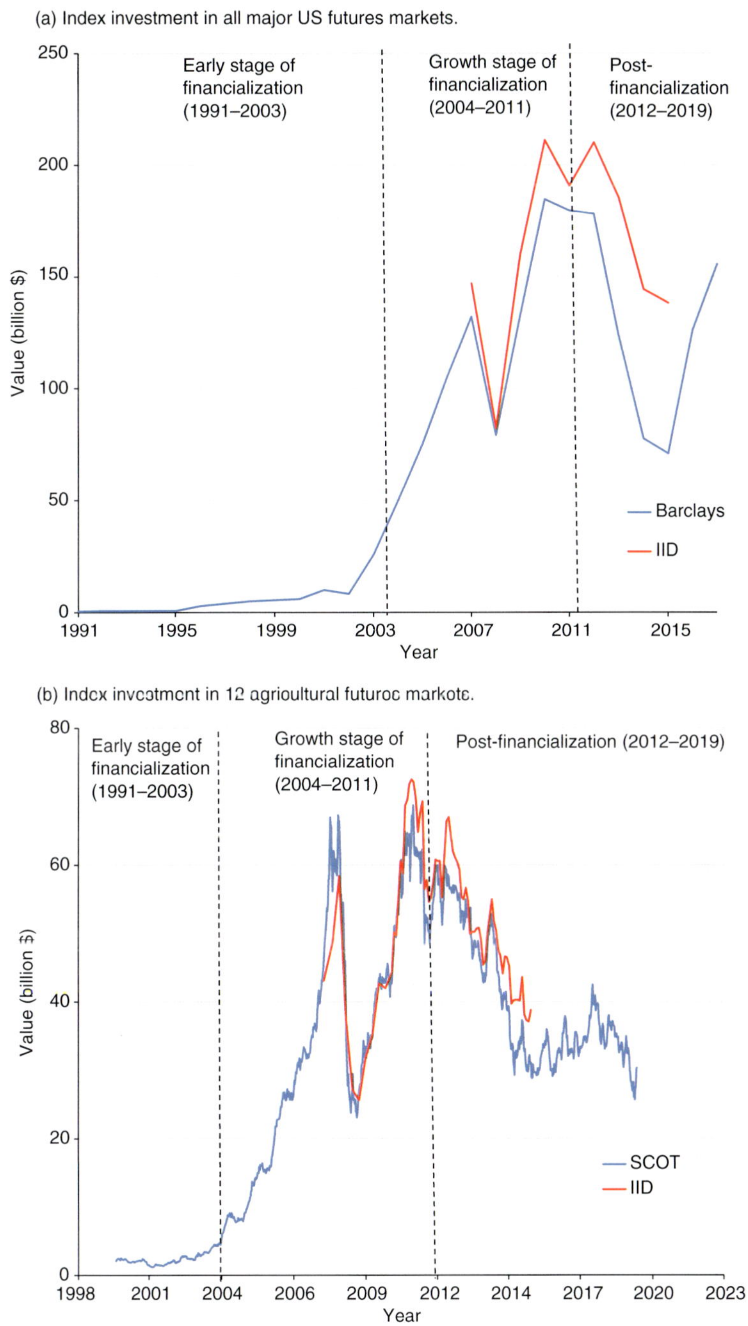

Fig. 12.2. Notional value of commodity index investment in (a) all major US futures markets; and (b) 12 agricultural futures markets. (Data for (a) provided by Barclays for 1991–2017; data for (b) calculated based on index positions from the Commodity Futures Trading Commission (CFTC) *Supplemental Commitments of Traders* (SCOT) report, which is publicly available for 2006–2019. Data for 2000–2005 from Aulerich *et al*. (2014). Index investment in all major futures markets is available from CFTC *Index Investment Data* (IID) reports from December 2007 to October 2015)

four stages represent the evolution of roll order flow demand from commodity index investors through time.

12.5 Estimated Spread Impacts

To estimate the market impact of roll trades in the event window, we examine daily price spreads between futures contracts with different maturities. Assuming that the daily settlement price of the nearby contract with a maturity date of T_1 is $F_t^{T_1}$ and the settlement price of the first deferred contract with a maturity date of T_2 is $F_t^{T_2}$, the spread at time *t* is

$$spread_t^{T_1,T_2} = \ln F_t^{T_1} - \ln F_t^{T_2} \qquad (12.1)$$

where *t* takes values from −10 to 14. The spread in this form measures the percentage difference in prices between the nearby and the first deferred contract.[11] In the presence of an impact, the spread is expected to be lower during the GR window because the nearby price will be pushed down by index-initiated sell orders and first deferred price will be pushed up by index-initiated buy orders. Otherwise, spreads will be roughly the same between pre-GR, GR, and post-GR windows.

We also measure the degree to which the spread moves during a time interval by calculating the change in spread from day t_1 to t_2 as

$$\Delta spread_{t_1,t_2} = spread_{t_2}^{T_1,T_2} - spread_{t_1}^{T_1,T_2} \qquad (12.2)$$

where t_1 and t_2 again take values from −10 to 14. For example, $\Delta spread_{0,4}$ corresponds to the change in spread over the GR window. The change in spread of a commodity for a given window (pre-GR, GR, or post-GR) is calculated based on Eqn 12.2. We examine the change in spread on a portfolio level by forming equally weighted portfolios of commodities.

In event studies, abnormal returns are computed to account for the impacts of other economic factors on asset prices. The abnormal return equals the raw return minus the expected return estimated from a multi-factor model. The use of abnormal returns is usually necessary when an event affects the price level (e.g. Henderson *et al.*, 2015). However, this adjustment is not necessary in our study because roll trades affect the price level of both the nearby and the first deferred contracts. The two commodity futures prices are tightly linked through storage arbitrage (e.g. Pindyck, 2001), so the impact of other economic factors is differenced out in the spread calculation.[12] In essence, the raw spread returns are abnormal returns, so there is no need to adjust returns by a multi-factor model.

Figure 12.3 presents the average spread between the nearby and first deferred contracts (nearby minus deferred contracts) of index commodities on each event day. Average spreads are calculated over three windows (pre-GR, GR, and post-GR) and four time periods (1980–1990, 1991–2003, 2004–2011, 2012–2019). For each monthly roll, spreads for a given window are averaged over commodities whose positions will be rolled. Since all individual monthly rolls are treated equally, this procedure generates the estimated spread impact for a randomly selected month. Monthly spreads are then averaged over months within a time period and reported in basis points. The spreads are normalized by subtracting day −10's spread to facilitate comparison across periods.

Figure 12.3 suggests several notable results. First, average spreads are close to zero over the event window during the pre-financialization stage (1980–1990), which is expected since this control period occurred before the development of major commodity indexes. Second, average spreads decline substantially during the early and growth stages of financialization (1991–2003 and 2004–2011, respectively). The maximum decline is roughly 30–40 basis points and this occurs during the GR window. While the decline is substantial in relative terms it is not large in absolute terms. For example, if the price of crude oil is $60/barrel, a 40-basis-point decline in the spread is only 24 cents/barrel. Third, there is a striking similarity in the pattern of average spreads during the early and growth stages to the pattern of observed roll behavior shown in Fig. 12.1. In both of these stages, there were substantial drops in average spreads prior to or during the GR window, followed by gradual rebounds to near zero in the following 2 weeks. The V-shaped patterns in Fig. 12.3 do not rule out a front-running impact as suggested by Mou (2011), but they are also consistent with the alternative explanation that some commodity index traders simply rolled 'early' or 'late' in order to avoid the congestion

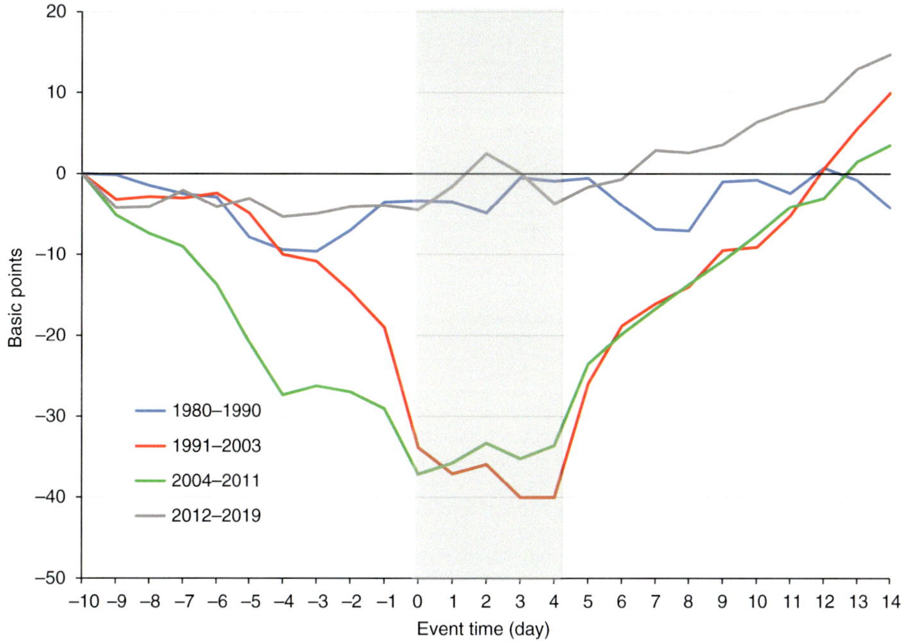

Notes: The spread is the difference in log prices between the nearby and first deferred contracts (nearby minus deferred) of the 19 index commodities. Each month, the spreads of each event day are normalized by subtracting day −10's spread, which are averaged over commodities whose positions will be rolled. The monthly normalized spreads are then averaged over months/rolls within a time period. The spreads are in basis points. Days 0–4 represent the GR window, ranging from the fifth through the ninth business day of each month (shaded area). Days −10 to −1 represent the 10 business days prior to the GR window and days 5–14 represent the 10 business days after the GR window. The sample period is from 1980 to 2019.

Fig. 12.3. Average spreads between the nearby and first deferred contracts of index commodities around the Goldman roll (GR) in four time periods.

and higher order flow costs during the GR. Fourth, the V-shaped patterns indicate that the order flow impacts from index rolling during the early and growth stages were temporary rather than permanent. Fifth, the V-shaped patterns also help to explain why previous time-series regression tests (e.g. Stoll and Whaley, 2010; Aulerich *et al.*, 2014; Sanders and Irwin, 2014, 2016) provide little evidence of order flow impacts from index rolling. It is clear from Fig. 12.3 that when there is an order flow impact it is highly non-linear. The linear regression specifications in previous studies are unlikely to pick up this form of an order flow impact. Sixth, average spreads are near zero in the post-financialization stage of 2012–2019, although there is a slight upward drift post-GR in this period.

We now turn to formal statistical analysis of the behavior of spreads during the roll event window. Table 12.2 presents the average change in spreads between the nearby and first deferred contracts of index commodities. The *t*-statistics for the null hypothesis that the mean change in spread is zero are computed based on Newey and West (1987) standard errors with four lags and are provided in parentheses. The average changes in spreads over the pre-GR, GR, and post-GR windows in the pre-financialization period of 1980–1990 are −1.8, 3.4, and −5.9 basis points, respectively, and are not significantly different from zero (*t*-statistics of −0.61, 0.91, and −0.78). By adding up values over the three windows, we obtain an average change of −5.3 basis points in spreads over the entire event window, which is insignificant at conventional levels. The finding that spreads stay nearly constant through the event window suggests that there are no significant roll impacts in the

Table 12.2. Average changes in spread between the nearby and first deferred contracts of index commodities.[a]

Period	No. of rolls	Pre-GR	GR	Post-GR	Full
1980–1990	132	−2.8	3.4	−5.9	−5.3
		(−0.61)	(0.91)	(−0.78)	(−0.55)
1991–2003	156	−16.5***	−22.1***	46.3***	7.6
		(−3.86)	(−5.69)	(5.54)	(0.76)
2004–2011	96	−32.1***	−5.6	36.6***	−1.1
		(−6.72)	(−1.24)	(7.75)	(−0.15)
2012–2019	96	−3.1	0.2	15.9***	13.0***
		(−0.90)	(0.06)	(4.11)	(2.57)
1991–2019	348	−17.1***	−11.4***	35.2***	6.7
		(−6.07)	(−4.25)	(8.07)	(1.29)

[a]The table presents the average changes in spread between the nearby and first deferred contracts (nearby minus deferred) of index commodities. The change in spread of a commodity for a given window (pre-GR, GR, post-GR) is calculated based on Equation (12.2). GR denotes the Goldman roll. The last column reports the total change in spread over the entire event window. For each month/roll, changes in spread are averaged over commodities whose positions will be rolled. The monthly changes in spread are then averaged over months within a time period. The GR window ranges from the fifth through the ninth business day of each month. The pre-GR window includes the 10 business days prior to the first roll date. The post-GR window includes the 10 business days after the last roll date. Changes in spread are in basis points. The *t*-statistics in parentheses are computed based on Newey and West (1987) standard errors with four lags. *, **, *** indicate significance at the levels of 10%, 5%, and 1%, respectively.

pre-financialization period, which is not surprising given that the S&P GSCI had not yet been created. As noted earlier, even without index investors, hedgers and traditional speculators often roll their positions from the nearby contract to the first deferred contract during the event window. Our results suggest that conventional rolling activity on the long and short side of the market is balanced and therefore, has very little impact on futures spreads. Consequently, this period serves as a valid control.

In the early financialization period of 1991–2003, the average change in spreads shows a completely different pattern over the event window (the third and fourth rows of Table 12.2). The spread decreases by 16.5 and 22.1 basis points over the pre-GR and GR windows, respectively, which are statistically significant with *t*-statistics of −3.86 and −5.69. The spread bounces back and increases by 46.3 basis points over the post-GR window. This V-shaped pattern clearly shows that the spreads decline as roll dates approach and rebound afterwards. Over the entire event window, the average change in the spread is 7.6 basis points and insignificant (*t*-statistic of 0.76), suggesting that the post-GR reversal fully offsets its decline in the pre-GR and GR windows. The full reversal of spreads over the post-GR window during 1991–2003 confirms that the identified roll impacts were temporary, which is consistent with uninformed order flows.

The results for the early stage of financialization imply that the S&P GSCI started to draw enough investment in the early 1990s and the roll order flow demand of index investors was surprisingly large relative to order flow supply. This is consistent with Aulerich *et al.*'s (2014) finding that the commodity index traders' proportion of long open interest averaged 8% in 12 agricultural futures markets as early as the year 2000.

The average change in spreads during the growth stage of financialization in 2004–2011 has a similar V-shaped pattern over the event window (the fifth and sixth rows of Table 12.2). The spreads decrease by 32.1 and 5.6 basis points over the pre-GR and GR windows, with the former estimate being statistically significant at the 1% level (*t*-statistic of −6.72). The decrease in spreads over the pre-GR window is consistent with Mou (2011), who finds that the spread decreases over a 15-business-day window over 2000–2010 with the last 5 days being GR roll dates in selected commodity markets. However, Mou only includes pre-GR and GR windows, while our extended results show that spreads rebound after the 5-day GR roll. Specifically, the spread increases by 36.6 basis points within the 2 weeks after the GR roll, which offsets the earlier decline and leaves a total change of −1.1 basis points. The full reversal again indicates temporary roll impacts and that roll trades are uninformed.

The most striking result in Table 12.2 is that the spreads no longer decrease significantly over the event window in the post-financialization period of 2012–2019 (the seventh and eighth rows of Table 12.2). The average changes in spreads over the pre-GR and GR windows are −3.1 and 0.2 basis points, respectively, and neither are statistically significant (t-statistics of −0.9 and 0.06), suggesting that the roll impacts found in earlier time periods have largely disappeared. The one anomalous finding during this time period is a statistically significant increase in spreads during the post-GR window. It is not clear why this occurs given the insignificant changes during the pre-GR and GR windows.

Overall, the results in Fig. 12.3 and Table 12.2 show that spreads in commodity futures markets decrease significantly as the S&P GSCI roll dates approach and reverse fully in the next 2 weeks – supporting the existence of temporary roll impacts. More importantly, the roll impacts emerge only in the early and growth stages of financialization (1991–2003 and 2004–2011, respectively) and then dissipate in the most recent post-financialization period (2012–2019).[13]

To further confirm that the V-shaped pattern in spreads is caused by the impacts of index roll trades, we examine spreads around the S&P GSCI roll dates for the group of out-of-index commodities. Presumably, the spreads for out-of-index commodities should show no differences between GR and non-GR windows since they are not covered by the index. Table 12.3 presents the average change in spreads between the nearby and first deferred contracts of out-of-index commodities. In the pre-financialization period of 1980–1990, the spreads show an odd downward trend over the GR and post-GR windows. Otherwise, the average changes in spread are quite small in magnitude and not significantly different from zero except for one case where the spread increases by 7.7 basis points over the GR window in 2004–2011. The pattern of changes in spread is also depicted in Fig. 12.4. There is no clear drop in the spread over the pre-GR or GR window in the early and growth stages of financialization, in contrast to the results for index commodities (Fig. 12.3 and Table 12.2). These results provide further confirmation that the pattern in average spreads observed for index commodities is related to rolling activity.

Table 12.3. Average changes in spread between the nearby and first deferred contracts of out-of-index commodities.[a]

Period	No. of rolls	Pre-GR	GR	Post-GR	Full
1980–1990	132	3.0	−10.9**	−23.2***	−30.7**
		(0.33)	(−1.98)	(−2.73)	(−2.08)
1991–2003	156	1.3	1.5	−0.4	2.4
		(0.22)	(0.34)	(−0.04)	(0.17)
2004–2011	96	−3.7	7.7*	13.1	17.2
		(−0.64)	(1.72)	(1.51)	(1.45)
2012–2019	96	0.9	8.3	−12.8	−3.3
		(0.12)	(1.06)	(−0.75)	(−0.12)
1991–2019	348	−0.1	5.1	−0.1	4.9
		(−0.05)	(1.58)	(−0.02)	(0.48)

[a]The table presents the average changes in spread between the nearby and first deferred contract (nearby minus deferred) of the 17 commodities that are not included in the S&P GSCI. The change in spread of a commodity for a given window (pre-GR, GR, post-GR) is calculated based on Eqn 12.2. GR denotes the Goldman roll. The last column reports the total change in spread over the entire event window. For each month/roll, changes in spread are averaged over commodities whose positions will be rolled. The monthly changes in spread are then averaged over months within each time period. The GR window ranges from the fifth through the ninth business day of each month. The pre-GR window includes the 10 business days prior to the first roll date. The post-GR window includes the 10 business days after the last roll date. Changes in spread are in basis points. The t-statistics in parentheses are computed based on Newey and West (1987) standard errors with four lags. *, **, *** indicate significance at the levels of 10%, 5%, and 1%, respectively.

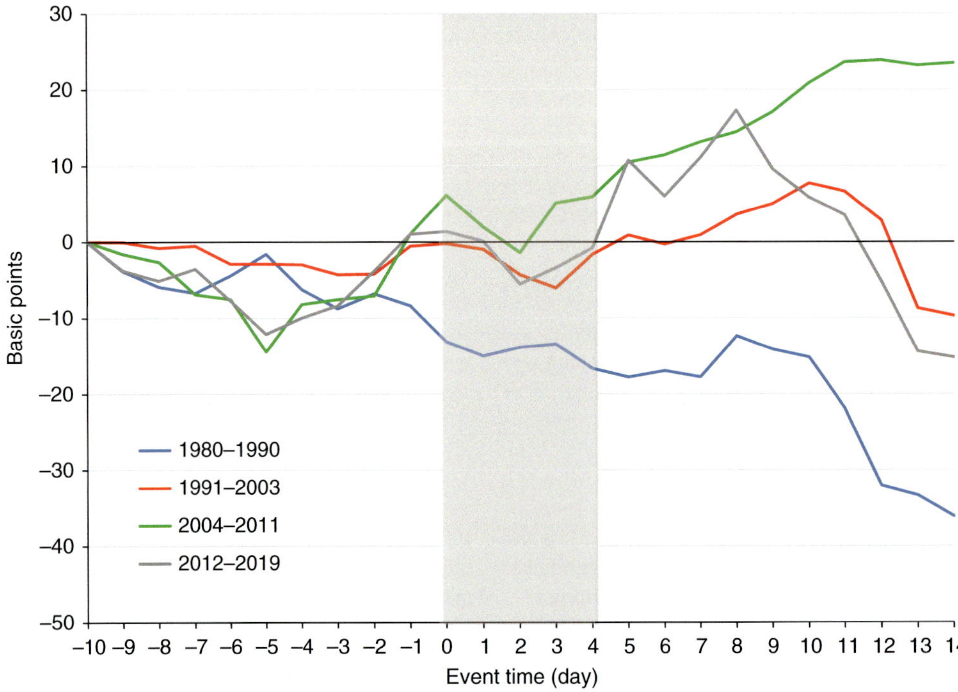

Notes: The spread is the difference in log prices between the nearby and first deferred contracts (nearby minus deferred) of the 17 out-of-index commodities. Each month, the spreads of each event day are normalized by subtracting day −10's spread, which are averaged over commodities whose positions will be rolled. The monthly normalized spreads are then averaged over months/rolls within a time period. The spreads are in basis points. Days 0–4 represent the GR window, ranging from the fifth through the ninth business day of each month (shaded area). Days −10 to −1 represent the 10 business days prior to the GR window and days 5–14 represent the 10 business days after the GR window. The sample period is from 1980 to 2019.

Fig. 12.4. Average spreads between the nearby and first deferred contracts of out-of-index commodities around the Goldman roll (GR) in four time periods.

As a final control, we examine spreads between the second and third deferred contracts of index commodities. We showed previously that the S&P GSCI rolling activity has an impact on the spread between the nearby and first deferred contracts. For other deferred contracts that are not held by the index, the spreads should not be affected. As shown in Table 12.4, the average changes in spread between the second and third deferred contracts are not statistically different from zero in the pre-GR, GR, or post-GR window in 1991–2003, providing no evidence of roll impacts. The spreads in 2004–2011 show a slight decline but do not reverse in the post-GR weeks. The slight drop in spreads cannot be associated with the S&P GSCI roll since the index does not hold those contracts; instead, it may be caused by the roll of other commodity indexes as discussed in the Online Appendix. The spreads in 2012–2019 show an even smaller decline compared with 2004–2011, suggesting that the impact, whatever the source, becomes smaller in recent years. We also examine the spreads between the third and fourth deferred contracts, the fourth and fifth deferred contracts, and the fifth and sixth deferred contracts. The results (reported in the Online Appendix) show that the changes in spread, despite being statistically significant in some cases, are generally small in magnitude (below 10 basis points) regardless of the window and time period. More important, the pattern in all deferred contracts is simply not consistent with the roll impacts displayed in Fig. 12.3.

Table 12.4. Average changes in spread between the second and third deferred contracts of index commodities.[a]

Period	No. of rolls	Pre-GR	GR	Post-GR	Full
1980–1990	132	−1.8	5.2**	−3.1	0.0
		(−0.58)	(2.12)	(−1.00)	(−0.01)
1991–2003	156	−1.6	3.4	−3.1	−1.4
		(−0.60)	(1.29)	(−0.98)	(−0.26)
2004–2011	96	−10.2***	−2.4	−6.0	−18.6***
		(−3.36)	(−1.19)	(−1.43)	(−4.06)
2012–2019	96	−7.5***	−2.4	−1.9	−11.9***
		(−3.09)	(−1.63)	(−0.59)	(−2.39)
1991–2019	348	−5.6***	0.1	−3.6*	−9.0***
		(−3.39)	(0.13)	(−1.74)	(−2.86)

[a]The table presents the average changes in spread between the second and third deferred contracts (second minus third) of index commodities. The change in spread of a commodity for a given window (pre-GR, GR, post-GR) is calculated based on Eqn 12.2. GR denotes the Goldman roll. The last column reports the total change in spread over the entire event window. For each month/roll, changes in spread are averaged over commodities whose positions will be rolled. The monthly changes in spread are then averaged over months within each time period. The GR window ranges from the fifth through the ninth business day of each month. The pre-GR window includes the 10 business days prior to the first roll date. The post-GR window includes the 10 business days after the last roll date. Changes in spread are in basis points. The *t*-statistics in parentheses are computed based on Newey and West (1987) standard errors with four lags. *, **, *** indicate significance at the levels of 10%, 5%, and 1%, respectively.

12.6 Order Flow Cost Estimates

From the perspective of index investors, order flow costs from roll trades arise implicitly in the form of a decrease in the price of the nearby contract and an increase in the price of the first deferred contract. It is common to measure such trading costs by comparing the price for a completed trade to a pre-trade price. Here, we assume the relevant price is the spread on day −10, which our previous results indicate has negligible impact from rolling activity by index investors. Any changes in spreads during the event window (after day −10) imply that index investors pay extra costs to complete their roll trades. To measure this trading cost, we compare the spread level on all event days with the starting spread value on day −10. Specifically, the trading cost of an event day is defined as day −10's spread minus the spread on that day.

Two issues are presented when estimating order flow costs of index investors as opposed to spread impacts. First, not all markets roll every month. For example, there are only five contracts per year in corn and therefore, only five rolls per year. Second, major commodity indexes are not equal-weighted across markets. As noted earlier, the S&P GSCI is a world-production-weighted index that reflects the value of commodities flowing through the global economy. This results in the index being concentrated in energy products. In order to reflect these characteristics of actual index investment, order flow costs are first summed for each year across all contract rolls for a commodity, which accounts for the differing number of rolls per year for some of the markets included in the S&P GSCI. Next, the order flow costs for commodities in a given year are averaged using the published weights for the S&P GSCI. We were able to obtain historical S&P GSCI weights for 2000–2019, which means that complete sets of actual weights are used during the 2004–2011 growth stage of financialization and the 2012–2019 post-financialization stage. During 1991–2003, the early financialization stage, actual weights are only available for 2000–2003; consequently, weights for the year 2000 are used to estimate weights over 1991–1999. This is not expected to have a material impact on the results for this period because the weights change slowly over time.

The average order flow costs caused by index rolling during the different stages of financialization are reported in Table 12.5. Since we are interested in estimating order flow costs of actual index investment, we do not include results for the pre-financialization control period of 1980–1990, as we did in the analysis of

Table 12.5. Average order flow costs due to commodity index rolling.[a]

Period	Pre-GR	GR	Post-GR	Full
1991–2003	0.31**	2.46***	0.45*	0.86***
	(2.33)	(4.50)	(1.86)	(4.20)
2004–2011	1.52***	4.36***	2.05**	2.32***
	(4.00)	(3.84)	(1.96)	(2.95)
2012–2019	0.29***	0.64**	0.17	0.30**
	(3.25)	(2.17)	(1.38)	(2.14)
1991–2019	0.64***	2.48***	0.81**	1.11***
	(2.72)	(3.58)	(1.99)	(2.89)

[a]The table presents the average trading costs of rolling positions around the Goldman roll (GR) dates in three time periods. The trading costs are in percent per year. For an event day, the trading cost is measured as day −10's spread minus the spread on that day. The trading cost of rolling positions over a window (pre-GR, GR, post-GR, and full) is an average of daily costs over days within the window. For a commodity, the trading cost is accumulated over rolls within each year to generate annual trading costs. We calculate a weighted average of annual trading costs over commodities with weights equal to the S&P GSCI weights each year and report their averages over years within each time period. The S&P GSCI weights are available for 2000–2019; the trading costs for 1991–1999 are estimated based on the year 2000's weights. The GR window ranges from the fifth through the ninth business day of each month, indexed by days 0–4. The pre-GR window includes the 10 business days prior to the first roll date, indexed by days −10 to −1. The post-GR window includes the 10 business days after the last roll date, indexed by days 5–14. 'Full' represents the full event window from days −10 to 14. The *t*-statistics in parentheses are computed based on Newey and West (1987) standard errors with four lags. *, **, *** indicate significance at the levels of 10%, 5%, and 1%, respectively.

spread impacts. In the early financialization period (1991–2003), the annual average trading costs are 0.31%, 2.46%, and 0.45% over the pre-GR, GR, and post-GR windows, respectively, with all three being statistically significant. These estimates imply that index investors, on average, paid almost 2.5% per year to roll their positions during the GR window, even with the relatively small size of aggregate index investment (see Fig. 12.2) during the early phase of financialization. The costs would have been negligible if trades were shifted to the pre-GR or post-GR window, which demonstrates the incentives for commodity index traders to widen the rolling window where possible. In the growth stage of financialization (2004–2011), annual average trading costs are much higher at 1.52%, 4.36%, and 2.05% over the pre-GR, GR, and post-GR windows, respectively, with all three again being statistically significant. It is noteworthy that the cost of rolling index positions during the GR window nearly doubled in the growth stage compared to the early stage. In the post-financialization period (2012–2019), annual average trading costs declined dramatically to 0.29%, 0.64% and 0.17% over the pre-GR, GR, and post-GR windows, respectively, with two of the three remaining statistically significant. Even with the decline in costs, a representative index fund following the standard 5-day GR rolling window would have absorbed 0.64% per year in roll costs, still a significant drag on performance considering the low expected (gross) returns on long-only commodity investments (Irwin *et al.*, 2020). The results for the full event window weight each day in the 25-day window equally and show an average trading cost of 0.86%, 2.32%, and 0.30% for the early, growth, and post-financialization stages, respectively. Surprisingly, order flow costs in the post-financialization stage dropped well below those in the early stage of financialization.

Some of the average costs shown in Table 12.5 may seem surprisingly high in light of our earlier point that the decline in spreads due to roll order flow trades is fairly small in absolute terms. The key is to recognize that the estimated spread impact was reported per monthly roll, while the observations in Table 12.5 are cumulated over all monthly rolls within a year for a commodity. In the earlier example, a 40-basis-point decline in the spread is only 24 cents/barrel, assuming the price of crude oil is $60/barrel. However, this cumulates to 4.8%, or $2.88/barrel, across the 12 monthly rolls crude oil futures within a year. This illustrates how relatively small order flow impacts per monthly roll can quickly accumulate to economically significant amounts.

We provide additional details on the evolution of order flow costs over time in Fig. 12.5,

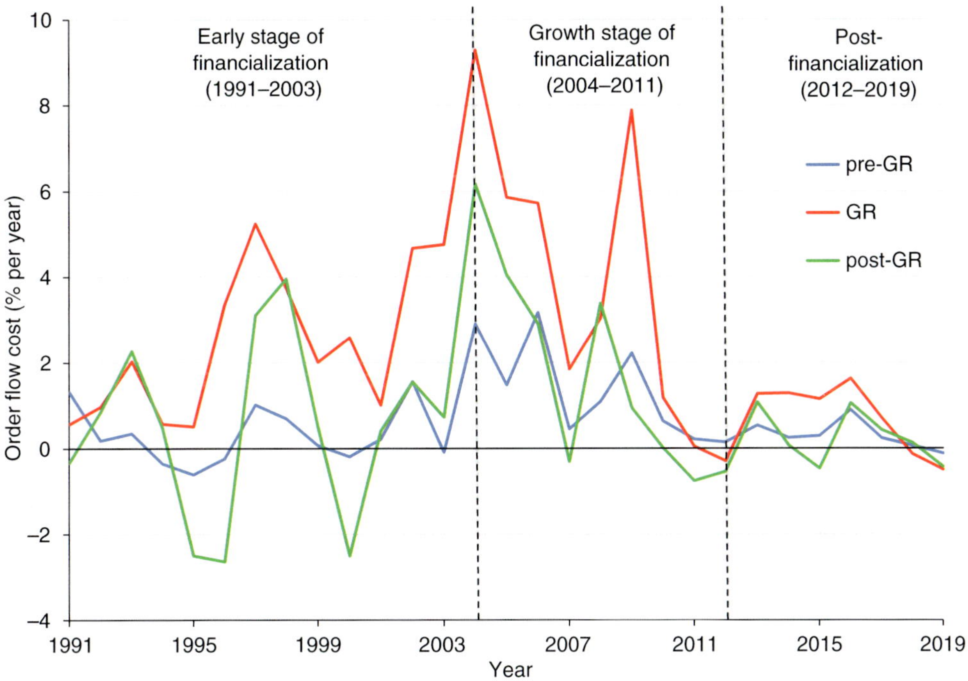

Notes: For each event day, order flow cost is measured as day −10's spread minus the spread on that day. The trading cost over a window (pre-GR, GR, or post-GR) is the average of daily cost over days within the window. For each commodity, the order flow cost is accumulated over rolls within each year to generate annual trading costs. Each year, a weighted average of the annual trading costs is calculated over commodities with weights equal to the S&P GSCI weights. These are the estimates shown in the figure. The S&P GSCI weights are available for 2000–2019; the trading costs for 1991–1999 are estimated based on the year 2000's weights'. The 5-day GR window ranges from the fifth through the ninth business day of each month. The pre-GR window includes the 10 business days prior to the first roll date. The post-GR window includes the 10 business days after the last roll date. The trading costs are in percentages per year and the sample period is 1991–2019.

Fig. 12.5. Annual order flow costs over the pre-Goldman roll, Goldman roll (GR), and post-Goldman roll periods.

which shows the estimated trading costs by year for the pre-GR, GR, and post-GR windows. The annual series track one another fairly closely, with cross-correlations ranging between 0.65 and 0.76. The most striking feature is the collapse in the level and volatility of order flow costs during the post-financialization period. Previously, order flow costs were much higher and much more volatile. For example, the standard deviation of annual order flow costs in all three windows during the post-financialization stage dropped about three-quarters compared to the growth stage. Another notable feature is the extremely high levels of order flow costs for some years in the GR window. Each year during the 5-year period from 2002 to 2006, estimated order flow costs exceeded 4% per annum and peaked at 9.3% in 2004. Costs were again very high during the GR in 2009 at 7.9%.

The next step in the analysis is to compute a weighted average of the annual pre-GR, GR, and post-GR costs shown in Fig. 12.5 to estimate the total order flow costs of index investors. This requires making an assumption about the amount of roll activity in each of the three roll windows. As discussed earlier, the only available data on the aggregate rolling behavior of index traders is the study by Aulerich *et al.* (2014), who used non-public CFTC data on index positions in 12 agricultural futures markets over 2004–2009. For their sample period, these authors found that an average of about 25%, 55%, and 20% of index roll trades were completed over the pre-GR, GR, and post-GR windows, respectively. We use these weights as they represent the best (and only) available data. This raises the possibility of inaccurate aggregate trading cost estimates if the weights change over time. Fortunately, Aulerich *et al.*'s sample period has a considerable overlap with the period of highest order flow costs shown in Fig. 12.5, so one can be reasonably confident about the weights and resulting aggregate cost estimates during this higher cost period. One has less confidence regarding the weights before and after the 2004–2009 period, but this is not as much of a limitation as it may first appear since the costs for all three roll windows are lower before 2004 and much lower after 2009. During these lower cost periods, aggregate cost estimates will be less sensitive to assumptions regarding the weights for each roll window. Nonetheless, it should be kept in mind that the aggregate costs estimates are conditional on the selected weighting.

Figure 12.6 presents the annual estimates of aggregate roll order flow costs for index investors based on the assumption of 25%, 55%, and 20% of index roll activity in the pre-GR, GR, and post-GR windows, respectively. The aggregate order flow cost from index rolling computed in this manner is positive in 24 out of 29 years over 1991–2019 and the average is 1.7% per year. The averages are 1.52%, 3.19%, and 046% per year for the pre-, growth, and post-financialization stages, respectively (1991–2003, 2004–2011, 2012–2019). Not surprisingly, total roll order flow costs are highest over 2002–2006, when costs average 4.4% per year. The peak in roll trading costs is 7.1% in 2004. It is noteworthy that roll order flow costs tended to decline sharply after 2006 (with the exception of 2009). For example, order flow costs in the 10 years from 1997–2006 average 3.2% compared to only 1.3% for the 10 years over 2007–2016, a nearly two-thirds decline. If we consider the period from 2011 to 2019, average costs are even lower at 0.4% per year. This pattern is particularly interesting since 2007–2008 was the period of greatest public policy concern about the impact of financialization in commodity futures markets. At least in terms of the impact on spreads in commodity futures markets and roll order flow costs, financialization had its biggest impact in the 5 years preceding 2007–2008.

It is clear from both Figs 12.5 and 12.6 that something dramatic happened to the order flow costs of index investors starting around 2007. Either the liquidity demand of index investors decreased, the supply of liquidity in commodity futures markets increased, or some combination of the two. We start with a discussion of possible changes in the liquidity demand of index investors. To begin, it makes sense that order flow demand pressure from index investors lessened during the post-financialization period. Figure 12.2(a) shows that total index investment declined rather sharply after 2011 before recovering somewhat starting in 2016. However, this is challenged by the fact that the downward trend in roll order flow costs actually began much earlier, around 2007, and near the beginning of the period of most rapid growth in index investment. Since the start of the downward trend in order flow costs preceded the decrease of index investment by several years, this suggests other demand or supply factors are likely at work. As a final point in this regard, order flow costs remained very low despite the rebound in total index investment that started in 2016.[14]

Another possible demand-side explanation is a change in the nature of order flow demand from index investors. Specifically, index investors may have reacted to the relatively high order flow costs before 2007 by altering roll strategies. This could have taken several forms, including a shift of rolling activity outside of the GR window to the pre- and post-GR windows, rolling less frequently, rolling outside of the 25-day window considered in this study, and adopting a less mechanical and more active rolling strategy. The latter change has been referred to as 'smart' rolling strategies associated with second and third generation commodity indexes (Miffre, 2012; Fethke and Prokopczuk, 2018). These newer commodity indexes base roll strategies on contango and backwardation in the term structure of futures prices. It is undoubtedly the case that changing roll strategies could have contributed to the downward trend in roll order costs that started in 2007. However, there are some limits to how much changing roll strategies could have impacted order flow costs, particularly in the post-financialization period. As shown in Fig. 12.5, roll order flow costs for all three roll windows (pre-GR, GR, and post-GR) fell precipitously and the range across the windows narrowed

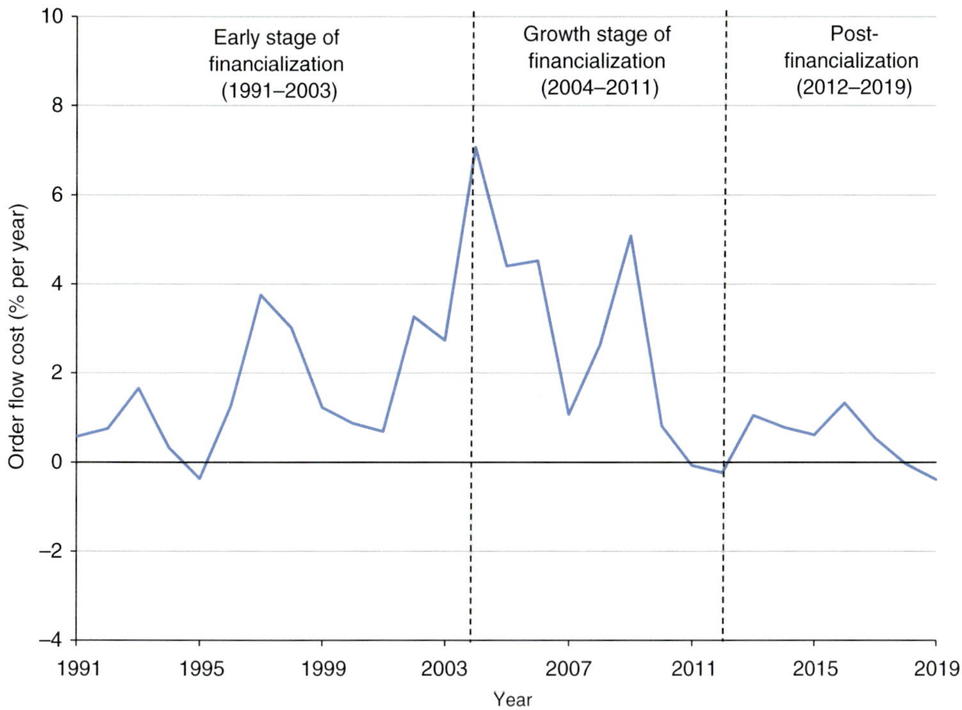

Notes: For each event day, order flow cost is measured as day −10's spread minus the spread on that day. The trading cost over a window (pre-GR, GR, or post-GR) is the average of daily cost over days within the window. GR denotes the Goldman roll. We assume that 55% of positions are rolled during the GR window, 25% during the pre-GR window, and 20% during the post-GR window. For each commodity, the order flow cost associated with a roll is a weighted average of the costs over the three windows and the costs are accumulated over rolls within each year to generate annual trading costs. Each year, a weighted average of the annual trading costs is calculated over commodities with weights equal to the S&P GSCI weights. These are the estimates shown in the figure. The S&P GSCI weights are available for 2000–2019; the trading costs for 1991–1999 are estimated based on the year 2000's weights. The 5-day GR window ranges from the fifth through the ninth business day of each month. The pre-GR window includes the 10 business days prior to the first roll date. The post-GR window includes the 10 business days after the last roll date. The trading costs are in percentages per year and the sample period is 1991–2019.

Fig. 12.6. Annual order flow costs due to commodity index rolling.

considerably during this stage. The implication is that there was a limited scope for reducing order flow costs by shifting rolling activities outside of the GR. There is also no publicly available data that documents widespread adoption of second and third generation commodity indexes.

We can now move on to consideration of possible supply-side explanations for the sharp decline in roll order flow costs starting in 2007. Figure 12.7 provides important context for this discussion. It shows the annual trading volume on GR days for nearby and first deferred contracts of index commodities, as well as the annual trading volume on all days in a year for the same nearby and first deferred contracts. The trends for the two series are similar and indicate an explosion in trading volume for index commodities that began in the mid–1000s. Both series more than quadrupled between 2005 and 2019. It is also interesting to observe that volume on GR days over the entire sample period increased somewhat faster than for other days. Volume on GR days represented 12% of trading on all days in 1991 and then rose to 22% in 2019.

The explosion in trading volume documented in Fig. 12.7 is suggestive of a massive increase in the supply of liquidity in commodity futures markets. Of course, identifying the cause

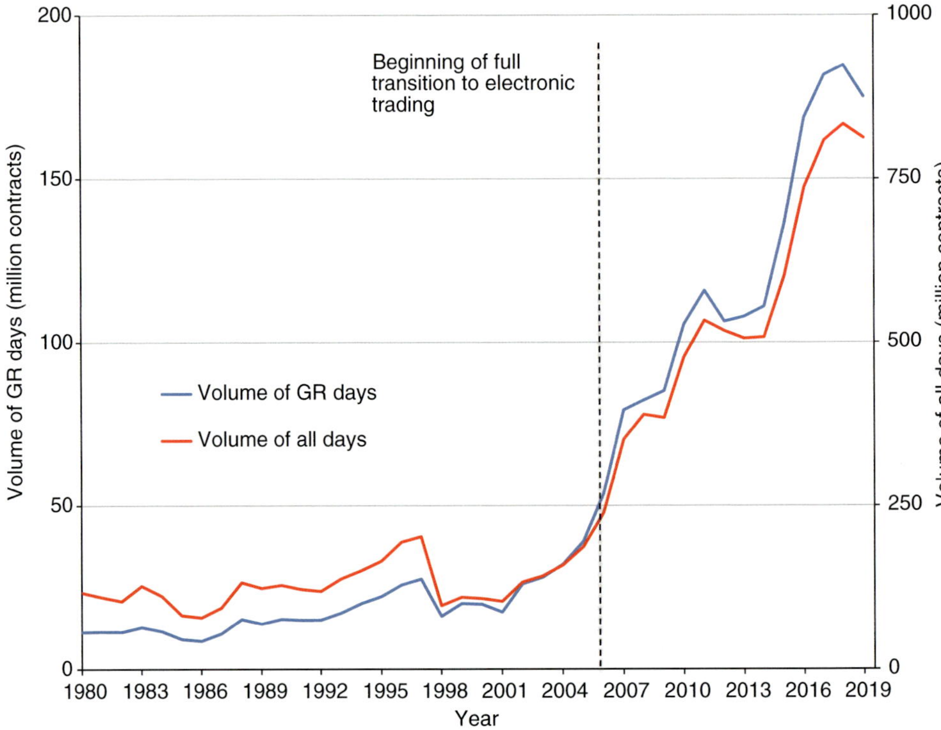

Notes: The annual trading volume of GR days is the sum of daily trading volumes of the nearby and first deferred contracts of index commodities during the monthly GR days of each year. The annual trading volume of all days is the sum of daily trading volumes of the nearby and first deferred contracts of index commodities during all trading days of each year. The dashed line indicates mid-2006, the beginning of the full transition to electronic trading. The sample period is 1980–2019.

Fig. 12.7. Annual trading volume of the nearby and first deferred contracts of index commodities during Goldman roll (GR) days and all days.

of the increase in trading volume requires information on the price (cost) of liquidity over the same period. Fortunately, there is ample evidence that liquidity costs in commodity futures markets declined substantially in the wake of the transition from open outcry pit trading to electronic trading that began in earnest during 2006 (e.g. Shah and Brorsen, 2011; Frank and Garcia, 2011; Boyd and Kurov, 2012; Wang *et al.*, 2014). The combination of increasing trading volume and declining liquidity costs is consistent with rightward shifts in the supply of liquidity in commodity futures markets. Based on this logic, we conclude that the transition to electronic trading was the main driver of the massive increase in trading volume shown in Fig. 12.7.

This discussion indicates that the transition to electronic trading in commodity futures markets led to a large increase in the supply of liquidity and a substantial decline in order execution costs for all market participants, including commodity index investors. Irwin and Sanders (2012) document the rapid transition from pit to electronic trading in agricultural futures markets, with over 80% of trading volume in the soybean futures market migrating to the electronic platform in the 18 months between July 2006 and December 2007. A similar pattern was observed in other commodity futures markets. For example, Boyd and Kurov (2012) show that about 60% of volume in four energy futures markets – WTI crude oil, heating oil, gasoline, and natural gas – had migrated to electronic trading by the end of 2006, and electronic trading volume exceeded 90% in 2008. The transition to electronic trading lines up with the start of the downward trend in order flow costs that began in 2007 (see Fig. 12.6). Another observation favouring this explanation is that the transition to lower roll order flow costs that began in 2007 was permanent, consistent with a major one-time structural change. In sum, it is likely that the increase in the supply of liquidity that accompanied the move to electronic trading in commodity futures markets was the primary contributor to the decline of roll order flow costs of index investors.

The transition to electronic trading is not the only possible supply-side explanation for the downward trend in roll order flow costs after 2007. Irwin and Sanders (2012) point out that market access increased at roughly the same time as the transition to electronic trading due to the internet revolution and innovations in financial technology. These structural changes also could have contributed to increasing the supply of liquidity.

The theory of sunshine trading (Admanti and Pfleiderer, 1991) offers an additional supply-side explanation. This theory asserts that predictable order flows, like those from index rolling, attract additional liquidity suppliers, and hence, have modest and temporary price effects. Based on this theory, Bessembinder (2015) argues that predictable order flows should have a declining price impact over time as they become better understood by market participants. Our previous observation of relatively larger volume increases on GR days compared to other days is consistent with the prediction of sunshine trading theory. However, the question of timing presents something of a challenge in this regard. While it makes sense that it would take some time for liquidity suppliers in commodity futures markets to understand and respond to the order flow demand from index investors, it is not clear why there would be a sudden response starting around 2007, given that index rolling had been much publicized and discussed in the marketplace for a number of years before this. Still, it seems reasonable to conclude that a sunshine trading effect contributed to the decline in roll order flow costs.

Finally, it is useful to point out that there is a logical supply-side explanation for the large upward spike in roll order flow costs during 2009 that is shown in Fig. 12.6. The Lehman Brothers financial crash in September 2008 and the Great Recession that followed represented a huge economic shock. Large economic shocks like this are often associated with a reduced supply of liquidity and higher order execution costs (e.g. Etula, 2013; Erdogan *et al.*, 2015). Hence, it is not surprising that order flow costs for index investors spiked upward in 2009. Notice that order flow costs in Fig. 12.6 quickly returned to their previous downward trend once the economic shock receded.

In sum, there is no single demand or supply of liquidity explanation that entirely accounts for the large decline in roll order flow costs of index investments that started in 2007. On the demand side, both the decline in the overall size of index investment and strategies to ameliorate demand pressure from index rolling likely had some impact. However, these same demand-side factors are confronted by challenges in terms of timing or scope. On the supply side, there is considerable evidence that the transition to electronic trading dramatically increased the supply of liquidity in commodity futures markets and this was the primary factor driving the decline in roll order flow costs. Other structural changes in market access and a sunshine trading effect also likely contributed to the increase in the supply of liquidity.

As a final step in the analysis, we follow Mou (2011) and combine information on estimated order flow costs with the size of index investment in commodity futures markets to provide an estimate of the total dollar size of the order flow cost associated with rolling. This is obtained by multiplying the annual average order flow costs (in percent and presented in Fig. 12.6) by the total dollar investment in commodity index funds for that same year (Barclays found in Fig. 12.2a). Since the Barclays data are not available for 2018 and 2019, for the purposes of this exercise we assume investment is the same for those years as in 2017. The result of this computation is that commodity index investors paid an estimated total of $29 billion in order flow costs during monthly rolls over 1991–2019. This total was heavily concentrated in the growth period of financialization when investment and percentage order flow costs were both large. The total order flow costs in the growth phase over 2004–2011 was an astounding $23 billion, or an average of $2.9 billion/year. By any reasonable standard, these costs represented a major drag on the performance of commodity index investments during this time period. With order flow costs of this magnitude, it is no wonder that supplying liquidity to index investors at the time was described in the financial press as 'taking candy from a baby' (Meyer and Cui, 2009). Average annual order flow costs declined to $474 million during the post-financialization years from 2012 to 2019. The over 80% decline in average dollar order flow costs from index rolling is indicative of a dramatic increase in the supply of liquidity in commodity futures markets driven primarily by the transition to electronic trading.

12.7 Conclusions

We examine the evolution of order flow costs caused by index rolling in commodity futures markets. The analysis focuses on the S&P GSCI because it is the first major investable commodity index and a benchmark for investment in commodity futures markets. Investors tracking the S&P GSCI sell the nearby futures contract and buy the next deferred contract over a 5-day window from the fifth through the ninth business day of each month, known as the Goldman roll (GR). Rolling of index positions is a major event in commodity futures markets because the entirety of index investment is rolled from one contract maturity to the next, and this can surpass a billion dollars of notional value in individual commodity futures markets (Stoll and Whaley, 2010).

The empirical analysis examines the daily spread in prices between the nearby and the next deferred month contract for 19 commodity futures included in the S&P GSCI over the 40-year period from 1980 to 2019. In the presence of an order flow impact, nearby prices should be pushed down by roll sell orders and next deferred contract prices should be pushed up by roll buy orders – resulting in a decrease in the spread between the nearby and the deferred contract (nearby minus deferred). We find that the spread between nearby and deferred futures prices decreases significantly in the early (1991–2003) and growth (2004–2011) phases of financialization, with the spreads reversing back over the post-GR period. The maximum decline is roughly 30–40 basis points during the GR window. The reversal shows that the order flow impact of index roll trades is temporary. Spread impacts decline substantially in the post-financialization period (2012–2019).

It is estimated that commodity index investors paid a total of $29 billion in order flow costs during monthly rolls over 1991–2019. This was heavily concentrated in the growth period of financialization over 2004–2011 when order flow costs totaled an astounding $23 billion, or an average of $2.9 billion/year. By any reasonable standard, these costs represented a major drag on the performance of commodity index investments during this time period. Average annual order flow costs declined to $474 million during the post-financialization years from 2012 to 2019. The more than 80% decline in average dollar order flow costs from index rolling is indicative of a dramatic increase in the supply of liquidity in commodity futures markets driven primarily by the transition to electronic trading.

Our results have three important implications for investors, traders, and policy makers. First, the performance of commodity index investments was disappointing through the early and growth stages of financialization (1991–2011). More to the point, Moran *et al.* (2020) show that index funds were the biggest losers in commodity futures markets during the 2000s. The high order flow cost of rolling during this period certainly was a major drag on returns and contributed to the poor investment performance. These same index funds may find performance somewhat closer to expectations as roll costs have diminished. Second, as the documented order flow cost of rolling has dissipated, so has the need to roll outside of the typical 5-day GR window. That is, with the increase in the supply of liquidity, funds that moved rolling outside of their benchmark index's official window may want to shift back. At present, they can do so at no additional rolling costs while minimizing tracking errors. Third, the results may have important ramifications for other commodity investors and traders. In particular, new investment vehicles – such as commodity indexes in the 1990s – may need to initially lower projected investment returns as order flow costs may exceed expectations until sufficient liquidity is available in the marketplace.

Acknowledgments

Scott H. Irwin is the Lawrence J. Norton Chair of Agricultural Marketing, Department of Agricultural and Consumer Economics, University of Illinois at Urbana-Champaign. Dwight R. Sanders is Professor, Department of Agribusiness Economics, Southern Illinois University Carbondale. Lei Yan is Statistician, School of Medicine, Yale University. Craig Pirrong provided helpful comments on an earlier draft. Correspondence can be directed to Scott Irwin: 344 Mumford Hall, 1301 W. Gregory Dr. University of Illinois at Urbana-Champaign, Urbana, IL 61801, phone: (217)–333–6087, email: sirwin@illinois.edu.

Notes

[1] Original citation: Irwin, S.H., Sanders, D.R. and Yan, L. (2022) The order flow cost of index rolling in commodity futures markets. *Applied Economic Perspectives and Policy*, first published online June 8 2022. Available at: https://doi.org/10.1002/aepp.13297 (accessed September 22 2022). Reprinted by permission of John Wiley and Sons and the Agricultural and Applied Economics Association. The Online Supplemental Appendix for the article is found here: https://onlinelibrary.wiley.com/doi/10.1002/aepp.13297 (accessed September 21 2022).

[2] We thank Michael Cohen, formerly of Barclays, for providing the data on commodity-linked investments.

[3] There is no small amount of confusion among market analysts and in the financial press about the difference between order flow impacts of roll trades and whether commodity investment returns are driven by the term structure of futures prices (contango or backwardation). Order flow costs are incurred regardless of the shape of the term structure of commodity futures prices at any particular point in time. In contrast, the argument that 'roll yield' (term structure) mechanically drives index investment returns is a myth (Sanders and Irwin, 2012; Bessembinder, 2018; Irwin *et al.*, 2020).

[4] Bessembinder *et al.* (2016) estimate the order flow costs of roll trades for the USO ETF (US Oil Exchange Traded Fund) to be about 3% per year, roughly the same magnitude as those estimated by Mou (2011). However, the USO ETF is not a commodity index investment since it is limited to a single futures market – WTI crude oil.

[5] See Table 12.A1 of the Online Appendix for a listing of the specific maturities of futures contracts held by the index at the beginning of each calendar month.

[6] The maturities of futures contracts traded on LME range from 1 day to 3 months consecutively. It is not clear from the S&P GSCI methodology which contracts are included in the index and how they are rolled forward. Some previous studies employ closing LME prices from the third Wednesday of each month as a proxy (e.g. Gorton and Rowenhorst, 2006; Rallis *et al.*, 2013).

[7] The source for these data is Aulerich *et al.* (2014). The positions are constructed using daily Large Trader Reporting Program (LTRS) data collected internally by the Commodity Futures Trading Commission (CFTC). We thank the authors for providing the data.

[8] The private fund replicates a proprietary commodity index. Further details can be found in Sanders and Irwin (2014, 2016).

[9] The event window is shorter than 25 business days in the WTI crude oil futures market because there are less than 10 business days between the last roll date (the ninth business day) and the last trading day of the nearby contract (the day 3 business day prior to the 25th calendar day). In this case, the event window ends on the last trading day of the nearby contract.

[10] The SCOT data are publicly available for 2006–2019. The source for the data over 2000–2005 is Aulerich *et al.* (2014). We thank the authors for providing these data.

[11] We follow market convention and define the spread as the price of the nearby contract minus the price of the deferred contract. This means the order flow pressure of index roll trades should *decrease* the spread. For example, consider a futures curve for a commodity that is in contango, so the deferred contract futures price is higher than the nearby price. The computed spread according to market convention is negative, and roll trades are expected to make the spread become even more negative; and hence, the spread will decrease. If one defines the spread in the opposite manner as the deferred contract futures price minus the nearby contract price, the expected impact of roll trades on the spread is simply reversed.

[12] In addition, the announcement of major news (e.g. US Federal Reserve Board policy, *World Agricultural Supply and Demand Estimates* (WASDE) reports, petroleum status reports, etc.) may overlap with roll

days, but their impact is expected to be minimal for similar reasons. First, the impacts from major news could be either positive or negative and is expected to offset each other, given that our sample includes a diverse group of commodities for a long period. This is one of the advantages of using an event study approach. Second, the impacts from major news, if not averaged out, should be on the level of futures prices rather than the spread between futures prices. Given that nearby and first deferred prices are tightly linked through storage arbitrage, the impact of major news should be differenced out in the spread calculation.

[13] We present results broken out by commodity sector in the Online Appendix. The analysis indicates that S&P GSCI rolling activity has a similar impact across commodity sectors, with the exception of metals, which show a limited impact.

[14] Because the actual quantity of funds tied directly to the S&P GSCI is unknown, we also checked the assets under management of the iShares S&P GSCI Commodity-Indexed Trust, the largest ETF that tracks the S&P GSCI. The fund's total assets decreased from $1.34 billion in early 2012 to $0.68 billion in early 2016 but bounced back to $1.48 billion in mid-2018. Hence, the pattern is similar to that reported in Fig. 12.2(a).

References

Admanti, A.R. and Pfleiderer, P. (1991) Sunshine trading and financial market equilibrium. *Review of Financial Studies* 4, 443–481.

Aulerich, N.M., Irwin, S.H. and Garcia, P. (2014) Bubbles, food prices, and speculation: evidence from the CFTC's daily large trader data files. In: Chavas, J.-P., Hummels, D. and Wright, B.D. (eds) *The Economics of Food Price Volatility*. University of Chicago Press, Chicago, Illinois, pp. 211–253.

Bessembinder, H. (2015) Predictable ETF order flows and market quality. *Journal of Trading* 10, 17–23.

Bessembinder, H. (2018) The 'roll yield' myth. *Financial Analysts Journal* 74, 41–53.

Bessembinder, H., Carrion, A., Tuttle, L. and Venkataraman, K. (2016) Liquidity, resiliency and market quality around predictable trades: theory and evidence. *Journal of Financial Economics* 121, 142–166.

Boyd, N.E. and Kurov, A. (2012) Trader survival: evidence from the energy futures markets. *Journal of Futures Markets* 32, 809–836.

Erdogan, O., Bennett, P. and Ozyildirim, C. (2015) Recession prediction using yield curve and stock market liquidity deviation measures. *Review of Finance* 29, 407–422.

Etula, E. (2013) Broker-dealer risk appetite and commodity returns. *Journal of Financial Econometrics* 11, 486–521.

Fethke, T. and Prokopczuk, M. (2018) Is commodity index investing profitable? *Journal of Index Investing* 9, 37–71.

Frank, J. and Garcia, P. (2011) Bid-ask spreads, volume, and volatility: evidence from livestock markets. *American Journal of Agricultural Economics* 93, 209–225.

Gorton, G.B. and Rouwenhorst, K.G. (2006) Facts and fantasies about commodity futures. *Financial Analysts Journal* 62, 47–68.

Gorton, G.B., Hayashi, F. and Rouwenhorst, K.G. (2012) The fundamentals of commodity futures returns. *Review of Finance* 17, 35–105.

Hamilton J.D. and Wu, J.C. (2014) Risk premia in crude oil futures prices. *Journal of International Money and Finance* 42, 9–37.

Henderson, B.J., Pearson, N.D. and Wang, L. (2015) New evidence on the financialization of commodity markets. *Review of Financial Studies* 28, 1285–1311.

Irwin, S.H. and Sanders, D.R. (2012) Financialization and structural change in commodity futures markets. *Journal of Agricultural and Applied Economics* 44, 371–396.

Irwin, S.H., Sanders, D.R., Smith, A. and Main, S. (2020) Investing in commodity futures: separating the wheat from the chaff. *Applied Economic Perspectives and Policy* 42, 583–610.

Kahn, V.M. (1997) For those who like coffee with their portfolio. *The New York Times*, August 31. Available at: https://www.nytimes.com/1997/08/31/business/for-those-who-like-coffee-with-their-portfolio.html?-searchResultPosition=9 (accessed September 21 2022).

Lehecka, G.V. (2015) Do hedging and speculative pressures drive commodity prices, or the other way round? *Empirical Economics* 49, 575–603.

Meyer, G. and Cui, C. (2009) U.S. Oil Fund finds itself at the mercy of traders. *The Wall Street Journal*, March 6. Available at: https://www.wsj.com/articles/SB123629874701846317 (accessed September 21 2022).

Miffre, J. (2012) Comparing first, second and third generation commodity indices. Working paper, EDHEC Business School, Paris.

Moran, N.M., Irwin S.H. and Garcia, P. (2020) Who wins and who loses? Trader returns and risk premiums in agricultural futures markets. *Applied Economic Perspectives and Policy* 42, 611–652.

Mou, Y. (2011) Limits to arbitrage and commodity index investment: front-running the Goldman roll. Working paper. DOI: 10.2139/ssrn.1716841.

Newey, W.K. and West, K.D. (1987) A simple, positive semi-sefinite, heteroskedasticity and autocorrelation consistent covariance matrix. *Econometrica* 55, 703–708.

Pindyck, R.S. (2001) The dynamics of commodity spot and futures markets: a primer. *Energy Journal* 22, 1–29.

Rallis, G., Miffre, J. and Fuertes, A.M. (2013) Strategic and tactical roles of enhanced commodity indices. *Journal of Futures Markets* 33, 965–992.

Ready, M. and Ready, R. (2022) Order flows and financial investor impacts in commodity futures markets. *Review of Financial Studies* 35, 4712–4755.

Sanders, D.R. and Irwin, S.H. (2012) A reappraisal of investing in commodity futures markets. *Applied Economic Perspectives and Policy* 34, 515–530.

Sanders, D.R. and Irwin, S.H. (2014) Energy futures prices and commodity index investment: new evidence from firm-level position data. *Energy Economics* 46, S57–S68.

Sanders, D.R. and Irwin, S.H. (2016) The 'necessity' of new position limits in agricultural futures markets: the verdict from daily firm level position data. *Applied Economic Perspectives and Policy* 38, 292–317.

Shah, S. and Brorsen, B.W. (2011) Electronic vs. open outcry: side-by-side trading of KCBT wheat futures. *Journal of Agricultural and Resource Economics* 36, 48–62.

Stoll, H.R. and Whaley, R.E. (2010) Commodity index investing and commodity futures prices. *Journal of Applied Finance* 20, 7–46.

Tang, K. and Xiong, W. (2012) Index investment and financialization of commodities. *Financial Analysts Journal* 68, 54–74.

Wang, X., Garcia, P. and Irwin, S.H. (2014) The behavior of bid-ask spreads in the electronically-traded corn futures market. *American Journal of Agricultural Economics* 96, 557–577.

13

Lessons Learned

The final challenge is to consider what we have learned after working on the topic of this book for over 15 years. In other words, after all the research and writing, what are the key lessons one can draw about the role of index funds in commodity futures markets? We believe there are three main lessons, and we share them below.

13.1 First Lesson: the Masters Hypothesis Is False

There are three basic tenets central to the Masters Hypothesis: (i) long-only index investors were directly responsible for driving commodity futures prices higher during price spikes, most notably in 2007–2008; (ii) the deviations from fundamental value during the price spikes were economically very large; and (iii) the impact was pervasive across commodity futures markets. If true, this would raise major questions about the efficiency of price discovery in commodity futures markets and the usefulness of the markets for managing risk. With such fundamental issues at stake, it is easy to understand why there was such a raging global controversy surrounding index funds in commodity futures markets.

As the chapters in this volume amply demonstrate, we tested the Masters Hypothesis with a dizzying array of empirical tests across numerous markets, time frames, and data sets. This exhaustive set of time-series and cross-sectional tests failed to find consistent temporal causality between index positions and commodity futures price movements. We could not find any evidence showing that positions held by index investors caused large changes in commodity futures prices. This body of work conclusively demonstrated that index speculation was not the main driver of the great commodity price spikes that occurred between 2007 and 2013. To be fair, there is a handful of other studies that purport to support the Masters Hypothesis. However, serious problems with the data or methods in these studies have been uncovered that render the results of questionable value. The bottom line is that there is no smoking gun anywhere in the data that would lead one to conclude that the Masters Hypothesis is even remotely close to the truth.

A colorful and insightful perspective on this issue was provided several years ago by Michael Cosgrove, a member of the Energy and Environmental Markets Advisory Committee (EEMAC) formed by the US Commodity Futures Trading Commission (CFTC) in the wake of the 2010 Dodd-Frank law. Among other things, this committee was charged with reviewing the empirical evidence about speculation in commodity futures markets and making recommendations about new regulations. After surveying the empirical evidence about ‘excessive speculation’ in

DOI:10.1079/9781800622104.0013

commodity futures markets, Cosgrove noted that 'Instead of being obvious, it is undetectable. If we claim that elephants were playing in the backyard then we would expect to see their footprints. The alleged excessive speculation, if it is taking place, is leaving no data footprints' (CFTC EEMAC, 2016, p.7). In other words, if the Masters Hypothesis is true, there should be an obvious and very large impact of index investment on commodity futures prices, and this impact should be easily detectable using the simplest of statistical tests. There is nothing in the data at all like the extremely large footprints expected under the Masters Hypothesis.

With apologies to Lester Telser, we submit that the Masters Hypothesis is false, and as such, should be respectfully interred and laid to rest.[1] This is important because market regulation is expensive in terms of both administrative burden and market friction. And this is multiplied many times over for regulation that relies on faulty reasoning and evidence.

13.2 Second Lesson: There Is Plenty Left to Do

While the Masters Hypothesis is demonstrably false, this does not mean that there aren't other interesting and unresolved questions about the role of index funds in commodity futures markets. By laying the Masters Hypothesis to rest, there is now more room for academic researchers to investigate other interesting economic impacts in commodity futures markets. Some of that work has been ongoing for a while, but it has been overshadowed by the all-consuming policy focus on the Masters Hypothesis.

From our perspective, the key is to think about interesting rational economic impacts that financialization may have in commodity futures markets. As we noted in Chapter 6 (this volume), a widely accepted definition of financialization is the large-scale buying of financial index investors in commodity futures markets. While large-scale buying by index investors is common to both financialization and the Masters Hypothesis, we think it is helpful to reserve the term 'financialization' for possible rational market impacts of the large-scale buying. By 'rational' we mean smaller but long-lasting impacts of index investment. We believe that Ready and Ready (2022, p. 4712) hit the nail on the head when they stated, 'Our results provide novel support for commodity financialization but highlight the importance of measuring the magnitude of financial investment, since even large financial flows have economically modest impacts on prices.'

There are several examples of potential rational impacts of financialization that deserve further attention. The first is the issue of market integration. Most notably, Tang and Xiong (2012) argued that commodity markets were not fully integrated with financial markets prior to the development of commodity index funds and that financialization better integrated financial and commodity markets with a concomitant increase in the correlation of financial and commodity prices. A number of studies have examined the correlation of financial and commodity prices, with some showing an increasing correlation and others finding a pattern of increasing and then decreasing correlation. Some find that the changes in correlations are driven by other factors. Further research is needed to help reach a clearer consensus on the issue.

A second issue is the impact of financialization on storage. As noted in Chapter 6 (this volume), classical economic writers argued that the introduction of futures trading in a commodity market flattens the supply of storage curve because the activity of futures speculators increases risk-bearing capacity. Hence, financialization may have shifted the supply-of-storage curve to the right so that the price of storage (or cost of carry) is lower and inventory higher. Others have argued that the impact is actually better thought of as a rightward shift in the demand for storage, which would increase the level of inventories and the price of storage. There is only a handful of studies that address these issues. In addition, there is the related and unresolved question of why the price of storage in the US wheat market went to such stratospheric levels in 2006–2010 (Garcia *et al.*, 2015). Wheat market participants remain adamant that index funds were the culprit. There is certainly plenty to work on here.

A third issue is the potential impact of financialization on the informational efficiency of commodity futures markets. In simplest terms, as the share of passive index positions grows in commodity futures markets, there may be fewer active traders that collect, analyze, and trade on new information. With fewer traders collecting and acting on new information, prices discovered in commodity futures markets may be less informative and market efficiency may decline. This is a very active research area in financial economics at the present time, which makes sense given the huge growth of index investment in the stock market. Important recent examples of work on this topic include Coles *et al.* (2022) and Goldstein and Yang (2022). Theoretical models predict anything from no effect, as active traders adjust to the presence of index positions, to permanent impacts on risk premiums and informational efficiency. Since price discovery is one of the two core functions of a commodity futures market, it will be important to sort out which of these two theoretical predictions is a better description of reality.

A fourth issue is the impact of financialization on the microstructure of commodity futures markets. This issue centers on the order flow impacts of index investor trades on the supply and demand for liquidity in these markets. Our contributions in this regard are reported in Chapters 11 and 12 (this volume), where we examined the impact of index rebalancing and roll trades in a broad range of commodity futures markets. One limitation of our work is the use of daily settlement prices. Additional insights can be gained with the use of high-frequency futures prices at sub-second intervals. Ready and Ready (2022) show the way in this regard. They develop regression models of order flow impacts that can be applied to a wide range of problems in commodity futures markets, including financialization impacts.

Regardless of the type of financialization impact investigated in the future, it is important to remember that one will not be looking for elephant footprints in the data but, rather, something closer to mouse tracks. This will require improved data and methods. There will be considerable challenges with regard to data. For example, the CFTC discontinued the *Index Investment Data* (IID) report in 2015, which means there has been no accurate data in the public domain on index positions in energy and metal futures markets since that time. It would certainly help if researchers had better access to position data from the non-public Large Trader Reporting System maintained by the CFTC. There are understandable legal restrictions on the use of these data, but they really are the best and only available data in many cases. In terms of methods, identification will take center-stage in the future because financialization price impacts, by definition, are small (or at least not large). Of course, it is always important to pay careful attention to causal identification in empirical analysis, but the detection of smaller market impacts places an even higher premium on strong identification.

13.3 Third Lesson: Attacks on Speculation Will Not End Anytime Soon

While we are completely confident that the Masters Hypothesis is false, we are not so naïve as to believe that this will be the end of controversies about speculation in commodity futures markets. There is a historical cycle to the attacks on speculation through the years, which is demonstrated in Fig. 13.1. When prices are low, natural sellers in the market, such as farmers, complain that speculators are the problem, and when prices are high, natural buyers in the market, such as consumers and processors, complain that speculators are the problem. Petzel (1981, p. 117) describes several instances of this cycle,

> In periods of rising prices (e.g. the early 1920s, the Korean War, inflation, and the 1970s) grain speculators have been accused of increasing the prices of agricultural commodities artificially. During the early 1930s when agricultural prices were low, grain speculators were accused of depressing prices.

We agree wholeheartedly with Petzel (1981, p. 126), that 'it is all too easy after suffering an economic loss to look for the villain in the piece.'

Commodity futures prices spiked once again in 2021–2022, and right on cue, concerns about speculation re-emerged. In a meeting of the EEMAC on September 20 2022, CFTC Commissioner Christy Goldsmith Romero delivered the latest salvo in the long-running battle over speculation:

> I propose that the Commission conduct a series of deep-dive studies in key commodities markets, starting with those that have been experiencing the most recent stress – natural gas, crude oil, and wheat. The CFTC has a significant amount of data and expertise at our disposal, and the objective of these reports would be supportive of

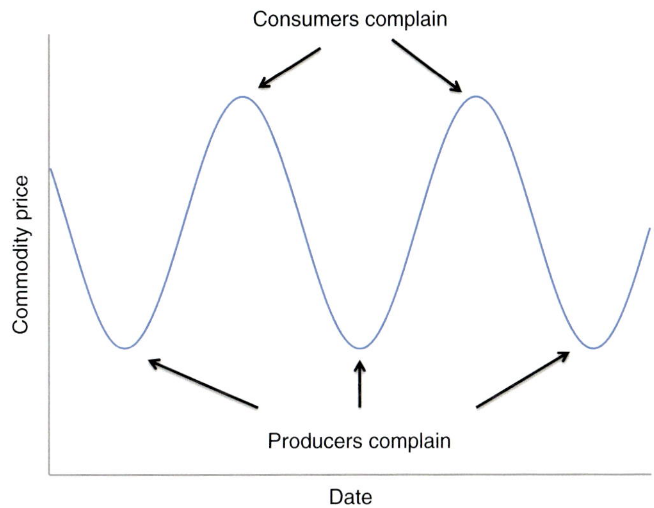

Fig. 13.1. The anti-speculation cycle in commodity futures markets.

> our core mission: To study whether prices are being determined by market fundamentals. We should not overlook any factor that might contribute to increased food, energy, and metals costs. And we should not assume that we can rely on previous conclusions without confirming those conclusions based on an independent review and analysis of the available data. The series of studies I propose would examine all material elements driving volatility and pricing in key commodities. One element of the studies should include an examination of the market presence of passive investment vehicles and other speculators, as well as dealers, to ensure they are providing useful liquidity or serving other useful functions and not distorting markets or otherwise undermining the price discovery process.
>
> (CFTC EEMAC, 2022)

One phrase in the Commissioner's statement is especially loaded considering the history of speculation controversies: '...include an examination of the market presence of passive investment vehicles and other speculators.' The mention of 'passive investment vehicles' makes it clear that, despite our best efforts, the Masters Hypothesis is alive and well, at least in the mind of one powerful market regulator. The mere mention of 'speculators' means that (sigh) the next phase of the anti-speculation cycle is upon us.

13.4 Closing Thoughts

In the introduction to this book (Chapter 1) we recounted the fortuitous events that caused our paths to intersect almost exactly 30 years ago. We literally had no idea that we would work together for so long and our work would have the impact that it did. There is an important lesson in that simple observation. Research opportunities are incredibly hard to predict, but you need to be ready when they present themselves. We have been very fortunate to have great opportunities presented to us, but at the same time, pursuing them has required a truly dogged determination over many years and the ability to weather lots and lots of rejection and controversy.

It also helps to be a little lucky. We have been lucky to work at great universities that have generously supported and encouraged our research. We also have a fortuitous complementarity in the skills we bring to the research enterprise. We also have to admit to being lucky in our choice of research areas. In that sense, the topic of speculation in commodity futures markets is the gift that keeps on giving. Since there is no sign that the controversy is going to be resolved anytime soon, we have no intention of slowing down or slacking off in the future.

Note

[1] The tongue-in-cheek apology is necessary because we borrowed the metaphor used here from Lester Telser. Surveying the literature on the risk premiums in commodity futures prices, Telser (2000, p. 551) remarked that 'It appears on this evidence that normal backwardation as a theory of futures should be respectfully interred.'

References

CFTC EEMAC (CFTC Energy and Environmental Markets Advisory Committee) (2016) Report on EEMAC's 2015 Review and Consideration of the CFTC's Proposed Rule on Position Limits. Commodity Futures Trading Commission (CFTC), February 25. Available at: https://www.findknowdo.com/sites/default/files/2016/02/25/00/Final%20Approved%20EEMAC%20Report%202-25-2016%20EMBARGOED.pdf (accessed October 6 2022.)

CFTC EEMAC (CFTC Energy and Environmental Markets Advisory Committee) (2022) Opening Statement of Commissioner Christy Goldsmith Romero before the Energy and Environmental Markets Advisory Committee. Commodity Futures Trading Commission (CFTC), September 20. Available at: https://www.cftc.gov/PressRoom/SpeechesTestimony/romerostatement092022 (accessed October 6 2022).

Coles, J.L., Heath, D. and Ringgenberg, M.C. (2022) On index investing. *Journal of Financial Economics* 145, 665–683.

Garcia, P., Irwin, S.H. and Smith, A. (2015) Futures market failure? *American Journal of Agricultural Economics* 97, 40–64.

Goldstein, I. and Yang, L. (2022) Commodity financialization and information transmission. *Journal of Finance* 77, 2613–2667.

Petzel, T.E. (1981) A new look at some old evidence: the wheat market scandal of 1925. *Food Research Institute Studies* 18, 117–128.

Ready, M. and Ready, R. (2022) Order flows and financial investor impacts in commodity futures markets. *Review of Financial Studies* 35, 4712–4755.

Tang, K. and Xiong, W. (2012) Index investment and financialization of commodities. *Financial Analysts Journal* 68, 54–74.

Telser, L.G.(2000). *Classic Futures: Lessons for the Past from the Electronic Age.* Risk Books, London.

Index

Note: page numbers in italic denote figures and tables, 'n' denotes a note.